Analysis and Optimisation of Stochastic Systems

The Institute of Mathematics and its Applications Conference Series

---

Optimization, *edited by* R. Fletcher
Combinatorial Mathematics and its Applications, *edited by* D. J. A. Welsh
Large Sparse Sets of Linear Equations, *edited by* J. K. Reid
Numerical Methods for Unconstrained Optimization, *edited by* W. Murray
The Mathematics of Finite Elements and Applications, *edited by* J. R. Whiteman
Software for Numerical Mathematics, *edited by* D. J. Evans
The Mathematical Theory of the Dynamics of Biological Populations, *edited by* M. S. Bartlett and R. W. Hiorns
Recent Mathematical Developments in Control, *edited by* D. J. Bell
Numerical Methods for Constrained Optimization, *edited by* P. E. Gill and W. Murray
Computational Methods and Problems in Aeronautical Fluid Dynamics, *edited by* B. L. Hewitt, C. R. Illingworth, R. C. Lock, K. W. Mangler, J. H. McDonnell, Catherine Richards and F. Walkden
Optimization in Action, *edited by* L. C. W. Dixon
The Mathematics of Finite Elements and Applications II, *edited by* J. R. Whiteman
The State of the Art in Numerical Analysis, *edited by* D. A. H. Jacobs
Fisheries Mathematics, *edited by* J. H. Steele
Numerical Software—Needs and Availability, *edited by* D. A. H. Jacobs
Recent Theoretical Developments in Control, *edited by* M. J. Gregson
The Mathematics of Hydrology and Water Resources, *edited by* E. H. Lloyd, T. O'Donnell and J. C. Wilkinson
Mathematical Aspects of Marine Traffic, *edited by* S. H. Hollingdale
Mathematical Modelling of Turbulent Diffusion in the Environment, *edited by* C. J. Harris
Mathematical Methods in Computer Graphics and Design, *edited by* K. W. Brodlie
Computational Techniques for Ordinary Differential Equations, *edited by* I. Gladwell and D. K. Sayers
Stochastic Programming, *edited by* M. A. H. Dempster
Analysis and Optimisation of Stochastic Systems, *edited by* O. L. R. Jacobs, M. H. A. Davis, M. A. H. Dempster, C. J. Harris and P. C. Parks

In Preparation

Numerical Methods in Applied Fluid Dynamics, *edited by* B. Hunt
Recent Developments in Markov Decision Processes, *edited by* D. J. White, R. Hartley and L. C. Thomas
Power from Sea Waves, *edited by* B. Count

Analysis and Optimisation of Stochastic Systems

Based on the proceedings of the International Conference on Analysis and Optimisation of Stochastic Systems held at the University of Oxford from 6–8 September, 1978, organised by The Institute of Mathematics and its Applications

Edited by

O.L.R. JACOBS
University of Oxford
Oxford, England

M.H.A. DAVIS
Imperial College
London, England

M.A.H. DEMPSTER
Balliol College
Oxford, England

C.J. HARRIS and **P.C. PARKS**
Royal Military College
Shrivenham, England

1980

ACADEMIC PRESS
A Subsidiary of Harcourt Brace Jovanovich, Publishers
London New York Toronto Sydney San Francisco

ACADEMIC PRESS INC. (LONDON) LTD.
24/28 Oval Road,
London NW1

United States Edition published by
ACADEMIC PRESS INC.
111 Fifth Avenue
New York, New York 10003

British Library Cataloguing in Publication Data
International Conference on Analysis and Optimisation
of Stochastic Systems, *University of Oxford, 1978*
Analysis and optimisation of stochastic systems.
1. System analysis—Congresses
2. Stochastic processes—Congresses
I. Jacobs, Oliver Louis Robert
II. Institute of Mathematics and its Applications
519.2 QA402 80-40224

ISBN 0-12-378680-0

Printed in Great Britain by
Whitstable Litho Ltd., Whitstable, Kent

PREFACE

This volume contains papers presented at an International Conference on Analysis and Optimisation of Stochastic Systems, held on 6th to 8th September 1978 at the University of Oxford, and organised by the Institute of Mathematics and its Applications. The Programme Committee of the conference are the Editors of this volume. The motivation for the conference was to provide an opportunity for reporting on recent advances and to identify areas for further research in mathematical methods for analysis and optimisation of stochastic systems. These methods provide the theoretical foundations of feedback control systems and of decision-making under uncertainty; subjects having applications in fields such as engineering, economics, operations research and biology. The theory is well developed for problems characterised by linear equations, quadratic costs and Gaussian random variables, but is otherwise incomplete, with the consequence that potentially interesting or valuable applications are frustrated for want of fundamental understanding.

The conference was to some extent a sequel to previous conferences on Stochastic Programming in July 1974 at Oxford and on Stochastic Systems in June 1975 at Lexington, Kentucky. It differed from the 1974 conference in having more material on stochastic control, and from the 1975 conference in also being slightly more biased towards applications.

All contributions submitted have been refereed; the resulting contents of this volume indicate the state of the subject in 1978. Eight topics emerged, so the papers are presented not in their order of presentation at the conference but in eight chapters: 1 Optimal stochastic control, 2 Stochastic optimisation, 3 Stochastic processes, 4 Algorithms, 5 Information, 6 State estimation, 7 Parameter estimation, 8 Applications. Some features of the indicated state of the subject are the increased amount of work on Estimation, the small amount of work on Adaptive Control and the emergence of Information as a separate topic.

Chapter 1 contains six papers on the theory of optimal control of stochastic systems. In paper 1.1 Beneš, Shepp and Witsenhausen discuss three problems in which a process having bounded variation tracks a Wiener process so as to minimize mean square tracking error. Elliott and Varaiya in 1.2 use a suitable stochastic maximum principle to give sufficient conditions for the optimal control of a partially observed diffusion. Paper 1.3 by Davis and Wan concerns the martingale theory for Markov control of a class of Markov jump processes and shows that Rishel's formulation of the "principle of optimality" holds for the class considered. The remaining three papers treat characterization of optimal controls for stochastic systems whose dynamics involve both diffusion and jump processes. In paper 1.4 Vermes uses the Bellman-Hamilton-Jacobi equation for the value-function of optimally controlled Markov processes to investigate detailed properties related to the "bang-bang" principle. Bismut in 1.5 discusses the optimal control of square-integrable semi-martingales invoking the duality theory of convex analysis to obtain necessary and sufficient conditions for the solvability of both original and dual problems. Semi-martingale systems are also investigated in paper 1.6 by Boel and Kohlmann, who obtain existence results and characterization results in terms of a stochastic maximum principle for optimal controls.

The four papers in Chapter 2 are a heterogeneous collection dealing with various types of stochastic optimisation problem. Bensoussan and Lions' paper 2.1 outlines some optimal stopping problems involving reflecting barriers and obtains the corresponding variational inequalities. In 2.2 Nisio constructs the nonlinear semigroup of operators corresponding to a Markovian system with both continuous control and controlled stopping. Optimal search problems are the subject of 2.3, in which Pursiheimo and Ruohonen give conditions for optimal deployment of search effort to locate a target which moves as a diffusion process. Sutherland in paper 2.4 discusses how the optimal control of a discrete-time system representing the growth of an economy may be formulated as an infinite dimensional linear program.

The papers in Chapter 3 treat some modelling and analysis questions for stochastic systems. In 3.1 Fleming outlines a method of obtaining Ventcel-Freidlin type estimates for exit probabilities for diffusion processes by analysing a related stochastic control problem. Virani and Sargent in 3.2 present a stochastic dynamical system whose sample functions, unlike those arising from diffusion processes, are "physically realisable" in the sense that they are absolutely continuous in time. Root's paper 3.3 presents an abstract framework for modelling families of stochastic systems in terms of input-output maps. In a more specific vein, Curtain demonstrates in 3.4 how a number of hyperbolic differential equations with random parameters arising in engineering and mathematical physics may be incorporated in the general model for infinite-dimensional stochastic systems developed by herself and her associates. In paper 3.5 Ansell, Bendell and Humble discuss (hierarchically) nested renewal processes and present some related optimisation problems arising in practice. In paper 3.6 Ichikawa uses semigroup theory to study stability questions and regulator problems concerning stochastic evolution equations in Hilbert space.

Chapter 4 contains three papers about algorithms. In 4.1 Haussmann considers a discrete-time approximation to optimal control and its convergence to the corresponding continuous-time problem. Dixon and James, in 4.2, are concerned with finding the minimum of a function corrupted by noise, with special reference to objective functions arising in parameter estimation for input-output systems. Gawthrop in 4.3, the only paper on adaptive control, discusses stability and convergence problems of self-tuning algorithms for identification and control of linear systems having unknown parameters.

Chapter 5 contains two contributions on the use of information in stochastic control and estimation. In 5.1 Witsenhausen demonstrates the importance of information structure in stochastic control, in particular he relates it to Shannon's Information Theory and to generalised rate-distortion theory. In less general but related terms Willems in 5.2 introduces the concept of information state for the control of uncertain systems; relationships with Kalman filtering and realisation theory are given in an applications section.

Chapter 6 contains seven contributions on state estimation; the first four papers, on linear systems, include some discussion of dual control problems. In 6.1 Brockett and Clark extend the early work of Brockett on Lie Algebras and Groups in control theory to the relationship between problems in estimation theory and the representations of Lie groups; they demonstrate that linear filtering has an intrinsic relationship with Heisenberg algebra. In 6.2 Grimble develops in the s-domain a time-invariant Kalman filter which generates correct finite-time state estimates at specified times: his approach allows for coloured noise as well as pure time-delays of the observed variable. In 6.3 Smith and Makov discuss Bayesian detection and estimation of jumps in discrete-time linear systems; computational aspects as well as approximation methods are considered. In 6.4 Pavon and Wets show that the well-known duality between estimation and control can be established by a stochastic variational approach. Balakrishnan in 6.5 reviews nonlinear filtering theory with "white" noise models. Some new algorithms for nonlinear state estimation via an equivalent two-point boundary value problem are given in 6.6 by Halme; Green's function and the corresponding integral equation are used rather than invariant imbedding. In 6.7 Marcus, Mitter and Ocone develop finite-dimensional state estimators for systems with nonlinearities which are analytic and expressable as finite Volterra series.

Chapter 7 contains six contributions on parameter estimation or, as some authors call it, system identification. In 7.1 Ljung summarises relationships between well-known recursive algorithms and also summarises his own results about their convergence. In 7.2 Larminat and Doncarli discuss suitable linear representations to be used with some of the algorithms. In 7.3, the only contribution about continuous-time systems in the chapter, Bagchi proves that continuous-time maximum likelihood estimates are consistent. In 7.4 Rissanen describes his procedure for determining the order of a system under identification by minimizing the amount of information (number of bits) needed to specify the estimated parameters together with the observed data. In 7.5 Zarrop describes a procedure for designing test signals so as to maximise obtainable information about parameters; the procedure requires initial estimates of the parameters but is believed to be convergent. In 7.6 Bittanti considers informational aspects of processes whose outputs have a periodic mean value; he suggests how discrete-time sample points might be distributed in time so as to maximise obtainable information. In 7.7 Goodrich and Caines discuss estimation of constant parameters from an ensemble of records of transient behaviour over a finite time (cross-sectional data); they prove consistency and asymptotic normality of maximum likelihood estimation as the number of records goes to infinity and give an example of the application of their results to interpretation of IQ data.

Chapter 8 contains four applications papers. In 8.1 Carmichael and Pritchard discuss scheduling of radiation doses to control the growth of cancer tumours. Uchida in 8.2 applies optimal control theory to statistical identification problems in economic system models. Tunnicliffe Wilson and Overton in 8.3 use differential dynamic programming techniques to maximise financial returns from a nuclear reactor. Finally in 8.4 Haslett considers stochastic problems arising in the storage of solar energy.

The Editors/Programme Committee are happy to acknowledge the contribution made by the staff of the Institute of Mathematics and its Applications who so efficiently organised the conference and prepared these proceedings for publication: in particular they are grateful to Mrs. C.A. Hinds for her part in the organisation and to Miss Janet Fulkes who typed these proceedings. All who attended the conference and all who read the volume are indebted to them and to the many referees for their generous help.

June 1980

The Institute thanks the authors of the papers, the executive editor, Dr. O. Jacobs, the editors Dr. M. Davies, Dr. M. Dempster, Professor C. Harris and Professor P. Parks and also Miss Janet Fulkes for typing the papers.

CONTENTS

CHAPTER 1

OPTIMAL STOCHASTIC CONTROL

1.1 SOME SOLVABLE STOCHASTIC CONTROL PROBLEMS

V. E. Beneš, L. A. Shepp and H. S. Witsenhausen

(Bell Laboratories, Murray Hill, New Jersey 07974, USA)

We solve explicitly some feedforward stochastic control problems that lead to free boundaries in one or two dimensions. All of them can be cast as optimally following a Wiener process $x + w_t$ started at x by a process ξ_t of bounded variation adapted to $x + w_t$, under various additional restrictions on ξ_t, so as to minimize the discounted mean square error criterion $L_\alpha(\xi) = E\int_0^\infty e^{-\alpha t}(x+w_t-\xi_t)^2 dt$ with discount rate $\alpha > 0$, or in one interesting case, the "finite horizon" version $L_T(\xi) = E\int_0^T (x+w_t-\xi_t)^2 dt$. It is helpful to think of ξ_t as the cumulative effect of control, and of its variation $\int_0^t |d\xi|$ as the "fuel used to date". The constraints on ξ will take the form of bounds on $|\dot{\xi}|$, or on the total available fuel.

Although much work [1-4] has been done earlier on similar problems in optimal control, a main result to be given here (Problem 3) is one of the few known cases in which a free boundary that is neither linear nor quadratic has been found explicitly. Applications of the problems considered here might include delta-modulation, dividend pricing, investment policy to meet demand, forecasting, or satellite control. The methods of solution are also of interest. The desired boundary is first derived or guessed by using Bellman's equation for the optimal value function (insofar as, and where, it applies), backed up by the heuristic principle that the optimal value function should be smooth to second order in space. Once the boundary is obtained, the corresponding policy is proved optimal by a "verification lemma" similar to those used in stochastic control. A final point of interest about several of these problems is this: they indicate that the local time of the controlled process on the boundary plays an explicit and natural role, and that the heretofore recondite concept of local time is a useful tool in a large class of applications (cf. Watanabe [5] or Stroock-Varadhan [6]).

The present account is limited to stating three problems and describing their solutions; theorems, proofs, and most of the heuristics will be published elsewhere.

Problem 1: Minimize $L_\alpha(\xi)$ over processes ξ_t adapted to $x + w_t$ under the constraints $\xi_0 = 0$, $-\infty < \theta_0 \leq \dot{\xi}_t \leq \theta_1 < \infty$; here ξ_t is a differentiable process whose rate of change is bounded between θ_0 and θ_1. In this case there is a number $\delta = \delta(\theta_0, \theta_1, \alpha) =$

$((\theta_0^2 + 2\alpha)^{\frac{1}{2}} - \theta_0)^{-1} - (\theta_1^2 + 2\alpha)^{\frac{1}{2}} - \theta_1)^{-1}$ such that the optimal process ξ_t is given by $\xi_0 = 0$ and

$$\dot{\xi}_t = \begin{cases} \theta_0 & \text{if } x + w_t - \xi_t < \delta \\ \theta_1 & \text{if } x + w_t - \xi_t \geq \delta. \end{cases}$$

The optimallly controlled process $x_t = x + w_t - \xi_t$ is a diffusion starting at x and solving the stochastic DE

$$dx_t = -\theta(x_t)dt + dw_t, \quad \theta(x) = \begin{cases} \theta_1 & x \geq \delta \\ \theta_0 & x < \delta \end{cases}$$

The optimal value function

$$L(x) = L_\alpha(\xi)_{opt} = E\left\{ \int_0^\infty e^{-\alpha t} x_t^2 dt \,\middle|\, x_0 = x \right\}$$

satisfies the backward equation $\alpha L(x) = x^2 + \frac{1}{2} L''(x) - \theta(x)L'(x)$, $x \neq \delta$, with the smoothness condition $L \in C^2$, which together with quadratic growth uniquely determine L as

$$L(x) = \begin{cases} \dfrac{x^2}{\alpha} + \dfrac{1}{\alpha^2} - \dfrac{2\theta_1 x}{\alpha^2} + \dfrac{2\theta_1^2}{\alpha^3} - \dfrac{2(\theta_1-\theta_0)(\bar\theta_1+\theta_1)e^{-(x-\delta)(\bar\theta_1-\theta_1)}}{\alpha^2(\bar\theta_0-\theta_0)(\bar\theta_0+\theta_0+\bar\theta_1-\theta_1)}, & x \geq \delta \\ \dfrac{x^2}{\alpha} + \dfrac{1}{\alpha^2} - \dfrac{2\theta_0 x}{\alpha^2} + \dfrac{2\theta_0^2}{\alpha^3} - \dfrac{2(\theta_1-\theta_0)(\bar\theta_0-\theta_0)e^{(x-\delta)(\bar\theta_0+\theta_0)}}{\alpha^2(\bar\theta_1+\theta_1)(\bar\theta_0+\theta_0+\bar\theta_1-\theta_1)}, & x<\delta \end{cases}$$

where $\bar\theta = (\theta^2+2\alpha)^{\frac{1}{2}}$ for $\theta = \theta_0, \theta_1$ and $\delta = (\bar\theta_0-\theta_0)^{-1} - (\bar\theta_1+\theta_1)^{-1}$. The first two terms represent the criterion value for $\xi \equiv 0$, i.e. if no control is undertaken.

Problem 2: Minimize the finite horizon loss

$$L_T(\xi) = E\int_0^T (x+w_t-\xi_t)^2 dt$$ over processes ξ_t adapted to $x + w_t$

under the constraints $\xi_0 = 0$, $\theta_0 \leq \dot\xi_t \leq \theta_1$. We can only solve this problem explicitly when $\theta_0 = 0$ and $\theta_1 = \infty$, i.e. when ξ_t must be nondecreasing. In this case ξ_t can be thought of as cumulative investment to meet the random demand $x + w_t$, with no option of disinvestment. It is intuitively reasonable that an optimal policy take the form of always meeting demand to within an amount f and not investing when within f of demand; the function f should decrease to 0 at T; it represents a "hedging" against possible future decreases in demand.

Indeed, it can be correctly guessed from a rescaling argument that the optimal policy is of the form

$$\dot\xi_t = \begin{cases} 0 \text{ if } x_t < \delta(T-t)^{\frac{1}{2}} \\ \infty \text{ if } x_t \geq \delta(T-t)^{\frac{1}{2}} \end{cases} \quad (\text{where } x_t = x+w_t-\xi_t),$$

that is, after possibly an initial jump at $t = 0$ to the boundary of the region $x \leq \delta(T-t)^{\frac{1}{2}}$, fuel is used ($\xi_t$ increases) only in order to stay in this shrinking region. The formal Bellman equation for $L(x,\tau) = \inf_{\xi} E\left\{ \int_{T-\tau}^{T} x_t^2 dt \,|\, x_{T-\tau} = x \right\}$ is

$$\frac{\partial L}{\partial \tau} = x^2 + \frac{1}{2}\frac{\partial^2 L}{\partial x^2} + \inf_{0 \leq u \leq \infty} \left| -u \frac{\partial L}{\partial x} \right|,$$

where τ = "time to go".

The control parameter u enters linearly and the infimum is at u = 0 or u = ∞. Thus we expect a problem with a free boundary $x = f(\tau)$ and conditions

$$\frac{\partial L}{\partial \tau} = x^2 + \frac{1}{2}\frac{\partial^2 L}{\partial x^2}, \quad x < f(\tau)$$

$$\frac{\partial L}{\partial x} = 0, \quad x \geq f(\tau)$$

for $L \in C^{2,1}$ of quadratic growth in x. Observing that the PDE becomes an ordinary DE under the rescaling

$L(x,\tau) = \tau^2 L(x\tau^{-\frac{1}{2}},1)$, we solve for L and f as $f(\tau) = \delta\tau^{\frac{1}{2}}$,

$$L(x,1) = \begin{cases} x^2 + \frac{1}{2} - 4\delta^3 e^{\delta^2/2}(x^4+6x^2+3)\displaystyle\int_{-x}^{\infty} \frac{e^{-u^2/2}}{(u^4+6u^2+3)^2}\,du, & x \leq \delta \\ \frac{1}{2}\delta^2, & x \geq \delta \end{cases}$$

with $\delta = 0.63883\ 32158\ldots$ the root of

$$1 = \frac{8\delta^3(\delta^4+6\delta^2+3)}{\delta^2+1} e^{\delta^2/2} \int_{-\delta}^{\infty} \frac{e^{-u^2/2}}{(u^4+6u^2+3)^2}\,du.$$

The increasing process ξ_t which achieves this value function can be defined (cf. Bather-Chernoff [1]) for $x \leq \delta T^{\frac{1}{2}}$ by

$$\xi_t = \max\left\{0,\ \max_{0\le s\le t}\left\{x+w_s-\delta(T-s)^{\frac{1}{2}}\right\}\right\},\ 0 \le t \le T. \tag{1}$$

(This leaves out the initial jump in ξ which is needed if $x>\delta T^{\frac{1}{2}}$.) It can be verified that ξ_t is the smallest increasing function starting at 0 that keeps the controlled process x_t in the region $x_t \leq \delta(T-t)^{\frac{1}{2}}$. It is natural, by analogy with Tanaka's formula, or Watanabe [5], to interpret ξ_t as the local time of the controlled process on the boundary.

Problem 3: Minimize $L_\alpha(\xi)$ over processes ξ_t adapted to $x + w_t$ under the constraints $\xi_0 = 0$, $\int_0^\infty |d\xi_t| \le v$ given. Here we broach one of the difficult "finite fuel" problems, for it is natural to regard the available variation as "fuel" to be used to increase or decrease the controlled process $x + w_t - \xi_t$ as the controller sees fit. Accordingly we formulate the problem in terms of an additional process $v_t = v - \int_0^t |d\xi|$, the fuel remaining at t. In this case the optimal control can be described as follows. There is a region $C = \{x,v: v<f(x), v>0\}$ defined by $f(x) = f(-x)$, $f(\pm u\sqrt{2\alpha}) = 0$, and

$$f'(x) = \left\{x^{-1}\tanh x - 1\right\}^{-1},\quad 0 < x \leq u\sqrt{2\alpha},$$

with $u = 1.119\ldots$ the root of $u \tanh u = 1$, such that no fuel is used if $x_t, v_t \in C$; fuel is used initially to jump to ∂C if starting from outside $\bar{C}$, and on $\partial C\ \{v>0\}$ in order to keep x_t, v_t within $\bar{C}$. It can be checked that C flares out with decreasing fuel, indicating that as fuel gets low, a larger error is needed to make us use it.

The process x_t, v_t is a kind of degenerate diffusion with 45° reflection on $\partial C \cap \{v>0\}$; when v_t reaches 0 it absorbs there, and then x_t becomes an uncontrolled Wiener process started at one of $\pm u\sqrt{2\alpha}$. The process x_t, v_t can be constructed explicitly out of the "driving" Brownian motion w_t by the formulas

$$x_t = x + w_t + a_t^- - a_t^+ , \quad a^{\pm}(0) = 0$$

$$v_t = v - a_t^+ - a_t^- ,$$

where $|d\xi| = da^+ + da^-$, i.e. a^+ and a^- are respectively the negative and positive variations of ξ, and are the unique nondecreasing solutions of the equations (cf. (1)):

$$a_t^+ = \max\left\{0, \max_{0\le s\le t}\left\{x + w_s + a_s^- - f^{-1}(v - a_s^+ - a_s^-)\right\}\right\}$$

$$a_t^- = \max\left\{0, \max_{0\le s\le t}\left\{-x - w_s + a_s^+ - f^{-1}(v - a_s^+ - a_s^-)\right\}\right\} ,$$

with f^{-1} the inverse of f on $v > 0$. Here it is natural to relate a^+ (resp. a^-) to the local time of the controlled process x_t on the boundary component $\partial C \cap \{x>0\}$ (resp. $\partial C \cap \{x<0\}$). When v_t reaches 0 we run out of fuel and the equations cease to apply; $v \equiv 0$ henceforth and x_t becomes a Brownian motion started from $\pm u\sqrt{2\alpha}$.

Inside $\bar{C}$ and on $v = 0$ the optimal value function has the form $L(x,v) = \frac{x^2}{\alpha} + \frac{1}{\alpha^2} + A(v)\cosh x\sqrt{2\alpha}$, where $A(0) = 0$ and

$$A'(v) = -\frac{1}{\alpha\sqrt{2\alpha}}(y \cosh y - \sinh y)$$

$$y = \sqrt{2\alpha}\, f^{-1}(v)$$

Note that the $A \leq 0$, so that term in $A(v)$ subtracts from the value $\frac{x^2}{\alpha} + \frac{1}{\alpha^2}$ incurred if no fuel is used. Inside C, $L(x,v)$ satisfies $\alpha L = x^2 + \frac{1}{2}\frac{\partial^2 L}{\partial x^2}$; on $\partial C \cap \{v>0\}$ it meets the boundary condition $L_x + (\operatorname{sgn} x)L_v = 0$ corresponding to reflection at a fixed 45° angle on each of the two components of $\partial C \cap \{v>0\}$; outside C it is constant on lines $\{v=x+f(x_0)-x_0,\ 0<x_0\leq u\sqrt{2\alpha}\}$ in $x > 0$, and on lines $\{v=-x+f(x_0)-x_0,\ u\sqrt{2\alpha}\leq x_0<0\}$ in $x < 0$; on $v = 0$, $L(x,v) = \frac{x^2}{\alpha} + \frac{1}{\alpha^2}$; if $x - v > u\sqrt{2\alpha}$, $L(x,v) = \frac{(x-v)^2}{\alpha} + \frac{1}{\alpha^2}$; if $x + v < - u\sqrt{2\alpha}$, $L(x,v) = \frac{(x+v)^2}{\alpha} + \frac{1}{\alpha^2}$.

We remark that Problem 3 involves a higher level of complication than the first two, in that no explicit representation like (1) for the optimal process ξ_t in terms of maxima is available, only a pair of functional equations that serve the same purpose, viz. they single out the smallest increasing functions whose addition to and subtraction from $x + w_t$ will keep the resulting sum in $\bar{C}$ until fuel funs out.

Other problems can be attacked by these methods. We conjecture extensions to controlling Bessel processes and other diffusions, under a finite fuel constraint. The boundaries are sometimes relatively straightforward to find, but construction of the process achieving the minimum can be a problem.

1 REFERENCES

1. Bather, J. and Chernoff, H. "Sequential decisions in the control of a spaceship", Proc. 5th Berkely Symp on Math. Stat. and Prob., University of California Press, (1966).

2. Bather, J. and Chernoff, H. "Sequential decisions in the control of a spaceship (finite fuel)", *J. Appl. Prob.*, **4**, 584-604, (1967).

3. Chernoff, H. "Optimal stochastic control", reprint, ca. (1968).

4. Heath, D. C. and Sudderth, W. D. "Continuous-time gambling problems", *Adv. Appl. Prob.*, **6**, 651-655, (1974).

5. Watanabe, S. "On stochastic differential equations for multidimensional diffusion processes with boundary conditions", *J. Math. Kyoto Univ.*, **11**, 169-180, (1971).

6. Stroock, D. W. and Varadhan, S. R. S. "Diffusion processes with boundary conditions", *Comms. on Pure and Appl. Math.*, **24**, 147-225, (1971).

1.2 A SUFFICIENT CONDITION FOR THE OPTIMAL CONTROL OF A PARTIALLY OBSERVED STOCHASTIC SYSTEM

R. J. Elliott

(Department of Mathematics, The University of Kentucky, Lexington, Kentucky 40506, USA)

and

(Department of Pure Mathematics, University of Hull, England)

P. P. Varaiya

(Department of Electrical Engineering and Computer Science, University of California, Berkely 94720, USA)

1 INTRODUCTION

The optimal control of a partially observed system is discussed, whose dynamics are described by a system of functional-differential equations. In a previous paper, [2], the first author has shown that if a control u*, which is adapted to the partially observed σ-fields, is optimal, then, almost surely, it minimizes the conditional expectation of a certain Hamiltonian with respect to the observed σ-fields when this expectation is taken with respect to the measure induced by u*. Because this result is in terms of a single Hamiltonian it is stronger than previous theorems of this kind.

In the present paper it is shown that a control u*, adapted to the partially observed σ-fields, is optimal if it minimizes the conditional expectations of a Hamiltonian, when these expectations are taken with respect to the measure induced by any other control. Finally, in the case when the cost is measurable with respect to the observed variables, an even simpler sufficient condition is given.

2 DYNAMICS

This paper is a sequel to [2], and we review the notation used there. The evolution of the system is described by a stochastic functional differential equation of the form

$$dx_t = f(t,x,u)dt + \sigma(t,x)dw_t, \tag{2.1}$$

where $t \in [0,1]$ and w is an m-dimensional Brownian motion. C denotes the space of continuous functions from $[0,1]$ to R^m, x denotes a member of C, and the value of x at time t is indicated by x_t. For simplicity we suppose that

$$x(0) = 0 \in R^m.$$

The control values u are chosen from a set U, which is a Borel subset of R^k with the Borel σ-field.

F_t is the σ-field on C generated by $\{x_s: x \in C,\ s \leq t\}$.

2.1 Definition

We suppose the $m \times m$ matrix $\sigma = (\sigma_{ij})$ satisfies

(i) for $1 \leq i,j \leq n$ $\sigma_{ij}(t,\cdot):C \to R$ is F_t measurable, and $\sigma_{ij}(\cdot,x):[0,1] \to R$ is Lebesgue measurable for each $x \in C$,

(ii) $\sigma(t,x)$ is nonsingular,

(iii) each σ_{ij} satisfies a uniform Lipschitz condition in x, when $x \in C$ is given the uniform norm

$$\|x\|_t = \sup_{0 \leq s \leq t} |x(t)|,$$

(iv) there is a constant $k_0 < \infty$ such that

$$\Sigma\ \sigma_{ij}^2 dt < k_0$$

a.s. P, where the measure P is defined below.

Suppose an m-dimensional Brownian motion B_t, defined on a probability space (Ω, A, μ), is given. Then from the above properties of σ there is a unique solution of the stochastic equation

$$dx_t = \sigma(t,x)dB_t,$$

$$x(0) = 0 \in R^m.$$

A measure P can, therefore, be defined on $(\mathcal{C},\mathcal{F}_1)$ by putting

$$P(A) = \mu\{\omega : x(\omega) \in A\},\ A \in \mathcal{F}_1.$$

Now suppose that $x_t \in R^m$ is written in terms of two components

$$x_t = (y_t, z_t)$$

where $y_t \in R^n$ and $z_t \in R^{m-n}$. f and σ similarly decompose as

$$f = (f_1, f_2),\ \sigma = \begin{pmatrix} \sigma_1 \\ \sigma_2 \end{pmatrix},$$

where f_1 (resp. f_2) is an n (resp. m-n) vector function, and $\sigma_1(t,x)$ (resp. $\sigma_2(t,x)$) is an $n\times m$ (resp. $(m-n)\times m$) matrix function.

We, therefore, have

$$dy_t = f_1(t,x,u)dt + \sigma_1(t,x)dw_t,$$

and the variables y represent the observations that are made of the process. The observation σ-fields $\mathcal{Y}_t$ of $\mathcal{C}$ are defined by

$$\mathcal{Y}_t = \sigma\{y_s : s \leq t\} \subset \mathcal{F}_t.$$

2.2 Definition

A partially observable control over $[s,t] \subset [0,1]$ is a measurable function

$$u : [s,t] \times \mathcal{C} \to U$$

such that:

(i) for each τ, $s \leq \tau \leq t$, $u(\tau,\cdot)$ is $\mathcal{Y}_t$ measurable and $E|u(\tau,\cdot)| < \infty$,

(ii) for each $x \in \mathcal{C}$, $u(\cdot,x)$ is Lebesgue measurable.

$\mathcal{N}_s^t$ will denote the set of partially observable controls over $[s,t]$, and we shall write $\mathcal{N}$ for $\mathcal{N}_0^1$.

2.3 Definition

We suppose the drift function f satisfies

(i) $f:[0,1]\times C\times U\to R^m$ is measurable,

(ii) for every $u \in N$ and every $t \in [0,1]$, $f^u(t,\cdot) = f(t,\cdot,u(t,\cdot))$ is F_t measurable,

(iii) for every $u \in N$ and every $x \in C$, $f^u(\cdot,x)$ is Lebesgue measurable,

(iv) for every $(t,x) \in [0,1]\times C$, $f(t,x,\cdot)$ is continuous on U,

(v) $|\sigma^{-1}(t,x)f^u(t,x)| \leq M(1 + \|x\|_t)$.

With the prime denoting a transpose, for $u \in N$ we define

$$\xi_s^t(f^u) = \int_s^t \{\sigma^{-1}(\tau,x)f^u(\tau,x)\}' \, dB_\tau - \tfrac{1}{2}\int_s^t |\sigma^{-1}(\tau,x)f^u(\tau,x)|^2 d\tau.$$

The linear growth condition 2.3(v) ensures that, if

$$\rho_s^t(u) = \exp(\xi^t(f^u)),$$

then $E[\rho_s^t(u)|F_s] = 1$ a.s. P.

Corresponding to any control $u \in N$, therefore, we can define a new probability measure P_u on (C,F_1) by putting

$$P_u(A) = \int_A \rho_0^1(u)\,dP, \text{ for } A \in F_1.$$

A measure μ_u is then induced on (Ω,A) by putting $\mu_u(B) = P_u(x(B))$, for $B \in A$. Girsanov's theorem, (see [3] and [1]), then states the following:

2.4 Theorem

Suppose, as above, that B_t is an m-dimensional Brownian motion on (Ω,A,μ) and x_t is the solution of

$$dx_t = \sigma(t,x)dB_t,\ x(0) = 0 \in R^m.$$

Then, under the measure P_u, w_t^u is a Brownian motion, where

$$dw_t^u = \sigma^{-1}(t,x)(dx_t - f^u(t,x)dt).$$

That is, from Lemma 6 of [3],

$$dx_t = f^u(t,x)dt + \sigma(t,x)dw_t^u \text{ and } x(0) = 0 \in R^m.$$

As is now well known, this change of measure method enables us to interpret solutions of the dynamical equations (2.1) under quite weak conditions on f.

2.5 Cost

For simplicity of exposition we suppose that just a terminal cost is associated with the process. This is of the form

$$g(x(1))$$

where g is bounded and F_1 measurable.

Writing E_u for the expectation with respect to P_u, if control $u \in N$ is used, the expected cost is

$$J(u) = E_u[g(x(1))].$$

The optimal control problem is to determine how $u \in N$ should be chosen so that J(u) is minimized.

3 A SUFFICIENT CONDITION FOR OPTIMALITY

In our previous paper [2] we proved that, if $u^* \in N$ is an optimal control then there is a predictable process g* and a set of zero measure $T \subset [0,1]$, such that if $t \notin T$

$$E_{u^*}[g^*\sigma^{-1}(f^u - f^{u^*})|Y_t] \geq 0 \text{ a.s. P., for any other control } u \in N.$$

We now establish the following sufficient condition for u* to be optimal.

3.1 Theorem

Suppose g* is the process defined below, and suppose that for every control $u \in N$, u* satisfies

$$E_u[g^*\sigma^{-1}(f^u - f^*)|Y_t] \geq 0 \text{ a.s. P.}$$

Then u* is an optimal control.

PROOF

For the measure P_{u^*} associated with the control u*

$$E_{u^*}[g(x(1)) | F_t]$$

is an F_t, P_{u^*} martingale. Because the Brownian motion $w_t^{u^*}$ is the innovations process for x_t, we can represent this martingale as a stochastic integral with respect to $w_t^{u^*}$, (see [1]). That is, there is a predictable process g* such that

$$E_{u^*}[g(x(1)) | F_t] = J + \int_0^t g^* dw^{u^*}.$$

Note that

$$E_{u^*}[g(x(1))] = J$$

and

$$g(x(1)) = J + \int_0^1 g^* dw^{u^*}.$$

Suppose $u \in N$ is any other control. Then the expected cost if u is used is

$$E_u[g(x(1))] = J + E_u\left[\int_0^1 g^* dw^{u^*}\right].$$

However,

$$E_u\left[\int_0^1 g^* dw^{u^*}\right] = E_u\left[\int_0^1 g^* dw^u\right] + E_u\left[\int_0^1 g^* \sigma^{-1}(f^u - f^{u^*}) dt\right].$$

As in [2],

$$\int_0^t g^* dw^u$$

is an F_t, P_u martingale so

$$E_u\left[\int_0^1 g^* dw^u\right] = 0$$

Also

$$E_u\left[\int_0^1 g^*\sigma^{-1}(f^u - f^{u^*})dt\right] = \int_0^1 E_u\left[g^*\sigma^{-1}(f^u - f^{u^*})dt\right]$$

≥ 0 by hypothesis.

Therefore,

$$J = E_{u^*}\left[g(x(1))\right] \leq E_u\left[g(x(1))\right]$$

so u* minimizes the expected cost and is optimal.

4 PARTIALLY OBSERVABLE COST

In this section we suppose that the cost g(x(1)) is measurable with respect to the observed σ-field $\mathcal{Y}_1$. This is a reasonable hypothesis, because if the controller notices the effect of the unobserved variables through the cost he is, in effect, obtaining some information about them. We first quote from [1] some results concerning the innovations process.

Recall that

$$dy_t = f_1(t,x,u)dt + \sigma_1(t,x)dw_t.$$

For any vector $\theta \in R^n$ write $\xi_t = \theta' y$. Then from Ito's differential formula

$$\int_0^t \theta' \sigma_1\sigma_1'\theta ds = \xi_t^2 - \xi_0^2 - 2\int_0^t \xi_s d\xi_s.$$

Therefore, $\sigma_1(t)\sigma_1'(t)$ is adapted to $\mathcal{Y}_t$, and so there are matrices Q_t and L_t adapted to $\mathcal{Y}_t$, with Q_t unitary and L_t diagonal, such that $\sigma_1(t)\sigma_1'(t) = Q_tL_tQ_t'$.

Write

$$\hat{f}_1^u = E_u[f^u | \mathcal{Y}_t],$$

$$T_t = (L_t)^{-\frac{1}{2}} Q_t.$$

4.1 Lemma

Define the process ν_t^u by

$$d\nu_t^u = T_t(dy_t - \hat{f}_1^u dt),$$

$$\nu_0^u = 0$$

Then $(\nu_t^u, \mathcal{Y}_t, P_u)$ is a Brownian motion, and if $(M_t, \mathcal{Y}_t, P_u)$ is a square integrable martingale then there exists a process ψ_t, adapted to $\mathcal{Y}_t$, such that

$$E_u\left[\int_0^1 |\psi_t|^2 dt\right] < \infty$$

and

$$M_t = M_0 + \int_0^t \psi_s d\nu_s.$$

The proof of these results can be found in Lemma 2.1 and Theorem 2.3 of [1].

For this partially observable cost we can state a slightly more explicit sufficient condition for a control $u^* \in \mathcal{N}$ to be optimal.

4.2 Theorem

Suppose $u^* \in \mathcal{N}$ has the property that for any other $u \in \mathcal{N}$, $\psi_t \cdot T_t \hat{f}_1^{u^*} \leq \psi_t \cdot T_t \hat{f}_1^u$ a.s., where ψ_t is the process defined below.

Then u is optimal.

PROOF

Again $E_{u*}[g(x(1))|\mathcal{Y}_t]$ is a square integrable $\mathcal{Y}_t$, P_{u*} martingale, so from Lemma 4.1

$$E_{u*}[g(s(1))|\mathcal{Y}_t] = J + \int_0^t \psi_t d\nu_t^{u*}.$$

Now

$$E_{u*}[g(x(1))] = J$$

and

$$E_{u*}[g(x(1))|\mathcal{Y}_1] = g(x(1)),$$

because $g(x(1))$ is $\mathcal{Y}_1$ measurable. Therefore, for any other control $u \in \mathcal{N}$

$$g(x(1)) = J + \int_0^1 \psi_s d\nu_s^u + \int_0^1 \psi_s T_s (\hat{f}_1^u - \hat{f}_1^{u*}) ds,$$

where

$$\hat{f}_1^u(s) = E_u[f_1^u(s)|\mathcal{Y}_s].$$

As in [2],

$$\int_0^t \psi_s d\nu_s^u$$

is a $\mathcal{Y}_t$, P_u martingale, so taking expectations with respect to P_u

$$E_u[g(x(1))] = J + \int_0^1 E_u[\psi_s T_s (\hat{f}_1^u - \hat{f}_1^{u*}) ds]$$

$$\geqslant J = E_{u*}[g(x(1))], \text{ by hypothesis.}$$

Therefore, $u^* \in \mathcal{N}$ is optimal.

5 REFERENCES

1. Davis, M. H. A. and Varaiya, P. P. "Dynamic programming conditions for partially observable stochastic systems", *SIAM Jour. Control,* **11**, 226-261, (1973).

2. Elliott, R. J. "The optimal control of a stochastic system", *SIAM Jour. Control,* **15**, 756-778, (1977).

3. Girsanov, I. V. "On transforming a certain class of stochastic processes by absolutely continuous substitution of measures", *Theory Prob. and Appl.,* **5**, 285-301, (1961).

1.3 THE PRINCIPLE OF OPTIMALITY FOR MARKOV JUMP PROCESSES

M. H. A. Davis and C. B. Wan
(Department of Computing and Control, Imperial College, London, UK)

ABSTRACT

One form of the "principle of optimality" in stochastic control states that the value function of dynamic programming is a martingale under an optimal control and a submartingale otherwise. This depends on the controls satisfying the "ε-lattice property" introduced by Striebel. While this property is immediate in the case of non-anticipative controls, it is less evident for Markovian controls. In this paper we show that it holds in the case of controlled Markov jump processes with bounded rates.

1 INTRODUCTION

Let (Ω,F,P) be a probability space, $(F_t)_{t\geq 0}$ an increasing family of sub-σ-fields of F, and (x_t) an F_t-adapted process. Suppose that control is exercised by choosing a process $u = \{u_t, t\geq 0\}$ from a family U of processes, with each of which is associated a measure P_u mutually absolutely continuous with respect to P. (This is the situation for the controlled jump processes described below or for the Girsanov formulation of stochastic differential equations discussed in [5].) Suppose the cost associated with $u\in U$ is $J(u) = E_u[g(x_{T_f})]$ where T_f is a fixed terminal time, g is a bounded function and E_u denotes expectation with respect to measure P_u. The *value function* for this problem is the F_t-adapted process

$$W_t = \bigwedge_{u\in U} E_u[g(x_{T_f})|F_t]$$

where "Λ" denotes the lattice infimum in $L_1(\Omega,F_t,P)$. Let $u\in U$ be a particular control and denote by $U(u,t+h)$ all controls that agree with u_s for $s \leqslant t+h$. Since this is a subclass of $\mathcal{U}$ and using the iteration property of conditional expectations we have

$$W_t \leqslant \bigwedge_{v\in\mathcal{U}(u,t+h)} E_u\{E_v[g(x_{T_f})|\mathcal{F}_{t+h}]|\mathcal{F}_t\} \qquad (1.1)$$

If we are allowed to interchange $\wedge$ and $E\{\cdot|\mathcal{F}_t\}$ then (1.1) gives the *submartingale inequality* for the value process W_t, and this is equivalent to Bellman's *principle of optimality* (see [3], [5]-[9]).

Striebel's contribution in [8] was to show that a sufficient condition for the legitimate interchange of infimum and conditional expectation is the so-called "ε-lattice property" (see § 2 below). This property is immediate in the case of "non-anticipative controls", as will be seen in Proposition 3.1 below. (This means, by the way, that the Byzantine complications of Lemma 3.1 in [5] were quite unnecessary.) On the other hand the ε-lattice property for "Markovian controls" is *not* immediate; the proof of Proposition 3.1 shows why. The purpose of this paper is to show that the ε-lattice property holds for systems of controlled Markovian jump processes with bounded rates, of the type considered in previous papers [6] [9]. Boel and Varaiya studied a similar problem in §5.5 of [3] and introduced an "assumption of approximation" under which the ε-lattice property holds; thus essentially the contribution of this paper is to state conditions under which this assumption is satisfied. It should be pointed out that no such results are, to the best of our knowledge, available for the stochastic differential equation case, so the proof of Lemma 6.1 of [5] is still incomplete.

The reason for studying Markovian controls in systems with a basically Markovian structure is that the value process for Markovian controls is then just a function of the current state, i.e. $W_t = W(t,x_t)$, and one easily concludes that non-anticipative controls cannot achieve lower cost. In an interesting paper on impulse control [7], Lepeltier and Marchal start with nonanticipating controls (for which the ε-lattice property is assured) and show directly that the value function depends only on the current state (and thus that Markovian controls are in fact minimizing). However the methods are specific to the impulse control set-up and it is not clear whether a similar approach could be used for other types of control problem.

In the next section we summarize Striebel's results on the ε-lattice property. In §3 we formulate the jump process control problem and state the main result, Theorem 3.3. The proof follows from two lemmas in §4, the second of which essentially states that under the given conditions Boel and Varaiya's Assumption A [3] is always satisfeid.

2 THE ε-LATTICE PROPERTY

Let (Ω,F,P) be a probability space and $\{f_\gamma : \gamma \in \Gamma\}$ be a subset $L_1(\Omega,F,P)$ such that $f_\gamma(\omega) \geq 0$ a.s. for each $\gamma \in \Gamma$. Denote by $\wedge$ the lattice infimum in L_1, and let λ_γ be the measure defined by $d\lambda_\gamma/dP = f_\gamma$. Striebel's results [8] are the following.

2.1 Lemma

$$\bigwedge_{\gamma\in\Gamma} f_\gamma = \frac{d\lambda}{dP}$$

Here λ is the measure defined by

$$\lambda(A) = \inf \sum_{i=1}^{n} \lambda_{\gamma_i}(A_i) \qquad A \in F$$

where the infimum is over n, $\gamma_i \ldots \gamma_n \in \Gamma$ and finite partitions $A_1 \ldots A_n$ of A.

2.2 Definition

Γ has the *ε-lattice property* for $\{f_\gamma\}$ if given $\gamma_1, \gamma_2 \in \Gamma$ and $\varepsilon > 0$ there exists $\gamma_0 \in \Gamma$ such that

$$f_{\gamma_0}(\omega) \leq f_{\gamma_1}(\omega) \wedge f_{\gamma_2}(\omega) + \varepsilon \qquad \text{a.s.}$$

2.3 Lemma

If Γ has the ε-lattice property then

$$\lambda(A) = \inf_{\gamma\in\Gamma} \lambda_\gamma(A)$$

2.4 Lemma

Let $\mathcal{B}$ be a sub-σ-field of $\mathcal{F}$ and suppose that Γ has the ε-lattice property for $\{E[f_\gamma|\mathcal{B}]\}$. Then

$$E\Big[\bigwedge_{\gamma\in\Gamma} f_\gamma \Big| \mathcal{B}\Big] = \bigwedge_{\gamma\in\Gamma} E[f_\gamma|\mathcal{B}] \quad \text{a.s.} \tag{2.5}$$

The proof of Lemma 2.4 runs as follows, given that γ has the ε-lattice property for $\{E[f_\gamma|\mathcal{B}]\}$.

Define

$$\lambda'_\gamma(A) = \int_A E[f_\gamma|\mathcal{B}]dP, \quad \lambda'(A) = \inf_{\gamma\in\Gamma} \lambda'_\gamma(A)$$

Then $\bigwedge_\gamma E[f_\gamma|\mathcal{B}] = d\lambda'/dP$ so that $d\lambda'/dP$ is $\mathcal{B}$-measurable. But for $B\in\mathcal{B}$, $\lambda'_\gamma(B) = \lambda_\gamma(B)$ and hence $\lambda'(B) = \lambda(B)$. Thus $d\lambda'/dP = E[d\lambda/dP|\mathcal{B}]$ and this is equivalent to (2.5).

3 FORMULATION OF CONTROLLED JUMP PROCESS

The jump process model considered here is a special case of the formulation given in [6], [9], and so we will only give brief details here. The jump process $\{x_t, t\geq 0\}$ takes values in a Blackwell space $(X,\mathcal{S})$ and is specified by a countable sequence of random variables $(T_0,Z_0,T_1,Z_1\ldots)$ defined on a probability space $(\Omega,\mathcal{F},P)$, where $\{T_i\}$ are the jump times and $\{Z_i\}$ the states, so that the sample path is given by

$$x_t = Z_i, \quad t \in [T_i, T_{i+1}[\qquad i=0,1\ldots$$

(It will always be the case that $0 = T_0 < T_1 \leqslant \ldots$ and $T_n \to \infty$ a.s.) Let $\mathcal{F}_t$ denote the generated family of σ-fields $\sigma\{x_s, s\leqslant t\}$. The fundamental family of counting processes associated with x_t is $\{p(.,A) : A \in \mathcal{S}\}$ defined by

$$p(t,A) = \sum_{\substack{s\leqslant t\\ x_s\neq x_{s-}}} I_{(x_s\in A)} = \sum_{T_i\leqslant t} I_{(Z_i\in A)}$$

Let $\lambda : R^+ \times X \times S \to [0,1]$ be a given function such that $\lambda(.,.,A)$ is measurable for each $A \in S$ and $\lambda(t,x,.)$ is a probability measure on (X,S) for each $(t,x) \in R^+ \times X$. We can then construct for each $x \in X$ a "base measure" P_x on (Ω, F_∞) such that $P_x[x_0=x] = 1$ and the "local description" of x_t under P_x is the pair $(\Lambda_t, \lambda_t(A)) = (t, \lambda(t,x_{t-},A))$, i.e. if we define

$$\tilde{p}(t,A) = \int_0^t \lambda(s,x_{s-},A)ds$$

then the process

$$q(t,A) = p(t,A) - \tilde{p}(t,A)$$

is an F_t-martingale for each $A \in S$. Under measure P_x, x_t is a regular step Markov process in the sense of [2, §1.12]. Note in particular the rate of the counting process $p(t,X) = \sum_i I_{(t \geq T_i)}$ is t so p(t,X) is a Poisson process by Watanabe's characterization [4].

Control of the jump process is effected by absolutely continuous change of measure.
Admissible controls U are F_t - predictable processes taking values in a given measurable space (U, B_u)
Markov controls U_m are measurable functions $u : R^+ \times X \to U$; each $u \in U_m$ defines an admissible control $\tilde{u} \in U$ by the recipe $\tilde{u}_t(\omega) = u(t, x_{t-}(\omega))$.

Let $\alpha : R^+ \times X \times U \to [c_1,K]$ and $\beta : R^+ \times X \times X \times U \to [c_1,K]$ be given measurable functions (here $0<c_1<K$) such that

$$\int_X \beta(t,x,y,u)\ \lambda(t,x,dy) = 1$$

for all $(t,x,u) \in R^+ \times X \times U$. Now take $T_f < \infty$ and $u \in U$ and define

$$L(u) = \prod_{\substack{x_s \neq x_{s-} \\ s \leq T_f}} \alpha(s,x_s,u_s)\beta(s,x_{s-},x_s,u_s)\exp\left(-\int_0^{T_f} (\alpha(s,x_{s-},u_s)-1)ds\right).$$

Then $L(u) > 0$ a.s. and it is shown in [6], [9] that $E_x L(u) = 1$; thus $L(u)$ defines a probability measure P_x^u on (Ω, F_{T_f}) by

$$\frac{dP_x^u}{dP_x} = L(u).$$

Under P_x^u, x_t is a jump process whose local description is the pair

$$\Lambda^u(t,\omega) = \int_0^t \alpha(s,x_{s-},u_s)ds$$

$$\lambda^u(t,\omega,A) = \int_A \beta(s,x_{s-},x,u_s)\lambda(s,x_{s-},dx),$$

i.e. $\left\{p(t,A) - \tilde{p}^u(t,A)\right\}$ is an (F_t, P_x^u) martingale, where

$$\tilde{p}^u(t,A) = \int_0^t \lambda^u(t,A)d\Lambda^u(t)$$

Thus the effect of the control is to produce an absolutely continuous change of local description, with $\alpha^u = d\Lambda^u/d\Lambda$, $\beta^u = d\lambda^u/d\lambda$. Note that if $u \in U_m$ then x_t is a regular step Markov process under measure P_x^u.

The cost $J(x,u)$ associated with a given starting point $x \in X$ and control $u \in U$ is defined as follows:
let $c : R^+ \times X \times U \to [0,K]$ be a given function and G_f a non-negative, F_{T_f} - measurable random variable, also bounded by K.
Then

$$J(x,u) = E_x^u\left[\int_0^{T_f} c(s,x_{s-},u_s)ds + G_f\right].$$

The objective is to choose a control so as to minimize the cost.

For $u \in U$ the *remaining cost process* is $\psi(u,t)$ defined by

$$\psi(u,t) = E_x^u\left[\int_t^{T_f} c(s,x_{s-},u_s)ds + G_f \middle| F_t\right]$$

It is shown in [6] that this only depends on u through the restriction of u to $[t,T_f]$. For fixed t, $\{\psi(u,t) : u \in U\}$ is a subset of $L_1(\Omega,F_t,P)$.

3.1 Proposition

$\{\psi(u,t) : u \in U\}$ has the O-lattice property.

Proof: Take $u^1, u^2 \in U$ and let

$A = \{\omega : \psi(u^1,t) \leqslant \psi(u^2,t)\}$

Now define†

$$u_s = \begin{cases} u_s^1 I_A + u_s^2 I_{A^c} & s \in [t,T_f] \\ \\ u_s^1 & s \in [0,t[\,. \end{cases} \tag{3.2}$$

Then clearly

$\psi(u,t) = \psi(u^1,t) \wedge \psi(u^2,t)$ a.s.

The above proof is included only to indicate why it does not work in the Markov case. Indeed, if $u^1, u^2 \in U_m$ then u_s defined by (3.2) is *not* in U_m. Thus some more subtle way of "mixing" u^1 and u^2 must be found.

For $u \in U_m$ and $t \in [0,T_f]$ there is, by the Markovian property, a function $\tilde{\psi}_{u,t}(y)$ such that

$\psi(u,t) = \tilde{\psi}_{u,t}(x_t)$ a.s.

---

† $A^c \equiv \Omega - A$ (the complement of A)

3.3 Theorem

For each $t \in [0,T_f]$ the set $\{\tilde{\psi}_{u,t} : u \in \mathcal{U}_m\}$ has the ε-lattice property.

Proof: It is clearly sufficient to establish the Theorem for the case t=0, and then

$$\tilde{\psi}_{u,t}(x) = J(x,u)$$

The fact that $\{J(.,u): u \in \mathcal{U}\}$ has the ε-lattice property follows directly from lemmas 4.1, 4.3 in the next section.

4 APPROXIMATING MARKOV CONTROLS

Fix a positive integer n and let $\Delta = T_f/2^n$. For $u \in \mathcal{U}_m$ let $\delta_n u$ be the control defined as follows

$$\delta_n u(t,\omega) = u(t, x_{k\Delta -}(\omega)), \quad t \in [k\Delta, (k+1)\Delta[$$

Then clearly $\delta_n u \in \mathcal{U}$ for $u \in \mathcal{U}_m$

4.1 Lemma

(cf. [3, Theorem 5.4]) $\{J(x,\delta_n u) : u \in \mathcal{U}_m\}$ has the 0-lattice property

Proof: This just uses discrete-time dynamic programming. Suppose $u^1, u^2 \in \mathcal{U}_m$ and for some k we have found a control $u \in \mathcal{U}_m$ such that

$$\psi(\delta_n u, (k+1)\Delta) \leqslant \psi(\delta_n u^1, (k+1)\Delta) \wedge \psi(\delta_n u^2, (k+1)\Delta) \quad \text{a.s.}$$

Now let $\tilde{u}^1, (\tilde{u}^2)$ be the concatenation of $\delta_n u^1 (\delta_n u^2)$ on $[k\Delta, (k+1)\Delta[$ with $\delta_n u$ on $[(k+1)\Delta, T_f]$. In view of the Markovian structure there are measurable functions $\eta^1(.)$, $\eta^2(.)$ such that

$$\psi(\tilde{u}^i, k\Delta) = \eta^i(x_{k\Delta}) \quad \text{a.s.,} \quad i = 1,2.$$

Let $B = \{x : \eta^1(x) \leq \eta^2(x)\}$ and define, for $t \in [k\Delta,(k+1)\Delta[$,

$u(t,x) = u^1(t,x)I_B(x) + u^2(t,x)I_{B^c}(x)$

Then $u \in \mathcal{U}_m$ and

$$\psi(\delta_n u,k\Delta) = \eta^1(x_{k\Delta}) \wedge \eta^2(x_{k\Delta}) \leq \psi(\delta_n u^1,k\Delta) \wedge \psi(\delta_n u^2,k\Delta) \quad \text{a.s.} \tag{4.2}$$

Thus by induction (4.2) holds for all k and taking k = 0 we see that there exists $u \in \mathcal{U}_m$ such that

$$\psi(\delta_n u,0) = J(x,\delta_n u) \leq J(x,\delta_n u^1) \wedge J(x,\delta_n u^2) \text{ for all } x \in X$$

This completes the proof.

In order to show that $\{J(x,u) : u \in \mathcal{U}_m\}$ has the ε-lattice property we need to show that the cost $J(x,\delta_n u)$ is close to $J(x,u)$ for large n. This was introduced as a hypothesis in [3] (Assumption A, p.110). The next lemma shows that it holds for our model, under the conditions stated.

4.3 Lemma

Take $u \in \mathcal{U}_m$ and $\varepsilon > 0$. Then there exists n' such that for $n \geq n'$

$$|J(x,u) - J(x,\delta_n u)| < \varepsilon \quad \text{for all } x \in X.$$

Proof: For $u \in \mathcal{U}_m$ denote

$$g(u) = \int_0^{T_f} c(s,x_{s-},u_s)ds.$$

Then

$$\begin{aligned} J(x,u) &= E_x^u[g(u) + G_f] \\ &= E_x[L(u)(g(u) + G_f)] \end{aligned}$$

so that

$$|J(x,u) - J(x,\delta_n u)|$$

$$= \left|E_x\left[L(u)(g(u) - g(\delta_n u))\right] + E_x\left[(L(u)-L(\delta_n u))(g(\delta_n u)+G_f)\right]\right|$$

$$\leq \left|E_x^u\left[g(u) - g(\delta_n u)\right]\right| + \left|E_x\left[(L(u) - L(\delta_n u))(g(\delta_n u) + G_f)\right]\right| \tag{4.4}$$

Now let

$$\chi_N = I_{(p(T_f,X)=N)}$$

(Recall that $p(T_f,X)$ is the total number of jumps in $[0,T_f]$.) Taking the first term on the right of (4.4) we have

$$g(u) - g(\delta_n u) = \int_0^{T_f}\left[c(s,x_{s-},u_s) - c(s,x_{s-},\delta_n u_s)\right]ds$$

and the integrand is non-zero only on intervals $[k\Delta, h(k+1)\Delta]$ in which jumps occur. Thus, on the set $(p(T_f,\chi) = N)$,

$$|g(u) - g(\delta_n u)| \leq K\Delta N$$

and hence

$$E_x^u|g(u) - g(\delta_n u)| = \sum_n E_x^u \chi_N|g(u) - g(\delta_n u)|$$

$$\leq \sum_N K\Delta N\, P_x^u\left[p(T_f,X) = N\right]$$

$$= K\Delta E_x^u\left[p(T_f,X)\right] \tag{4.5}$$

$$= K\Delta E_x^u\left[\tilde{p}^u(T_f,X)\right]$$

$$\leq K^2 T_f \Delta.$$

For the second term in 4.4, denote $Z = g(\delta_n u) + G_f$. Then $\chi_N Z$ is a.s. equal to an F_{T_N} - measurable random variable and hence

$$E_x \chi_N Z L(u) = E_x \chi_N Z L^{(N)}(u)$$

where $L^{(N)}$ is the restriction of $L(u)$ to $F_{T_f \wedge T_N}$, namely

$$L^{(N)}(u) = \pi(u) \exp(-\int_0^{T_f \wedge T_N} (\alpha(u)-1)ds)$$

where

$$\pi(u) = \prod_{\substack{x_s \neq x_{s-} \\ s \leq T_N \wedge T_f}} \alpha(s,x_{s-},u_s)\beta(s,x_{s-},x_s,u_s)$$

and

$$\alpha(u) = \alpha(s,u(s,x_{s-}),x_{s-}).$$

We thus have, using the fact that Z is bounded by $K(1+T_f)$,

$$\begin{aligned} |E_x(L(u) - L(\delta_n u))Z| &\leq \sum_N |E_x \chi_N Z(L^{(N)}(u) - L^{(N)}(\delta_n u))| \\ &\leq K(1+T_f) \sum_N E_x \chi_N |L^{(N)}(u) - L^{(N)}(\delta_n u)| \end{aligned} \tag{4.6}$$

Now let $A_{N,n}$ be the event that $p(T_f,x) = N$ and that no two of the jumps of x_t take place in the same time interval $[k\Delta,(k+1)\Delta[$. Note that, on $A_{N,u}$, $\pi(u) = \pi(\delta_n u)$ and that $\pi(u)$ is bounded by K^{2N}. Thus for $\omega \in A_{N,n}$,

$$\begin{aligned} |L^{(N)}(u) - L^{(N)}(\delta_n u)| &= \pi(u)e^{T_f}|\exp(\int_0^{T_f}\alpha(u)ds) - \exp(\int_0^{T_f}\alpha(\delta_n u)ds)| \\ &= \pi(u)e^{T_f}\exp(\tfrac{1}{2}\int_0^{T_f}(\alpha(u) + \alpha(\delta_n u))ds|\sinh(\tfrac{1}{2}\int_0^{T_f}(\alpha(u) - \alpha(\delta_u n))ds)| \\ &\leq K^{2N} e^{T_f} e^{K\Delta N} \sinh K\Delta N . \end{aligned}$$

Also $L^{(N)}(u)$ is itself bounded by some constant K', and hence

$$\sum_N E_x \chi_N |L^{(N)}(u) - L^{(N)}(\delta_n u)|$$

$$= \sum_N E_x \chi_N I_{A_{N,n}} |L^{(N)}(n) - L^{(N)}(\delta_n u)|$$

$$+ \sum_N E_x \chi_N I_{A^c_{N,n}} |L^{(N)}(n) - L^{(N)}(\delta_n u)| \tag{4.7}$$

$$\leqslant e^{T_f} E_x\left[K^{2p} e^{K\Delta p} \sinh(K\Delta p)\right] + K' \sum_N E_x\left[\chi_N I_{A^c_{N,n}}\right]$$

where $p := p(T_f,X)$. Recalling that $p(t,X)$ is (under measure P_x) a Poisson process we see that the first term in (4.7) is finite for any Δ and converges to 0 as $\Delta \to 0$. For the second term we have

$$\sum_N E_x\left[\chi_N I_{A^c_{N,n}}\right] = \sum_N P_x\left[A^c_{N,n} | p=N\right] P_x\left[p=N\right] \tag{4.8}$$

Since $p(t,x)$ is a Poisson process the jump times $T_1 \ldots T_N$ are, conditional on $[p=N]$, uniformly distributed on $[0,T_f]$. Thus

$$P_x\left[A_{N,n} | p=N\right] = (1-\Delta)(1-2\Delta)\ldots(1-N\Delta)$$

$$= 1 - \Delta(1+2+\ldots N) + o(\Delta)$$

$$= 1 - \tfrac{1}{2}\Delta N(N+1) + o(\Delta)$$

Therefore, for small Δ,

$$P\left[A^c_{N,n} | p=N\right] \leqslant \Delta N(N+1)$$

so that, from (4.8),

$$\sum_N E_x\left[\chi_N I_{A^c_{N,n}}\right] \leqslant \Delta E_x\left[p(p+1)\right] \tag{4.9}$$

Combining (4.4)-(4.9) we see that $|J(x,u) - J(x,\delta_n u)| \to 0$ as $n \to \infty$, uniformly in x. This completes the proof.

5 REFERENCES

1. Bismut, J. M. "Control of jump processes and applications", *Bull. Soc. Math. France,* **106**, 25-60, (1978).

2. Blumenthal, R. M. and Getoor, R. K. "Markov processes and potential theory", Academic Press, NY, (1968).

3. Boel, R. and Varaiya, P. "Optimal control of jump processes", *SIAM J. Control and Opt.,* **15**, 92-119, (1977).

4. Brémaud, P. "An extension of Watanabe's theorem of characterization of Poisson processes", *J. Appl. Prob.,* **12**, 396-399, (1975).

5. Davis, M. H. A. and Varaiya, P. "Dynamic programming conditions for partially - observable stochastic systems", *SIAM J. Control,* **11**, 226-261, (1973).

6. Davis, M. H. A. and Elliott, R. J. "Optimal control of a jump process", *Z. Wahrsch'theorie verw. Geb.,* **40**, 183-202, (1977).

7. Lepeltier, J. P. and Marchal, B. "Techniques probabilistes dans le control impulsionnel", *Stochastics,* **2**, 243-286, (1979).

8. Striebel, C. "Martingale conditions for the optimal control of continuous-time stochastic systems", Symposium on Nonlinear Estimation, San Diego, (1974).

9. Wan, C. B. and Davis, M. H. A. "Existence of optimal controls for stochastic jump processes", *SIAM J. Control and Opt.,* **17**, 511-522, (1979).

1.4 EXTREMALITY PROPERTIES OF THE OPTIMAL STRATEGY IN MARKOVIAN CONTROL PROBLEMS

D. Vermes

(Bolyai Institute, University of Szeged, Hungary)

1 INTRODUCTION

The aim of the present paper is to establish some necessary optimality conditions for Markovian control problems with continuous time.

The underlying controllable objects are described by Markov processes, their state evolution can (but need not) include diffusion, drift and jump components. We consider Markov strategies, i.e. the value of the control depends on the completely observable state only. We want to minimize the expected hitting time of a fixed target set. As we are interested in necessary optimality conditions, we assume the existence of an optimal strategy.

The starting point of our approach will be the Bellman equation together with some regularity properties of the optimal cost function. It is known ([7], [9], [10], [11], [12]) that for broad classes of Markovian control problems the Bellman equation is a necessary and sufficient condition of optimality and hence it is often regarded as the ace of stochastic control theory. But the determination of the optimal strategy via solution of the Bellman equation turns out to be extremely difficult, and can be carried out only in the most simple situations. The point is that the minimization in every state over the whole action space is an enormously laborious undertaking.

The aim of this paper is to prove some necessary optimality conditions, which essentially restrict the set of actions in which the value of the optimal strategy can lie. For processes with coefficients depending linearly on control the first result states that the optimal strategy cannot take values from the interior of the convex hull of the action space.

According to the second result the optimal strategy is either a continuous function of the state or it jumps between "neighbouring" points of the action space. These theorems are analogues of Weierstrass' extremality principle and the Weierstrass-Erdman jump conditions of the calculus of variations. As a consequence one can find a correspondence between the shape of the action space and the continuity of the optimal strategy.

The last result of the paper will be a stochastic bang-bang principle stating that the values of the optimal strategy can always be chosen from the extreme points of the action space. An immediate consequence of this theorem is that the optimum in the broader class of randomised strategies is not better than the optimum in the class of pure strategies. We prove all results for general non-linear problems. The validity of the bang-bang principle for non-linear stochastic systems is surprising, since it does not hold in the deterministic case. Notice also that the set of extreme points is neither closed nor evolves continuously in space even if the convex hull does so. Since the action space is in general infinite dimensional, it is important that we can show the bang-bang principle for the set of extreme points and not for its closure.

As the Bellman equation together with regularity properties of the cost function is a necessary and sufficient optimality criterion, all other necessary conditions can be deduced from it. We shall use this method. Another possible approach to similar extremity properties would proceed via the maximum principle of Kushner and Bismut (see the survey [2]). Notice that in our approach the regularity of the cost function plays a role not less important than the actual form of the equation itself. Papers which deduce the necessity of the Bellman equation assuming a priori these regularity (e.g. differentiability) properties could not serve as a basis for our discussion. Therefore we systematically refer to our previous paper [12].

A stochastic bang-bang principle was proved by Benes [1] for one-dimensional diffusions with controllable drift and by Johansen [8] for Markov chains with finitely many states. Time-optimal control was considered also by Brémaud [3] and Haussmann et. al. [7].

2 DIFFUSIONS

To make the fundamental ideas clear in this section we shall deal with only the time-optimal control of non-degenerate diffusions. For this widely investigated class not all results will be new; some of them follow from the maximum principle of Bismut or intersect with the results of Benes, Davis or Haussmann. But our approach which deduces the extremality conditions from the Bellman equation is new, and it can be applied practically without change for general Markovian control problems. This will be done in the third section.

Let the state space E be a bounded Lipschitzian domain in R^n, the action space Y a compact separable metric space. We assume that all strategies $u \in U$ are Borel measurable mappings from E into Y. A point $x \in E$ is called a continuity point of the strategy $u \in U$ if u is continuous in a neighbourhood of x. We call a strategy piecewise continuous if its continuity points have full Lebesgue measure in E, and we assume that all piecewise continuous strategies are admissible.

We consider the Markov process $\left(x_t^u, P_x^u\right)$ given on E as solution of the equation

$$dx_t^u = b(x_t, u(x_t))dt + a(x_t)dw_t, \quad x_0^u = x \tag{1}$$

and killed at the first exit time τ^u from E [5]. The coefficients $a: E \to R^{n\times n}$ and $b: E \times Y \to R^n$ are supposed to be bounded, Lipschitz-continuous functions, while w_t is an n-dimensional Brownian motion.

A strategy v is said to be optimal if it minimizes the expected exit time τ for any initial state, i.e. if $\psi(x) := E_x^v\tau^v \leq \inf_{u \in U} E_x^u \tau^u$. (Here E_x^u denotes the expectation w.r.t. P_x^u. If no confusion can occur we shall omit the superscript u,v from the random variables x_t^u, τ^v and denote the control-dependence only at the distributions e.g. $E_x^v\tau$, $P_x^v(x_t \in A)$.) As we are interested in necessary optimality conditions, we assume the existence of a piecewise continuous optimal strategy $v \in U$, and suppose that $E_x^v\tau^v < \infty$ for any $x \in E$. (The letter v will be reserved for the optimal strategy.)

Our starting point will be the fact that under the above assumptions (c.f. [6], [9], [12])

$$(*)\begin{cases}\text{the optimal cost function } \psi \text{ is twice continuously differentiable, } \psi(x) = 0 \text{ if } x \in \partial E \text{ and } \psi \text{ satisfies the Bellman equation} \\ \sum_{ij} a_{ij}(x)\psi_{x_i x_j}(x) + \sum_i b_i(x,v(x))\psi_{x_i}(x) + 1 = \\ \sum_{ij} a_{ij}(x)\psi_{x_i x_j}(x) + \min_{u(x) Y} \sum_i b_i(x,u(x))\psi_{x_i}(x) + 1 = 0 \\ \text{for any continuity point } x \text{ of } v.\end{cases}$$

Moreover if w is a strategy such that together with $\phi(x) := E_x^w \tau^w$ it possesses the same properties as v and ψ in (*) then w is optimal as well. For notational convenience we introduce the operators $A = \Sigma a_{ij}(x)\partial^2/\partial x_i \partial x_j$, $B^u = \Sigma b_i(x,u(x))\partial/\partial x_i$ and $L = A + B^u$. Then the Bellman equation can be written as $L^v\psi(x) + 1 = \min_{u(x)\in Y} L^u\psi(x) + 1 = 0$. We remark that in the proofs we use a bit less than (*). We do not need that $\psi \in C^2$; only that ψ is in the domain of the C-closure of $A|C^2$.

We want to deduce some necessary optimality conditions from (*). In nonlinear problems there is no hope of a direct connection between the optimality of the strategy and the geometrical properties of Y. Therefore we introduce some auxiliary notions. The subset $I(x) := \{b(x,y) : y \in Y\}$ of R^n will be called the indicatrix at point $x \in E$ c.f. [4]. The points of $I(x)$, denoted by $i^y(x)$ or $i(x,y)$, and will be referred to as "velocities". Keeping generalisations in mind in the present section we will not use the consequences of the fact that $I(x)$ is finite dimensional, we shall regard it as a subset of a Polish space. Hence hyperplanes will not be given by their normal vectors but by the corresponding continuous linear functionals. Of course for the problems of this section $i(x,y) \equiv b(x,y)$ and for any linear functional ℓ there is a $\xi \in R^n$ such that $\ell(w) = \langle \xi, w \rangle$. If the function $i(x,y)$ depends linearly on y, then Y and $I(x)$ are linearly isomorphic and in all subsequent statements the words "indicatrix" and "velocity" can be substituted by "action space" and "decision" respectively.

We call a point $x \in E$ trivial if at x every decision $y \in Y$ is equally effective, i.e. $L^y \psi(x) = -1$ for all $y \in Y$. Our first result shows that the optimal velocities necessarily take values from the boundary of the indicatrix. Theorem 1 can be regarded as the stochastic analog of Weierstrass' extremality principle from the calculus of variations [4].

Theorem 1. If x is a non-trivial continuity point of the optimal strategy then $i^v(x)$ cannot lie in the interior of the convex hull of $I(x)$.

Proof. Assume $i^v(x) \in \text{int co} I(x)$. This implies that $I(x)$ has points on both sides of every closed hyperplane going through $i^v(x) = i(x, v(x))$. More precisely for every non-trivial continuous linear functional ℓ defined on the linear hull of $I(x)$ with $\ell(i^v(x)) = c$ there exist velocities $w', w'' \in I(x)$ such that $\ell(w') < c$ and $\ell(w'') > c$. Consider the linear functional

$$\ell^*(w) := \sum_{i=1}^{m} w_i \psi_{x_i}(x). \tag{2}$$

Since x is a non-trivial point, ℓ^* is also non-trivial and of course continuous. $\ell^*(i^v(x)) = -1 - A\psi(x)$ by (*) and the existence of an action $y \in Y$ with $\ell^*(i(x,y)) < -1 - A\psi(x)$ would contradict the Bellman equation and hence the optimality of v. Q.e.d.

The following theorem will characterise the behaviour of the optimal strategy at discontinuity points as well. Let us call two points of a subset of a linear topological space S "neighbouring" if they have a common closed supporting hyperplane in S. Our result states that the optimal velocities either evolve continuously or perform jumps between neighbouring points of the indicatrix. In the calculus of variations the analogous result is known as the Weierstrass-Erdman jump condition [4].

Theorem 2. Let G be an arbitrary subset of E containing x_0 and denote by S the linear hull of $\cup_{x \in G} I(x)$. Consider all sequences $(x_n) \subset G$ of continuity points of v tending to x_0. If

x_0 is non-trivial then all limit points of the sequences $\{i^v(x_n);\ x_n \to x_0\}$ lie on the same closed hyperplane of S supporting $I(x_0)$.

Proof. We have to show the existence of a non-trivial continuous linear functional ℓ on S such that $\ell(w) \geq \lim_{x_n \to x_0} \ell(i^v(x_n))$ holds for all $w \in I(x_0)$ and that the limit on the right-hand side is independent of the sequence $x_n \to x_0$. Consider the functional ℓ^* defined by (2), and let (x_n) and (x'_m) be two sequences of continuity points tending to x_0. With the notation $\psi' = \text{grad}\psi$ we have

$$\ell^*(i^v(x_n) - i^v(x'_m)) = \langle b^v(x_n) - b^v(x'_m);\ \psi'(x)\rangle = \langle b^v(x_n);\ \psi'(x_n)\rangle -$$

$$-\langle b^v(x'_m);\ \psi'(x'_m)\rangle + \langle b^v(x_n);\ \psi'(x) - \psi'(x_n)\rangle - \langle b^v(x'_m);\ \psi'(x) - \psi'(x'_m)\rangle .$$

The difference of the first two terms is $A\psi(x'_m) - A\psi(x_n)$ and it tends to zero as $|x_n - x'_m| \to 0$, since $A\psi \in C$. By the continuity of $\psi'(x)$ and the boundedness of b the last two terms tend to zero too. Hence we have shown that all limit points lie on the same hyperplane. The assertion that ℓ^* supports $I(x_0)$ follows from the Bellman equation.

Theorems 1 and 2 have some direct consequences regarding the continuity properties of the optimal strategy. A set H is called strictly convex if every supporting hyperplane has only one common point with H. The set H is called strictly non-convex if its extreme points are all isolated from one another and H has supporting hyperplanes only in these extreme points.

Corollary. If in a region G of E the indicatrix $I(x)$ is strictly convex for all $x \in G$ then (except trivial points) the optimal velocity function i^v is continuous in G. If Y is strictly non-convex and $i(x,y)$ is linear in y then (except trivial points) the optimal strategy is piecewise constant.

One of the fundamental theorems of deterministic control theory is the bang-bang principle which states that the optimal strategy in linear time-optimal control problems takes a.e. values from the extreme points of Y. When it holds, this result is stronger than Weierstrass' extremality principle, but unfortunately it cannot be generalised to non-linear problems. In fact there are known examples where the value of i^v is always from the non-extremal boundary points of the indicatrix. All the more surprising then is the next theorem stating that in stochastic control theory the bang-bang principle holds also for non-linear problems.

Theorem 3. There exists a Lebesgue-measurable strategy not worse than the optimum v, such that $i^u(x)\in \mathrm{extr}\,I(x)$ for every $x\in E$.

Proof. Denote K a compact subset of the linear hull S of $\cup_{x\in E} I(x)$ such that $\cup_{x\in E} I(x)\subset K$. Since K is a separable metrisable space (recall that S is Polish) there exists a sequence (f_n) of continuous convex functions separating the points of K such that $\sup_{x,n}|f_n(x)|\leq 1$. Set $F_0:=\Sigma_1^\infty 2^{-n} f_n$ then F_0 is continuous and strictly convex. Define $F(x,w)$ on $E\times K$ by $F(x,w):=F_0(w)$ if $w\in I(x)$ and $F(x,w)=-\infty$ otherwise. Since $I(x)$ is compact for any $x\in E$ and depends continuously on x, function F will be upper semicontinuous. Denote $\mathcal{S}$ the set of continuous functions g on $E\times K$ such that for every fixed $x\in E$ $g(x,\cdot)$ is a concave function of the second variable. Define $G(x,w):=\inf\{g(x,w):\ g\in\mathcal{S},\ g\geq F\}$, then G is upper semicontinuous and $\{w\in K: G(x,w)=F(x,w)\neq -\infty\}=\mathrm{extr}\,I(x)$ holds true.

Consider the set

$$H:=\{(x,y)\in E\times Y: L^y\psi(x)=L^v\psi(x);G(x,i^y(x))=F(x,i^y(x))\neq-\infty\}.$$

Since all functions appearing in the definition of H are upper semicontinuous, H is the union of countably many compacts of ExY and the measurable choice theroem is applicable ([6],p.199). To show that the set

$$H_x:=\{y\in Y:(x,y)\in H\}=\{y:L^y\psi(x)=L^v\psi(x);\ i(x,y)\in \mathrm{extr}\,I(x)\}$$

is non-empty for every $x\in E$, observe that the continuous linear functional ℓ^* defined by (2) attains its infimum $L^v\psi(x)-A\psi(x)$ in an extreme point of $I(x)$. Finally the application of the measurable choice theorem to the set H shows the existence of a Lebesgue-measurable strategy u such that $L^u\psi(x)=L^v\psi(x)=-1$ and $i^u(x)\in\text{extr } I(x)$ for every $x\in E$. Since u is equally effective as v the theorem is proved.

From Theorem 3 we immediately see that if the optimal strategy is unique then it is necessarily extremal in the sense that $i^v(x)\in\text{extr} I(x)$.

Of the several consequences of Theorem 3 we formulate here only the following one. If i(x,y) depends linearly on y and Y is a polyhedron then the optimal strategy is necessarily piecewise constant in any region of E where it is unique.

3 GENERAL PROCESSES

First we recall in abstract form what we have used in the discussions of Section 2.

(I) (*) states that the C-infinitesimal operator L^u of the process x_t^u can be decomposed into the sum of a control-independent operator A and a control dependent operator B^u. The optimal cost function ψ belongs to the domain of L^u for any continuous strategy $u\in U$ and also for $u\equiv v$. Moreover ψ belongs to the domain of A too, consequently $A\psi\in C$ and $L^u\psi(x)=L^{u(x)}\psi(x)=A\psi(x)+B^{u(x)}\psi(x)$ for a.e. $x\in E$. Finally the Bellman equation holds in the form

$$L^v\psi(x)+1=\min_{u(x)\in Y} L^{u(x)}\psi(x)+1=0 \tag{3}$$

in every continuity point x of v.

(II) For any $x\in E$ there exists a continuous mapping i(x,y) of Y into a compact set $I(x)$ of a Polish space S, and a continuous linear functional ℓ_x^* on S such that $\ell_x^*(i(x,y))=B^y\psi(x)$ holds true. Moreover $\ell_x^*(w)$ depends continuously on x for any fixed $w\in S$. (The latter fact is used only in the proof of Theorem 2.)

If (I) and (II) hold true then Theorems 1,2 and 3 remain valid, their proofs can be repeated word for word.

We formulate now the general Markovian control problem treated in [12] for which (I) holds true (Theorem 1 of [12]).

Let the state space $E \subset R^n$ be a three-fold Cartesian product $E = E_1 \times E_2 \times E_3$. $E_1 \subset R^{n_1}$ ($n_1 = 0$ or 1) is a left-closed right-open interval if $n_1 = 1$ or a one-point set if $n_1 = 0$. E_1 corresponds to the at most one-dimensional deterministic state component. $E_2 \subset R^{n_2}$ ($0 \leq n_2 \leq n$) is a bounded Lipschitzian domain and serves as state space for the non-degenerate diffusion (or stable) part. Finally $E_3 \subset R^{n_3}$ with $0 \leq n_3 \leq n$ and $n_1 + n_2 + n_3 = n$ corresponds to the pure jump component of the process. In addition to [12] we assume here that E_3 contains only points of R^{n_3} with integer coordinates. We consider E provided with the topology induced from R^n. We denote the one-point compactification of E by E_Δ and the new "point at infinity" by Δ. C_0 denotes the set of continuous functions on E tending to zero as $x \to \Delta$, while C^k denotes the set of functions having k continuous partial derivatives w.r.t. the first $n_1 + n_2$ coordinates. We say that a property holds "a.e. in E" if for any $\xi \in E_3$ the intersection of the exceptional set with $R^{n_1+n_2} \times \{\xi\}$ has $n_1 + n_2$ dimensional Lebesgue-measure zero. The action space Y is a compact Polish space.

For $f \in C^\infty$ the characteristic operator L^y of the process x_t^y is of the form

$$L^y f(x) = \sum_{i,j=n_1+1}^{n_1+n_2} a_{ij}(x) f_{x_i x_j}(x) + \sum_{i=1}^{n_1+n_2} b_i(x,y) f_{x_i}(x) +$$

$$+ \int_E \left[f(x+z) - f(x)\right] Q_x^y(dz) + \int_G \left[f(x+z) - f(x) - \sum_{i=n_1+1}^{n_1+n_2} z_i f_{x_i}(x)\right] \tilde{\Pi}_x^y(dz) + \tag{4}$$

$$+ \int_{\overline{G}} \left[f(x+z) - f(x)\right] \tilde{\Pi}_x^y(dz)$$

with $\tilde{\Pi}_x^y(dz)=c(x,z)dz/|z|^{n_2+\alpha}+\Pi_x^y(dz)$ for some $0<\alpha<2$ and $G=E_2\cap\{z:|z|<1\}$, $\bar{G}=E_2\setminus G$. The objects defining L^y underly the following conditions.

(i) If $n_2\neq 0$ then one of the following cases hold true: $(a_{ij}(x))$ is a uniformly positive definite n_2-dimensional matrix (case "$\alpha=2$") or $c(x,z)$ is a uniformly positive real valued function (case "$\alpha<2$").

(ii) a_{ij}, b_i, c are bounded Lipschitz continuous functions of their arguments. If $n_1=1$ then b_1 is uniformly positive and does not depend on y. If $a=0$ and $\alpha\leq 1$ then $b_i\equiv 0$ for $i>n_1$.

(iii) Denote $M_\rho(H)$ the set of measures μ on $H\subset E$ such that $\int_H \phi_\rho(x)d\mu(x)<\infty$ with $\phi_\rho(x):=|x|_\rho^\delta/(1+|x|_\rho^\delta)$. We say that μ_n tend weakly to μ in $M_\rho(H)$ if $\mu_n(f)\to\mu(f)$ for any $f\in C_0$ such that f/ϕ_ρ remains bounded. With this notation $Q_x^y\in M_0(E)$, the set $\{Q_x^y: x\in E,\ y\in Y\}$ is contained in a weakly compact subset of $M_0(E)$. Further $\Pi_y^x\in M_\rho(E_2)$ for all $x\in E$, $y\in Y$ with the same $0<\rho<\alpha$, and $\sup_{x,y}\Pi_y^x(dz)\leq K\cdot dz/|z|^{n_2+\rho}$. Moreover both Q_x^y and $\tilde{\Pi}_x^y$ are weakly continuous functions of x and y in $M_0(E)$, $M_\rho(E_2)$ respectively, and $\{Q_x^y(f): y\in Y\}$, $\{\Pi_x^y(f\phi_\alpha): y\in Y\}$ are uniformly continuous families for any uniformly continuous function f.

Regarding the strategy space and the performance criterion we make the same assumptions as in Section 2. Under conditions (i), (ii), (iii) to every $u\in U$ there corresponds a unique Markov process x_t^u on E; we consider x_t^u as killed at the first exit time τ^u from E.

We define the mapping $i: E\times Y\to R^{n_2}\times M_0(E)\times M_\rho(E_2)$ by

$$i(x,y):=(b_{n_1+1}(x,y),\ldots,\ b_{n_1+n_2}(x,y),\ Q_x^y,\ \Pi_x^y) \tag{5}$$

In view of assumptions (ii), (iii) function $i(x,y)$ is continuous and the indicatrix $I(x):=\{i(x,y): y\in Y\}$ is compact in the metrisable product (weak) topology of $R^{n_2}\times M_0(E)\times M_\rho(E_2)$. In order to show (II) we prove the following lemma.

Lemma. For any $x\in E$

$$\ell_x^*(b,Q,\Pi) := \sum_{k=n_1+1}^{n_1+n_2} b_k\psi_{x_k}(x)+\int_E [\psi(x+z)-\psi(x)]Q(dz)+ \tag{6}$$
$$+\int_G [\psi(x+z)-\psi(x)-\sum_{k=n_1+1}^{n_1+n_2} z_k\psi_{x_k}(x)]\Pi(dz)+\int_{\bar{G}} [\psi(x+z)-\psi(x)]\Pi(dz)$$

defines a continuous linear functional on $S=R^{n_2}\times M_0(E)\times M_\rho(E_2)$. Moreover $\ell_x^*(w)$ is a continuous function of x for any fixed $w=(b,Q,\Pi)\in S$. (Derivatives ψ_{x_i} are to be interpreted in the generalised sense.)

Proof. We show that all the four terms of (6) define continuous linear functionals on the corresponding subspaces of S. First of all notice that $\psi\in C_0$ since the process $\overset{v}{x}_t$ is Fellerian and killed at Δ. Denote A the C-closure of the operator A on C^∞, consisting of the terms from (4) corresponding to (a_{ij}), b_1 (if $n_1=1$) and $c(x,z)/|z|^{n_2+\alpha}$. Theorem 1 of [12] states that $\psi\in\mathcal{D}(A)$. This implies that ψ is Hölderian with any exponent $\gamma<\alpha$. (Recall that we set $\alpha=2$ if (a_{ij}) is non-degenerate.) Especially if $\alpha>1$ then $\psi\in C^1$ and the lemma is proved for the first term of (6). To prove it for the third term too, we have only to remark that $\rho<\alpha$ and if ψ is ρ-Hölderian then $[\psi(x+z)-\psi(x)-\Sigma z_k\psi_{x_k}(x)]/\phi_\rho(z)$ is bounded. Since ψ is continuous in the (Polish) topology of E (recall that all the at most countably many points of E_3 are isolated) the second and fourth terms of (6) define continuous linear functionals on $M_0(E)$ and $M_\rho(\bar{G})=M_0(\bar{G})$ respectively. The continuity of $\ell_x^*(w)$ in x is obvious. Q.e.d.

As (I) is implied by Theorem 1 of [12] and (II) follows from the preceding lemma we have shown that Theorems 1, 2 and 3 of Section 2 remain valid for the more general Markovian control problem described in this section.

4 RANDOMISED STRATEGIES

Until now we were dealing with so called pure (or simple) strategies u mapping E into Y. In many problems it is meaningful to consider the broader class of mixed or randomised strategies. A mapping μ of E into the space of probability measures on Y is called a randomised strategy if for every Borel-set $\Gamma \subset Y$ the function $\mu_x(\Gamma)$ is measurable in x. With each randomised strategy μ we associate the Markov process determined by the characteristic operator L^μ of the form (4) but with functions $\hat{b}_i^\mu(x) := \int_Y \mu_x(dy) b_i(x,y)$, $\hat{Q}_x^\mu(\Gamma) := \int_Y \mu_x(dy) Q_x^y(\Gamma)$ and $\hat{\Pi}_x^\mu(\Gamma) := \int_Y \mu_x(dy) \Pi_x^y(\Gamma)$ instead of b_i, Q and Π respectively.

Having introduced the broader class of randomised strategies there arises the question, whether there exists a randomised strategy which is better than all simple strategies. The answer follows immediately from the bang-bang principle.

Theorem 4. There exists a Lebesgue measurable simple strategy which is as good as the optimal randomised strategy.
Proof. Suppose μ is the optimal randomised strategy. Then the corresponding velocities $i^\mu(x) = (\hat{b}^\mu(x), \hat{Q}_x^\mu, \hat{\Pi}_x^\mu)$ lie in $\mathrm{co}I(x)$. By Theorem 3 there exists a strategy ν not worse than μ such that i^ν takes its values from extr $\mathrm{co}I(x)$. But since the compact set $I(x)$ contains all extreme points of its convex hull, ν is simple. Q.e.d.

5 REFERENCES

1. Benes, V. E. "Girsanov functionals and optimal bang-bang laws for final value stochastic control", *Stoch. Proc. Appl.*, **2**, 127-140, (1973).

2. Bismut, J. M. "An introductory approach to duality in optimal stochastic control", *SIAM Review*, **20**, 62-78, (1978).

3. Brémaud, P. "Bang-bang control for point processes", *Advances in Appl. Probability*, **8**, 385-394, (1976).

4. Caratheodory, C. "Variationsrechnung", Leipzig, Berlin, (1935).

5. Dynkin, E. B. "Markov Processes", Springer Verlag, Berlin, (1965).

6. Fleming, W. H. and Rishel, R. W. "Deterministic and Stochastic Optimal Control", Springer Verlag, Berlin, (1975).

7. Haussmann, U. G., Anderson, W. F., Boyarsky, A. "A new stochastic time optimal control problem", *SIAM J. Control Optimization,* **16**, 1-15, (1978).

8. Johansen, S. "The bang-bang problem for stochastic matrices", *Z. Warsch. theorie und Verw. Geb.*, **26**, 191-195, (1973).

9. Krylov, N. V. "Controlled Processes of Diffusion Type", (In Russian), Nauka, Moscow, (1977).

10. Pragarauskas, H. "On Bellman equation for controlled discontinuous stochastic process". In"Proceedings of the Internat. Symp. on Stochastic Differential Equations, Vilnius, 1978", Lecture Notes in Mathematics, Springer Verlag, (1979).

11. Vermes, D. "A necessary and sufficient condition of optimality for Markovian control problems", *Acta. Sci. Math. (Szeged)*, **34**, 401-413, (1973).

12. Vermes, D. "On the semigroup theory of stochastic control", In "Proceedings of the Internat. Symp. on Stochastic Differential Equations, Vilnius, 1978", Lecture Notes in Mathematics, Springer Verlag, (1979).

1.5 DUALITY METHODS IN THE CONTROL OF SEMI-MARTINGALES

J-M. Bismut

(Université Paris-Sud Orsay, France)

1 INTRODUCTION

The purpose of this paper is to give a general convex analysis framework for the control of square-integrable semi-martingales. It is an extension of our previous work [1], where we were only concerned with the control of semi-martingales for which the martingale term is a stochastic integral with respect to a Brownian motion.

More precisely, let us consider a not necessarily continuous square-integrable martingale $w=(w_1,\ldots,w_m)$, such that its quadratic variation in the sense of [8, p. 267] satisfies $d\langle w_i,w_j\rangle=\delta_{ij}dt$. In the controlled stochastic differential equation

$$\begin{aligned} dx &= f(\omega,t,x^-,u)dt + \sigma(\omega,t,x^-,u)dw \\ x(0) &= x \end{aligned} \tag{1.1}$$

the distinction between predictable and optional controls becomes crucial. To take both cases simultaneously into account, we assume that J is a σ-field containing the predictable σ-field and included in the optional σ-field, and that u is required to be J-measurable. We want then to minimize

$$E\{\int_0^T L(\omega,t,x^-,u)\left(\frac{dt+d[w_1,w_1]+\ldots+d[w_m,w_m]}{m+1}\right)+\Psi(x_T)\} \tag{1.2}$$

$d[w_i,w_i]$ is the second quadratic variation of w [8, p.267] which allows the criterion to take account of the control of jumps.

As in [1], where we followed the general line of reasoning of Rockafellar [11] for deterministic control, we define a general problem of control of semi-martingales, which under convexity assumptions includes as a special case the control problem associated with (1.1-1.2).

In Section 2, the main notation and concepts of the paper are introduced. In Section 3, the control problem is defined in a way formally inspired by [11], and very similar to [1]. A dual problem is then defined. In Section 4, perturbation methods are introduced, which allow us to prove that both problems are dual to each other in the sense of Rockafellar [12, Chapter VI]. In Section 5, necessary and sufficient conditions are given for the solution of both problems. In Section 6, various applications are given, including the treatment of (1.1-1.2). A general Pontryagin maximum principle is obtained, which extends the maximum principle obtained in Theorem V.I of [1] and [2]. An application is given to the control of probability measure densities, which generalizes the control of Girsanov densities [2-6].

Since the paper is very technical in nature, the reader is advised to consult Section 6 for a justification of most of the technical assumptions. Reference to [4], where the linear-quadratic case is fully developed, could also be useful.

2 NOTATION

(Ω,F,P) is a complete probability space. $\{F_t\}_{t\geq 0}$ is an increasing right-continuous sequence of complete sub σ-fields of F. The symbol O (resp. P) denotes the optional (resp. predictable) σ-field on $\Omega\times R^+$ [7, IV-D61]. J is a σ-field of $\Omega\times R^+$ such that $P\subset J\subset O$ (in general J is equal to P or to O).

$w = (w_1,w_2,\ldots w_m)$ is a m-dimensional square-integrable martingale adapted to $\{F_t\}_{t\geq 0}$ such that, in the sense of [8, p. 267], we have

$$d\langle w_i,w_j\rangle = \delta_{ij}dt \tag{2.1}$$

$$\text{if } J \neq P,\ d[w_j,w_i]=0 \text{ when } i \neq j.$$

Expression (2.1) implies that w is quasi-left continuous.

V is a finite-dimensional Euclidian space. All the stochastic processes considered here are with values in V.

T is a positive constant.

For $0 \leq s \leq T$, L_2^s is the space of the F_s-measurable square-integrable random variables. L is the set of the square-integrable martingales null at time zero and stopped at time T. L is identified with a closed subspace of L_2^T, with the induced norm.

C_2^T is the space of the right-continuous processes x with left-hand limits, which are adapted and such that $E(\sup_{0 \leq t \leq T} |x_t|^2) < +\infty$.

dS is the measure on $\Omega \times R^+$ defined by

$$dS = \frac{1}{m+1} (dt + \sum_1^m d\,[w_i, w_i]). \qquad (2.2)$$

${}^J L^i_{22}$ (resp. ${}^J L_{22}$) is the quotient Hilbert space of the J-measurable processes H such that

$$\|H\|_i = (E \int_0^T |H|^2 \, d[w_i, w_i])^{1/2} < +\infty \qquad (2.3)$$

(resp.

$$\|H\| = (E \int_0^T |H|^2 \, dS)^{1/2} < +\infty) \qquad (2.3')$$

If J = P, dS may be replaced by dt in (2.3').

${}^J W$ is the subspace of L which consists of the martingales M which may be written

$$M_t = \sum_1^m \int_0^t H_i \, dw_i \qquad (2.4)$$

with $(H_1,\ldots,H_m)\in\prod_1^m {}^J L_{22}^{\,i}$. Because of (2.1), it follows that

$$\|M\|_{L_2^T} = (\sum_1^m \|H_i\|_i^2)^{1/2}. \tag{2.5}$$

${}^J W$ is a stable space in the sense of [8, p. 262], because $P\subset J$.

${}^J W^\perp$ is the orthogonal space to ${}^J W$ in L in the sense of [8, p. 262]. If J = O, by Proposition III.1 of [4], ${}^O W^\perp$ is the set of martingales $M\in L$ such that for $i = 1,\ldots,m$, $d[M,w_i] = 0$. We assume that ${}^J W^\perp$ is decomposed into the sum of two weakly orthogonal subspaces W_1 and W_2 (practically, W_1 will be either $\{0\}$ or ${}^J W$).

For $i = 0,\ldots,m$, d_i is the optional density

$$d_0 = \frac{dt}{dS}, \quad i = 0$$
$$d_i = \frac{d[w_i,w_i]}{dS}, \quad i \neq 0 \tag{2.6}$$

${}^j{}_i(.)$ is the conditional expectation operator for the measure dt when $i = 0$, and for the measure $d[w_i,w_i]$ when $i \neq 0$, relative to the σ-field J. ${}^j(.)$ is the conditional expectation operator for the measure dS relative to the σ-field J. ${}^j d_i$ is the expectation of the J-measurable density of (2.6). Note that ${}^P d_i = 1$.

${}^J L_{21}$ is the quotient space of the J-measurable processes u such that

$$\|u\|_{21} = (E(\int_0^T |u|\, dS)^2)^{1/2} < +\infty \tag{2.7}$$

${}^{J}L_{2_\infty}$ is the quotient space of the J-measurable processes y such that

$$\|y\|_{2_\infty} = (E\,(S.\underset{0 \leqslant t \leqslant T}{\text{ess sup}} |y_t|^2))^{1/2} < +\infty \tag{2.8}$$

where $S.\underset{0 \leqslant t \leqslant T}{\text{ess sup}} |y_t|^2$ is the random variable which associates to each ω the essential supremum of $|y_t|^2$ for the measure dS_ω.

If J = P, ${}^{P}L_{21}$ and ${}^{P}L_{2_\infty}$ may be replaced in the whole text by ${}^{P}L'_{21}$ and ${}^{P}L'_{2_\infty}$, where (2.7) and (2.8) are replaced by

$$\|u\|'_{21} = (E\,(\int_0^T |u|\,dt)^2)^{1/2} < +\infty \tag{2.7'}$$

$$\|y\|'_{2_\infty} = (E\,(\underset{0 \leqslant t \leqslant T}{\text{ess sup}} |y_t|^2))^{1/2} < +\infty \tag{2.8'}$$

and the ess sup in (2.8') is taken with respect to the measure dt.

Duality brackets are defined between ${}^{J}L_{21}$ and ${}^{J}L_{2_\infty}$ by

$$\langle u, y\rangle = E \int_0^T \langle u_t, y_t\rangle dS \tag{2.9}$$

and between ${}^{P}L'_{21}$ and ${}^{P}L_{2_\infty}$ by

$$\langle u, y\rangle = E \int_0^T \langle u_t, y_t\rangle \, dt \tag{2.9'}$$

All the Hilbert spaces considered are taken to be in duality with themselves by means of the scalar product.

Finally, if $c=(c_1 \ldots c_m)$ are constants and $v=(v_1 \ldots v_m)$ are vectors, cv is the family of vectors $(c_1 v_1 \ldots c_m v_m)$.

3 THE CONTROL PROBLEM

In this section, we define the control problem and its dual. The definitions will be very formal. We refer to Section 6 for an interpretation of the various conditions.

Let L be a normal convex integrand [9] on $(\Omega \times R^+) \times V \times V \times V^m$, where $\Omega \times R^+$ is endowed with the σ-field J, i.e. a function with values in $R \cup \{+\infty\}$, such that:-

1. For each $(\omega,t) \in \Omega \times R^+$, $L(\omega,t,.)$ is convex, lower semi-continuous and not identically $+\infty$.
2. There exists a countable collection C of J-measurable functions with values in $V \times V \times V^m$ such that, if $v \in C$, $L(\omega,t,v(\omega,t))$ is J-measurable, and moreover for each (ω,t), $C(\omega,t) \cap D(\omega,t)$ is dense in $D(\omega,t)$, where

$$D(\omega,t) = \{x \in V \times V \times V^m;\ L(\omega,t,x) < \infty\} \text{ and } C(\omega,t) = \{v(\omega,t); v \in C\}. \tag{3.1}$$

By [9], when u is measurable (resp. J-measurable) on $\Omega \times R^+$, $L(\omega,t,u(\omega,t))$ is also a measurable (resp. J-measurable) function.

Let L^* be the dual integrand of L, i.e.

$$L^*(\omega,t,v') = \sup_{v \in V \times V \times V^m} \langle v',v\rangle - L(\omega,t,v). \tag{3.2}$$

M is then defined by

$$M(\omega,t,p,s,H') = L^*(\omega,t,s,p,H'). \tag{3.3}$$

We make the following assumptions on L and M:-

1. For $(\omega,t)\in\Omega \times R^+$, if for a family $I\subset 1,..,m$, ${}^j d_i = 0$ when $i\in I$, then for any $(x,y,H)\in V \times V \times V^m$ with $H = (H_1,..,H_m)$,

if H^I is the same family of vectors as H except that the vectors with indices $i\in I$ have been replaced by zero, then

$$L(\omega,t,x,y,H^I) \leqslant L(\omega,t,x,y,H). \tag{3.4}$$

2. There exists (x_0,y_0,H_0) and (p_0,s_0,H'_0) in ${}^J L_{2_\infty} \times {}^J L_{21} \times \prod_1^m {}^J L^i_{22}$ such that

$$E\int_0^T L(\omega,t,x_0,y_0,\sqrt{{}^j d}\,H_0)\,ds < +\infty$$
$$E\int_0^T M(\omega,t,p_0,s_0,\sqrt{{}^j d}\,H'_0)\,ds < +\infty. \tag{3.5}$$

Notice that condition 1 is then also verified for M. In fact

$$M(\omega,t,p,s,H') = \sup <s,x>+<p,y> + <H',H> - L(\omega,t,x,y,H)$$
$$\geqslant \sup <s,x> + <p,y> + <H',H^I> - L(\omega,t,x,y,H^I)$$
$$\geqslant \sup <s,x> + <p,y> + <H'^I,H> - L(\omega,t,x,y,H) \tag{3.6}$$
$$= M(\omega,t,p,s,H'^I)$$

Condition 1 is a purely technical assumption which is verifiable in all practical cases (see Section 6).

Observe that all the spaces of measurable processes which are considered here are <u>decomposable</u> in the sense of Rockafellar [9], i.e. if we take ${}^J L_{21}$ as an example, then:-

(a) ${}^J L_{\infty\infty} \subset {}^J L_{21}$.

(b) If $x \in L_{21}$, $A \in J$, then $1_A x \in {}^J L_{21}$.

It is then possible to apply all the known results on normal convex integrals and duality given by Rockafellar in [9] and [10].

ℓ_O (resp. ℓ_T) is a lower semi-continuous convex function defined on L_2^O (resp. L_2^T) with values in $R\cup\{+\infty\}$, and not identically $+\infty$. Let ℓ_O^* and ℓ_T^* be their duals. We define ℓ and m on $L_2^O \times L_2^T$ by

$$\ell(c_O, c_T) = \ell_O(c_O) + \ell_T(c_T)$$
$$m(c_O, c_T) = \ell_O^*(c_O) + \ell_T^*(-c_T). \tag{3.7}$$

We define the following spaces

$$\mathcal{J}_R = L_2^O \times \mathcal{J}_{L_{21}} \times \prod_1^m \mathcal{J}_{L_{22}^i} \times \mathcal{J}_W^{\perp}$$

$$\mathcal{J}_{R_1} = L_2^O \times \mathcal{J}_{L_{21}} \times \prod_1^m \mathcal{J}_{L_{22}^i} \times W_1$$
$$\mathcal{J}_{R_2} = L_2^O \times \mathcal{J}_{L_{21}} \times \prod_1^m \mathcal{J}_{L_{22}^i} \times W_2 \tag{3.8}$$

To any $X = (x_O, \dot{x}, H, M) \in \mathcal{J}_R$, we associate the process x defined by

$$x_t = x_O + \int_O^t p\,\dot{x}\,ds + \int_O^t H\,dw + M_t. \tag{3.9}$$

Now x is in C_2^T. Indeed:-

(a) By Doob's inequality [14, VI, Remark 2], $\int_O^t H\,dw + M_t$ is in C_2^T.

b) It is easily checked that if $\dot{x} \in {}^J L_{21}$, $\int_0^t {}^p|\dot{x}|\,ds$ is the predictable compensator of $\int_0^t |\dot{x}|\,ds$. By [14, VII, T60-Remark b], we know that $\int_0^T {}^p|\dot{x}|\,dt$ is square-integrable. Knowing that $|{}^p\dot{x}| < {}^p|\dot{x}|$, it follows that $\int_0^t {}^p\dot{x}\,ds$ is also in C_2^T. In particular, if $\dot{x} \in {}^P L_{21}$, then $\dot{x} \in {}^P L'_{21}$.

The correspondence $X \longrightarrow x$ is one to one if $J = P$.

For simplicity of notation, we drop the dependence on ω of L, x, etc. in the sequel.

We now introduce the functionals $\Phi_{\ell,L}$ and $\Phi_{m,M}$.

<u>Definition 3.1</u>: $\Phi_{\ell,L}$ is the functional defined on ${}^J R$ by

$$X = (x_0,\dot{x},H,M) \longrightarrow \begin{cases} \ell(x_0,x_T) + E\int_0^T L(t,x^-,\dot{x},\sqrt{{}^J dH})\,ds & \text{if } X \in {}^J R_1 \\ +\infty \text{ otherwise.} \end{cases} \tag{3.10}$$

$\Phi_{m,M}$ is the functional obtained as $\Phi_{\ell,L}$ in (3.10) by replacing ℓ by m, L by M and ${}^J R_1$ by ${}^J R_2$.

The integral in (3.10) is well defined by (3.5) and [9].

$\Phi_{\ell,L}$ and $\Phi_{m,M}$ are convex functions on ${}^J R$.

We now define the control problem

<u>Definition 3.2</u>: Problem (P) is the minimization of $\Phi_{\ell,L}$ on ${}^J R$. Its dual problem (P') is the minimization of $\Phi_{m,M}$ on ${}^J R$.

It is easily verified that the dual problem of problem (P') is problem (P).

The introduction of the normalization factor $\sqrt{{}^{j}d}$ in the integrals can be justified on the grounds that for H to be in ${}^{J}L^{i}_{22}$, it is necessary and sufficient that $\sqrt{{}^{j}d}\,H$ is in ${}^{J}L_{22}$. H_i being an equivalence class for the measure $d[w_i,w_i]$, it would not make sense to introduce it directly in the integrand because dS and $d[w_i,w_i]$ are generally not equivalent. If $J = P$, $\sqrt{{}^{P}d_i} = 1$.

If the dependence of L on (ω,t) is optional, it is possible to introduce the conditional expectation ${}^{j}L$ of the integrand L with respect to J for the measure dS as in [13], and to come back to the previous formulation of the problem. However, in [13], ${}^{J}L$ is defined only if O and J are completed with the null sets of O for the measure dS. ${}^{j}L(t,x^-,\dot{x}\sqrt{{}^{j}d}\,H)$ would then be measurable relative to the augmented σ-field J^*. If $J = P$, one must be careful in formally writing that

$$E\int_0^T L(t,x^-,\dot{x},\sqrt{{}^{P}d}\,H)\,dS = E\int_0^t {}^{P}L(t,x^-,\dot{x},\sqrt{{}^{P}d}\,H)\,dt \tag{3.11}$$

because dS and dt do not have the same null sets but only the same P-measurable null sets. Equality makes sense only if a P-measurable representive of ${}^{P}L(t,x^-,\dot{x},\sqrt{{}^{P}d}\,H)$ is taken.

4 PERTURBATION METHODS

As in [1] and [11], we introduce a perturbation in the problem (P) which will make problem (P') the dual of Problem (P) in the sense of Rockafellar [12, Chapter VI].

Let ${}^{J}R'$ be the space

$${}^{J}R' = {}^{J}L_{2_\infty} \times L_2^T \tag{4.1}$$

A duality is defined between ${}^{J}R$ and ${}^{J}R'$ by

$$\langle X, (y,b)\rangle = E\int_0^T \langle \dot{x},y\rangle\, dS + E\langle x_T, b\rangle. \tag{4.2}$$

If J = P and if ${}^P L_{21}$ and ${}^P L_{2_\infty}$ are replaced by ${}^P L'_{21}$ and ${}^P L'_{2_\infty}$ in (4.2), dS is replaced by dt. This duality defines locally convex topologies on R and R' which are Hausdorff. Indeed:-

(a) If for any $(y,b)\in {}^J R'$, $\langle X,(y,b)\rangle = 0$, then $\dot{x} = 0$, $x_T = 0$. The martingale x_t is then equal to 0, and x_0, H, M are all zero.

(b) If for any X in ${}^J R$, $\langle X, (y,b)\rangle = 0$, then y = 0 and, using the decomposition of the square integrable martingale $E^{F_t} b$ given in [8, p. 262],

$$E^{F_t} b = b_0 + N + N' \quad (b_0,N,N')\in L_2^0 \times {}^J W \times {}^J W^{\perp}, \tag{4.3}$$

from which it is easily seen that b = 0.

<u>Definition 4.1</u>: For $(y,b)\in {}^J R'$, the functional $\Phi^{y,b}_{\ell,L}$ is defined on ${}^J R$ by

$$X = (x_0,\dot{x},H,M) \longrightarrow \begin{cases} \ell(x_0,x_T-b) + E\int_0^T L(t,x^-+y,\dot{x},\sqrt{j}\,d\,H)\,dS & \text{if } X\in {}^J R_1 \\ +\infty & \text{otherwise.} \end{cases} \tag{4.4}$$

$\Phi^{y,b}_{m,M}$ is defined similarly with the modifications indicated previously in Definition 3.1.

The functionals $\phi_{\ell,L}$ and $\phi_{m,M}$ are defined by

$$\begin{aligned} \phi_{\ell,L}(y,b) &= \inf_{X\in {}^J R} \Phi^{y,b}_{\ell,L}(X) \\ \phi_{m,M}(y,b) &= \inf_{P\in {}^J R} \Phi^{y,b}_{m,M}(P). \end{aligned} \tag{4.5}$$

We have then the fundamental result:

Theorem 4.1: $\phi_{\ell,L}$ and $\phi_{m,M}$ are convex functions on ${}^{J}R'$, and their duals on ${}^{J}R$ are given by

$$\phi^{*}_{\ell,L} = \Phi_{m,M}$$
$$\phi^{*}_{m,M} = \Phi_{\ell,L}. \tag{4.6}$$

Further, one has

$$\Phi^{*}_{\ell,L}(y,b) = \liminf_{(y',b')\to(y,b)} \phi_{m,M}(y',b') \tag{4.7}$$

except in the case where $\Phi_{\ell,L}$ is identically $+\infty$, and $\phi_{m,M}$ is equal to $+\infty$ on a neighbourhood of (y,b) (the topology is any locally convex topology compatible with the duality ${}^{J}R$, ${}^{J}R'$). The corresponding result also holds for $\Phi^{*}_{m,M}$.

Proof: The proof follows that of Theorem III.1 of [1]. For $P = (p_O,\dot{p},H',M')\in {}^{J}R$, we have

$$\Phi^{*}_{1,L}(P) = \sup_{\substack{X\in {}^{J}R_1 \\ (y,b)\in {}^{J}R'}} E\int_O^T \langle\dot{p},y\rangle\, dS + E\langle p_T,b\rangle - E\int_O^T L(t,x^{-}+y,\dot{x},\sqrt{{}^{j}d\, H})\, dS$$
$$-\ell_O(x_O) - \ell_T(x_T-b). \tag{4.8}$$

But, since x^{-} defines an element of ${}^{J}L_{2_\infty}$ (or of ${}^{p}L'_{2_\infty}$), it follows that

$$\phi^{*}_{\ell,L}(P) = \sup_{\substack{X\in {}^{J}R_1 \\ (z,b')\in {}^{J}R'}} E\int_O^T \langle\dot{p},z\rangle dS - E\int_O^T \langle\dot{p},x^{-}\rangle dS - E\langle p_T,b'\rangle + E\langle p_T,x_T\rangle$$
$$- E\int_O^T L(t,z,\dot{x},\sqrt{{}^{j}d\, H})\, dS - \ell_O(x_O)-\ell_T(b'). \tag{4.9}$$

x and p are given by

$$x_t = x_o + \int_o^t {}^p\dot{x}\, ds + \int_o^t H\, dw + M_{1t} \qquad M_1 \in W_1$$

$$p_t = p_o + \int_o^t {}^p\dot{p}\, ds + \int_o^t H'\, dw + M'_{1t} + M'_{2t} \quad (M'_1, M'_2) \in W_1 \times W_2. \tag{4.10}$$

(a) In the general case, where J is not necessarily equal to P, by Proposition III.2 of [4], we have

$$E < p_T, x_T> = E(<p_o, x_o> + \int_o^T <{}^p\dot{p}, x^- > ds + \int_o^T <p^-, {}^p\dot{x}>ds$$

$$+ \int_o^T <H,H'>d[w,w] + < M_{1T}, M'_{1T}>). \tag{4.11}$$

From (4.11) we see that

$$\phi^*_{\ell,L}(P) = \sup_{M_1 \in W_1} < M'_{1T}, M_{1T}> + \sup_{(\dot{x},H,z)\in {}^J L_{21} \times \prod_1^m {}^J L_{22}^i \times {}^J L_{2_\infty}} E(\int_o^T <\dot{p},z> dS + \int_o^T <p^-, \dot{x}>dS +$$

$$+ \int_o^T {}^j d <H,H'>dS) - E\int_o^T L(t,z,\dot{x},\sqrt{{}^j d}\, H)\, dS$$

$$+ \sup_{(x_o,b')\in L_2^O \times L_2^T} E < p_o, x_o> - E<p_T, b' > - \ell_o(x_o) - \ell_T(b'). \tag{4.12}$$

It is clear that

$$\sup_{M_1 \in W_1} \langle M'_{1T}, M_{1T} \rangle = \begin{cases} 0 & \text{if } M'_1 = 0 \\ +\infty & \text{otherwise.} \end{cases} \tag{4.13}$$

Moreover in (4.12), we may replace $\prod_1^m {}^{J}L_{22}^{i}$ by the set of the dS equivalence classes of processes $(H_1, \ldots H_m)$ such that

$$\sum_1^m E \int_0^T d_i |H_i|^2 \, dS < +\infty, \tag{4.14}$$

and this last space is clearly decomposable.

Let $\tilde{L}$ be the convex integrand

$$\tilde{L}(t,x,y,H) = L(t,x,y,\sqrt{{}^j d}\, H). \tag{4.14}$$

If ${}^j d_i \neq 0$ for every $i \in \{1..m\}$, then

$$\tilde{L}^*(t,s,p,{}^j dH') = L^*(t,s,p,\sqrt{{}^j d}\, H'). \tag{4.15}$$

If ${}^j d_i = 0$ for $i \in I$ then by (3.4)

$$\tilde{L}^*(t,s,p,{}^j dH') = \sup_{(x,y,H) \in V \times V \times V^m} \langle s,x\rangle + \langle p,y\rangle + \langle {}^j dH'^I, H\rangle - L(t,x,y,\sqrt{{}^j d}\, H^I)$$

$$= \sup_{(x,y,H) \in V \times V \times V^m} \langle s,x\rangle + \langle p,y\rangle + \langle \sqrt{{}^j d}\, H', H\rangle - L(t,x,y,H) \tag{4.16}$$

$$= L^*(t,s,p,\sqrt{{}^j d}\, H').$$

Then (4.15) holds in every case. It follows from [9] that the second sup in (4.12) is equal to

$$E \int_0^T M(t,p^-,\dot{p},\sqrt{{}^j d}\, H') \, dS. \tag{4.17}$$

The third sup in (4.12) is trivially equal to

$$\ell_O^*(p_O) + \ell_T^*(-p_T). \tag{4.18}$$

(b) In the case where $J = P$ and where $d[w_i, w_j]$ is not necessarily equal to O for $i \neq j$, we may apply Proposition I.1 of [4] to obtain, instead of (4.11)

$$E\langle p_T, x_T\rangle = E(\langle p_O, x_O\rangle + \int_O^T \langle \dot{p}, x^-\rangle\, ds + \int_O^T \langle p^-, \dot{x}\rangle\, ds + \int_O^T \langle H, H'\rangle\, ds$$

$$+ \langle M_{1T}, M'_{1T}\rangle). \tag{4.11'}$$

The proof continues as in the previous case. □

5 DUALITY OF INFIMA

When $X \in {}^J R_1$ and $P \in {}^J R_2$, we have

$$\ell_O(x_O) + \ell_T(x_T) + \ell_O^*(p_O) + \ell_T^*(-p_T) \geq E(\langle p_O, x_O\rangle - \langle p_T, x_T\rangle). \tag{5.1}$$

Moreover

$$E\int_O^T L(t, x^-, \dot{x}, \sqrt{}^j d\ H)\, dS + E\int_O^T M(t, p^-, \dot{p}, \sqrt{}^j d\ H')\, ds \geq E(\int_O^T \langle x^-, \dot{p}\rangle\, dS$$

$$+ \int_O^T \langle \dot{x}, p^-\rangle\, dS + \int_O^T d\langle H, H'\rangle\, dS) = E(\langle p_T, x_T\rangle - \langle p_O, x_O\rangle). \tag{5.2}$$

We thus have

<u>Proposition 5.1</u>: For any $(X, P) \in {}^J R \times {}^J R$,

$$\Phi_{\ell, L}(X) + \Phi_{m, M}(P) \geq O. \tag{5.3}$$

In particular

$$\inf_{X \in {}^J R} \Phi_{\ell,L}(X) \geqslant -\inf_{P \in {}^J R} \Phi_{m,M}(P). \tag{5.4}$$

Theorem 5.1: If $\Phi_{\ell,L}$ and $\Phi_{m,M}$ are not identically $+\infty$, the following conditions are equivalent:-

(a) $\inf_{X \in {}^J R} \Phi_{\ell,L}(X) = -\inf_{P \in {}^J R} \Phi_{m,M}(P)$.

(b) $\inf_{X \in {}^J R} \Phi_{\ell,L}(X) = \liminf_{(y,b)\to 0} \inf_{X \in {}^J R} \Phi^{y,b}_{\ell,L}(X)$. (5.5)

(c) $\inf_{P \in {}^J R} \Phi_{m,M}(P) = \liminf_{(y,b)\to 0} \inf_{P \in {}^J R} \Phi^{y,b}_{m,M}(P)$.

Proof: By Theorem 4.1, each of the stated relations is equivalent to the lower semi-continuity at zero of $\phi_{\ell,L}$ and $\phi_{m,M}$ (see [11] and [12]). □

We then give the following definition of coextremality:

Definition 5.1: $X \in {}^J R_1$ and $P \in {}^J R_2$ will be said to be coextremal if †

$$\begin{aligned} &(\dot{p}, p^-, \sqrt{j}_d H') \in \partial L(t, x^-, \dot{x}, \sqrt{j}_d H) \text{ a.e. } (dS) \\ &p_O \in \partial \ell_O(x_O), \; -p_T \in \partial \ell_T(x_T). \end{aligned} \tag{5.6}$$

The conditions (5.6) may be written

$$\begin{aligned} &(\dot{x}, x^-, \sqrt{j}_D H) \in \partial M(t, p^-, \dot{p}, \sqrt{j}_d H') \text{ a.e. } (dS) \\ &x_O \in \partial \ell^*_O(p_O), \quad x_T \in \partial \ell^*_T(-p_T). \end{aligned} \tag{5.6'}$$

† (∂L is the subdifferential of L).

Coextremality is then symmetrical in X and P.

Theorem 5.2: The following assertions are equivalent:-

(a) X and P are coextremal.

(b) X minimizes $\Phi_{\ell,L}$, P minimizes $\Phi_{m,M}$ and the equivalent conditions of Theorem 5.1 are verified.

Proof: This is obvious because equality in (5.1) and (5.2) is equivalent to each of the stated properties. □

6 APPLICATIONS

We assume that the assumptions of Section 2 hold.

(a) The linear quadratic case [2,3,4]

(A,B,C,D,M) are matrix-valued J-measurable bounded processes. N is a matrix-valued J-measurable bounded process with symmetric values such that

$$\langle Nu,u\rangle \geqslant \lambda |u|^2 \tag{6.1}$$

for some $\lambda > 0$.

M_1 is a matrix-valued F_T-measurable bounded random variable. We consider the following equation

$$\begin{aligned} dx &= {}^{P}(Ax^- + Cu)\,dt + (Bx^- + Du)\,dw \\ x(0) &= x_0 \end{aligned} \tag{6.2}$$

with $x_0 \in L_i$.

We want to minimize

$$\frac{1}{2}\left(E\int_0^T (|Mx^-|^2 + \langle Nu,u\rangle)\,ds + E|M_1 x_T|^2\right), \tag{6.3}$$

when u is in ${}^{J}L_{22}$ (u takes its values in a vector space U).

Let L be defined by

$$L(t,x,y,H) = \inf \frac{1}{2} (|Mx|^2 + \langle Nu,u\rangle)$$

$$\begin{cases} u \in U \\ y = Ax + Cu \\ H = \sqrt{{}^j d}\,(Bx + Du). \end{cases} \tag{6.4}$$

It is easily checked that L is a normal convex integrand. Moreover (3.4) is trivially verified, and L(t,O) and L*(t,O) are both zero. (3.5) is then also verified.

ℓ_O and ℓ_T are then defined by:

$$\ell_O(x) = \begin{cases} O \text{ if } x = x_O \\ + \infty \text{ otherwise,} \end{cases} \tag{6.5}$$

$$\ell_T(x) = \frac{1}{2} E|M_1 x|^2.$$

By Theorem III.1 of [4], if $u \in {}^J L_{22}$, (6.2) has a unique solution and x is in C_2^T. The minimization of (6.3) is then equivalent to the minimization of $\Phi_{\ell,L}$ with $W_1 = \{O\}$.

If M and M_1 have inverses, the dual problem consists of the minimization of

$$E\langle p_O, x_O\rangle + \frac{1}{2} \Big(E\int_O^T (|M^{*-1}(\dot{p} + A^* p^- + {}^j d\, B^* H)|^2 + \langle N^{-1}(C^* p^- + {}^j dD^* H), \tag{6.6}$$

$$C^* p^- + {}^j dD^* H\rangle)\, ds + E|M^{*-1} p_T{}^2|\Big)$$

where p may be written as

$$p_t = p_O + \int_O^t p_{\dot{p}}\, ds + \int_O^t H\, dw + M_t' \tag{6.7}$$

$$\text{with } (p_O, \dot{p}, H, M') \in L_2^O \times {}^J L_{21} \times \prod_1^m {}^J L_{22}^i \times {}^J W^{\perp}. \qquad (6.7)$$

X corresponding to u in (6.2) and $P \in {}^J R_2$ will be coextremal if:-

(1) (6.2) is verified.

(2) The following relations hold

$$dp = {}^P(M^* M x^- - A^* p^- - {}^j d\, B^* H)\,dt + H\,dw + dM'$$

$$p_T = -M_1^* M_1 x_T \quad \text{a.e.} \qquad (6.8)$$

$$Nu = C^* p^- + {}^j d\, D^* H \quad \text{a.e. } (dS).$$

By noticing that

$$^P({}^j d\, B^* H) = {}^P(dB^* H) = \sum_1^m {}^P i\, B_i\, H_i,$$

we see that if M is P-measurable, (6.8) is equivalent to (3.36) in [4].

More generally, if K is a positive finite convex integrand on $(\Omega \times R^+, J) \times V \times U$, such that $K(t,x,u) \leqslant h(x)$, where h is a fixed finite function on V, we want to minimize

$$E\int_O^T K(t,x^-,u)\,dS, \qquad (6.9)$$

where x satisfies (6.2) and where u takes its values in a compact convex set U_d included in U.

To put the problem in its standard form, we define again L as in (6.4), while adding the constraint $u \in U_d$. Moreover, if $x \in V$, $u \in U_d$, $L(t,x,Ax + Cu, Bx + Du)$ is bounded, and $L^*(t,O)$ is $\leqslant$ O. (3.5) is then verified.

Let K' be the integrand

$$K'(t,x,u) = K(t,x,u) + \ell_{U_d}(u) \tag{6.10}$$

where ℓ_{U_d} is O on U_d and $+\infty$ elsewhere.

The dual problem consists then of the minimization of

$$E\langle p_O, x_O\rangle + E\int_O^T K'^*(t,k,\ C^*p^- + {}^j dD^*H)\,dS \tag{6.11}$$

where $k \in {}^J L_{21}$ and p is the unique solution in the sense of Theorem III.2 of [4] of

$$\begin{aligned} dp &= {}^p(k - A^*p^- - dB^*H)dt + H\,dw + dM' \\ p_T &= O. \end{aligned} \tag{6.12}$$

The coextremality conditions may be written as

$$(k, C^*p^- + {}^j d\ D^*H) \in \partial\ K'(t,x^-,u) \text{ a.e. } (dS). \tag{6.12}$$

(b) <u>The maximum principle</u>

U is now a compact metrizable space. Let us consider the stochastic differential equation

$$\begin{aligned} dx &= {}^p f(\omega,t,x^-,u)dt + \sigma(\omega,t,x^-,u)dw \\ x(O) &= x_O \in L_2^O \end{aligned} \tag{6.13}$$

where b and σ are J-measurable in (ω,t), uniformly Lipschitz in x^-, continuous in u, and satisfy a growth condition

$$(|f| + |\sigma|)(\omega,t,x,u) \leqslant k(1 + |x|) \tag{6.14}$$

When u is O-measurable, it is easily seen that (6.13) has a unique solution by using standard approximation techniques, as in [4, Theorem III.1].

K is a nonnegative function on $\Omega \times R^+ \times V \times U$, which is J-measurable in (ω,t), continuous in (x,u), and such that $K(\omega,t,x,u) \leqq k(1+|x|^r)$, $r \geqq 1$. Finally Ψ is a convex function on V, which is finite everywhere.

We want to minimize

$$E\int_0^T K(t,x^-,u)\,dS + E\,\Psi(x_T) \tag{6.15}$$

when u is a J-measurable control.

Let us define L by

$$L(\omega,t,x,y,H)=\inf K(\omega,t,x,u)$$

$$\begin{cases} u\in U \\ y=f(\omega,t,x,u) \\ H=\sqrt{{}^{j}d}\;\sigma\;(\omega,t,x,u). \end{cases} \tag{6.16}$$

If L is convex, L is a normal convex integrand, because L is l.s.c., moreover, if (x_n,u_n) is a countable dense family in V x U, for each (ω,t), $\{x_n,f(\omega,t,x_n,u_n),\sqrt{{}^{j}d}\;\sigma(\omega,t,x_n,u_n)\}$ is dense in $D(\omega,t)$.L and M then satisfy (3.5) as previously indicated, and (3.4) is trivially verified.

By defining ℓ_0 as in (6.5) and $\ell_T(x)$ as $E\,\Psi(x)$, this problem is then equivalent to the minimization of $\Phi_{\ell,L}$, with $W_1 = \{0\}$.

Let us write the coextremality conditions between $X\in{}^{J}R_1$ and $P\in{}^{J}R_2$ as

$$\begin{aligned} &\langle x^-,\dot{p}\rangle+\langle\dot{x},p^-\rangle + {}^{j}d\langle H,H'\rangle-L(t,x^-,\dot{x},\sqrt{{}^{j}d}\;H)=L^*(t,\dot{p},p^-,\sqrt{{}^{j}d}\;H') \\ &p_T\in-\frac{\partial\Psi}{\partial x}(x_T). \end{aligned} \tag{6.17}$$

Let u_0 be a J-measurable process such that equality holds a.e. (dS) in (6.16) (i.e. u_0 corresponds to the optimum X); such a process exists by Theorem 2 of [10], see [1, p.393].

Then

$$L^*(t,\dot{p},p^-,\sqrt{\;}^j d\,H') = \sup_{(x,u)\in U\times V} <x,\dot{p}> + <f(t,x,u),p^-> + d< H',$$

(6.18)

$\sigma(t,x,u)>-K(t,x,u)$.

We have then the following essential result:

Theorem 6.1: Under the previous assumptions, if b,σ,K are differentiable in x^-, let H be the random function

$$H = <f(t,x^-,u),p^-> + {}^j d<\sigma(t,x^-,u),H'> - K(t,x^-,u). \qquad (6.19)$$

Then for $x\in {}^J R_1$ and $p\in {}^J R_2$ to be coextremal, it is necessary that

$$dx = \frac{{}^p\partial H}{\partial p^-}\,dt + \frac{\partial H}{\partial {}^j dH'}\,dw$$

$$x(0) = x_0$$

(6.20)

$$dp = -\frac{{}^p\partial H}{\partial x^-}\,dt + \frac{\partial H}{\partial {}^j d\sigma}\,dw + dM'$$

$$p_T = -\frac{\partial \Psi}{\partial x_T}(x_T)$$

H is maximum at u_0 a.e. (dS).

Proof: This follows from (6.18).□

Remark: If J=P, (6.20) is equivalent ot Theorem V.I of [1].

(c) The control of densities

We assume here that $(w_1,\ldots,w_{m-1})$ are continuous, and that w_m is a pure jump martingale.

Let K be a J-measurable set-valued mapping [10] defined on $\Omega \times R^+$ with non-empty compact convex values in R^m. If b is a J-measurable selection of K, let Z be given by

$$dZ = Z^- < b, dw>$$
$$Z(0) = 1. \tag{6.21}$$

If $A_T \in L_2^T$, we want to find b minimizing $E(A_T Z_T)$.

By reasoning as in [6], this problem may be put in the standard form of Definition 3.2, and the convexity assumption may be dropped. Let us apply Theorem 6.1. H is given by

$$H = \left(\frac{m+1}{m} \sum_1^{m-1} H_i b_i^j 1(_{\Delta w_m = 0}) + (m+1) H_m b_m^j (1_{\Delta w_m \neq 0}) \cdot \right) Z^- \tag{6.22}$$

Then (6.20) may be written

$$dp = - \left(\sum_1^{m-1} H_i b_i + {}^p m(H_m b_m) \right) dt + H\, dw + dM'$$

$$p_T = -A_T \tag{6.23}$$

$$\max H \quad \text{a.e. } (dS)$$
$$b \in K(\omega, t) \; .$$

This principle is useful for the control of jump diffusions [5]. To prove the existence of optimal X and P in the general case, the same methods as in [15, Section D] may be applied. For the full treatment of the linear-quadratic case, we refer to [4].

7 REFERENCES

1. Bismut, J.-M "Conjugate convex functions in optimal stochastic control", *J. Math. Anal. and Appl.*, **44**, 384-404, (1973).

2. Bismut, J.-M. "An introductory approach to duality in optimal stochastic control", *SIAM Review*, **20**, 62-78, (1978).

3. Bismut, J.-M. "Linear quadratic optimal stochastic control with random coefficients", *SIAM J. of Control and Opt.*, **14**, 419-444, (1976).

4. Bismut, J.-M. Controle des systemes lineaires quadratiques: applications de l'integrale stochastique", Seminaire de Probabilités no 12, Lecture notes in mathematics No 649, Berlin-Heidelberg-New-York: Springer, 180-264, (1978).

5. Bismut, J.-M. "Control of jump processes and Applications", *Bul. Soc, Math. France,* to appear (1978).

6. Bismut, J.-M. "Duality methods in the control of densities", *SIAM J. of Control and Optimization,* to appear.

7. Dellacherie, C., Meyer, P.A. "Probabilites et Potentiels", 2° edition, Paris: Hermann (1975).

8. Meyer, P.A. "Cours sur les integrales stochastiques", Seminaire de Probabilités no 10, Lecture notes in mathematics no 511, Berlin-Heidelberg-New-York:Springer, 245-400, (1976).

9. Rockafellar, R.T. "Integrals which are convex functionals", *Pacific J. of Math.*, **24**, 525-539, (1968).

10. Rockafellar, R.T. "Measurable dependence of convex sets and functions on parameters", *J. Math. Anal. and Appl.*, **28**, 4-25, (1969).

11. Rockafellar, R.T. "Conjugate convex functions in optimal control and the calculus of variations", *J. Math. Anal. and Appl.*, **32**, 174-222, (1970).

12. Rockafellar, R.T. "Convex Analysis", Princeton: Princeton University Press, (1970).

13. Bismut, J.-M. "Integrales convexes et Probabilites", *J. Math. Anal. and Appl.*, **42**, 639-673, (1973).

14. Meyer, P.A. "Probabilites et Potentiels", Paris:Hermann (1966).

15. Bismut, J.-M. "Analyse convexe et Probabilites, These, Universite Paris VI, (1973).

1.6 STOCHASTIC OPTIMAL CONTROL OVER DOUBLE MARTINGALES

R. Boel

(N.F.W.O. (Belgian Foundation for Scientific Research) presently on leave at Bell Telephone Laboratories, Murray Hill, New Jersey 07974, U.S.A.)

and

M. Kohlmann*

(Institut für Angewandte Mathematik der Universität, D-5300 Bonn, FRG)

ABSTRACT

This paper considers stochastic systems perturbed both by Wiener noise and Poisson processes. The dynamics are described by a change of probability measure. Two abstract existence results and an optimality criterion are treated.

1 INTRODUCTION

In this paper we will discuss the optimal control of stochastic dynamical systems influenced both by Wiener noise and (generally interdependent) Poisson processes. To this specific example, we will apply the abstract existence theorems and optimality criteria of [3]. Proofs will be given only when they are significantly easier than those in [3] since the purpose of this paper is to clarify some of the abstract issues raised there.

Following Beneš' [1] and Davis and Varaiya's [5] approach to systems with Wiener noise, we let different control laws correspond to different probability measures on a fixed probability space. Several authors (Boel and Varaiya [3], Davis and Elliott [4], Kohlmann [10]) have used the same approach for systems with jump process noise. Here we combine the results of the above mentioned papers. In § 3 we explain

*The author was partly supported by the Deutsche Forschungsgemeinschaft as a member of the SFB 72, Bonn.

the model by defining the class of admissible controls, how they influence the likelihood ratios between different probability measures, and how they influence the cost. In § 4 we state two existence theorems, assuming the likelihood ratios to be respectively, square integrable, and uniformly integrable. The second theorem holds for partially observable systems. This is obtained at the cost of strong topological assumptions. In §5 we give, assuming existence of at least one optimal control, a necessary condition for optimality. This is written as a maximum principle, following the approach of Elliot [7].

In §2 we discuss the properties of double martingales, i.e. sums of stochastic integrals over Brownian motion and compensated Poisson processes. For all details the reader is referred to [3] and the references therein.

A similar control system, with simultaneous continuous and discountinuous noise, has recently been treated by Gertner and Rapaport [8].

2 DOUBLE MARTINGALES

The basic probability space throughout this paper is $(\Omega, F, F_t, P_O)_{t\in[0,1]}$, where $\Omega=\Omega'\times\Omega''$, $F_t=F'_t\otimes F''_t$, $F=F_1$, $P_O=P'_O\times P''_O$. $(\Omega', F', F'_t, P'_O)$ is the usual Wiener space, i.e. $W^O_t(\omega) = W^O_t(\omega', \omega'') = \omega'_t$ is a Brownian motion. $(\omega'', F'', F''_t, P''_O)$ is a space carrying K Poisson processes $N^i_t(\omega) = N^i_t(\omega', \omega''_1, \ldots, \omega''_K) = \omega''_{i,t}$.

We assume P''_O such that $N^i_t - \int_O^t \lambda^i_s(\omega)\,ds$ is an F_t-local martingale.

The rate process λ^i_t is P_O-independent of F'_t, i.e. of the Brownian motion. Besides the complete information family of σ-algebras F_t, we also fix an increasing family (G_t) of sub-σ-algebras $(G_t \subset F_t)$, describing the observed history of the system (e.g. $G_t = F'_t\times\{\phi,\Omega''\}$ or $\{\phi,\Omega'\}\times F''_t$).

A process $A_t(\omega)$ is called $F_t(G_t)$-predictable, roughly, if it is the limit of a sequence of left-continuous, $F_t(G_t)$-adapted processes. For any locally square integrable martingale

M_t ($M_t \in M^2_{loc}(F_t, P_0)$, i.e. $\exists$ sequence of F_t-stopping times $T_n, T_n \uparrow 1$, such that $\sup_t E\, M^2_{t\wedge T_n} < \infty$ and $M_{t\wedge T_N}$ a martingale) we can define the predictable quadratic variation $\langle M\rangle_t$ so that $M_t^2 - \langle M\rangle_t$ is a local martingale. Then we define (i=1,2)

$$L^i(M, F_t) = \{g_t \mid g_t \text{ is } F_t\text{-predictable, } \|g\|^M_i = E_0 \int_0^1 |g_s|^i d\langle M\rangle_s < \infty\}.$$

g_t is said to be in $L^i_{loc}(M, F_t)$ if there exists an increasing sequence of F_t-stopping times T_n, such that $T_n \uparrow 1$ and $g_{t\wedge T_n} \in L^i(M, F_t)$. Similar definitions can be given for G_t-martingales. For further details, see [3] or [2,9,10].

Elliott [6] has shown that for any (F_t, P_0)-local martingale M_t, there exist F_t-predictable processes $g_t \in L^2_{loc}(W^0, F_t)$, $h^i_t \in L^1_{loc}(N^i_t, F_t)$ such that

$$M_t = \int_0^t g_s \, d\, W^0_s + \sum_{i=1}^K \int_0^t h^i_s (dN^i_s - \lambda^i_s ds). \tag{1}$$

M_t is called a double martingale.

For $\phi \in L^1_{loc}(W^0, F_t)$ and $\psi^i \in L^1_{loc}(N^i, F_t)$, define the martingale (exponential formula):

$$\varepsilon(\phi, \Psi)_t = \varepsilon(\phi)\varepsilon(\psi^1, \ldots, \psi^K)_t$$

$$= \exp\left[\int_0^t \phi_s dW^0_s - \frac{1}{2}\int_0^t \phi_s^2 ds - \sum_{i=1}^K \int_0^t \psi^i_s \ \lambda^i_s \ ds\right] \tag{2}$$

$$\times \prod_{i=1}^K \prod_{s\leq t} \left[1 + \psi^i_s \, I_{\{N^i_s \neq N^i_{s-}\}}\right].$$

Assuming $E_0\varepsilon(\phi,\Psi^i)_1 = 1$, and $\varepsilon(\phi,\Psi^i)_1 \geq 0$ (by martingale property $\varepsilon(\phi,\Psi^i)_t \geq 0$) one can define a new probability measure P on (Ω,F) by

$$\frac{dP}{dP_0} = \varepsilon(\phi,\Psi^i)_1. \tag{3}$$

By the translation theorem of van Schuppen and Wong [12], one easily verifies that, under P, $W_t = W_t^0 - \int_0^t \phi_s ds$ is a Brownian motion, and N_t^i a Poisson process with rate $(1+\psi_s^i)\lambda_s^i$. Note that W_t and N_t^i are P-dependent in general.

Jacod and Mémin [9, Thm. 8.3] have shown that martingale representation results, such as (1), are preserved under absolutely continuous change of measure. Hence, any (F_t,P) - local martingale M_t can be written as

$$M_t = \int_0^t g_s(dW_s^0 - \phi_s ds) + \sum_{i=1}^K \int_0^t h_s^i(dN_s^i - (1+\psi_s^i)\lambda_s^i ds) \tag{4}$$

with $g_t \varepsilon L^2_{loc}(W)$, $h_t^i \varepsilon L^1_{loc}(N^i)$ (now P-integrable).

Remark: All jump times are assumed totally inaccessible. A double martingale can then be decomposed into a continuous part of unbounded variation and a bounded variation part.

3 OPTIMAL CONTROL MODEL

Consider the following simple example (K=2). Let N_t^1 count repairs of a machine, while N_t^2 counts breakdowns of the machine; they are independent Poisson processes with rates λ^1 and λ^2, also independent of a Brownian motion W_t^0 influencing the output X_t of the machine. The dynamics are classically described by $dX_t = f(t,X_t,N_t^1 - N_t^2,v_t)dt + dW_t$ and control of the repair process is described by changing the rate to $u_t\lambda^1$. The controls u_t and v_t are chosen to minimize some function of X_t. In our model the dynamics would be described by the basic space $(\Omega,\mathcal{F},P_0)$ carrying W_t^0, N_t^1, N_t^2 and a change of probability measure

$$\frac{dP_{u,v}}{dP_0} = \exp\left[\int_0^1 f(s,W_s^0,N_s^1-N_s^2,v_s)\,dW_s^0 - \int_0^1 f_s^2(\cdot)\,ds - \int_0^1 u_s\,\lambda^1\,ds\right] \times \prod_{t\leq 1}\left[1+u_t\,I_{\{\Delta N_t^1=1\}}\right].$$

More generally, we suppose that $(\Omega,\mathcal{F},\mathcal{F}_t,P_0)$ and $(\mathcal{G}_t)$ as in §2 are given. Also specified is a class of control laws $\mathcal{U}$. Each $(u_t)\in\mathcal{U}$ is a $\mathcal{G}_t$-adapted stochastic process taking values in a bounded subset U of $\mathbb{R}^n$. $\mathcal{U}$ is assumed closed under concatenation (see [2,3]). Assume that u determines $\mathcal{F}_t$-predictable processes $\phi_s(u_s)\in L^2_{loc}(W^0)$, $\psi_s^i(u_s)\in L^1_{loc}(N^i)$ such that $\varepsilon(\phi_s(u),\psi_s^i(u))_t>0$ and $E_0\varepsilon(\phi_s(u_s),\psi_s^i(u_s))_1 = 1$. With each $u\in\mathcal{U}$ there corresponds a probability measure P_u defined by

$$\frac{dP_u}{dP_0} = \varepsilon(\phi_u,\psi_u^i)_1, \tag{5}$$

i.e. under P_u, $W_t^0 - \int_0^t \phi_s(u_s)\,ds$ is a Brownian motion, while N_t^i is a point process with rate $(1+\psi_s^i(u_s))\lambda_s^i$.

To simplify the statement of the theorems later on, we introduce some new notation:

$$\Phi := \{(\phi_s(u_s),\ \psi_s^i(u_s),\ i=1,\dots,K)\ |\ u\in\mathcal{U}\}$$

$$\mathcal{D}(\Phi) := \{\varepsilon(\phi,\psi^i)_1\,|\,(\phi,\psi^i)\in\Phi\}\subset L_1(\Omega,\mathcal{F},P_0).$$

To identify results here with those in [3], note that the control martingale n_t^u used in [3] is :

$$n_t^u = \int_0^t \phi_s(u_s)\,dW_s^O + \sum_{i=1}^{K}\int_0^t (1+\psi_s^i(u_s))(dN_s^i - \lambda_s^i ds).$$

Finally, we define the cost to be minimized. Let Y_t^u be a real valued, (F_t,P_u)-semimartingale, i.e. the sum of an (F_t,P_u)-martingale and an F_t-predictable process. Then a control $u\in U$ is optimal if it achieves the minimum in:

$$J(u^*) = J^* = \inf_{u\in U} J(u) \tag{6}$$

where

$$\begin{aligned} J(u) &= J(\phi_s(u),\psi_s^i(u_s)) = E_u(Y_1{}^u). \\ &= E_O(\varepsilon(\phi(u),\psi^i(u))_1 . Y_1{}^u) \end{aligned} \tag{7}$$

In [1,3,10] it is explained how one can often choose $Y_1^u = Y_1$, independent of the control law, by expanding the probability space.

Remark: In the complete information case, $F_t = G_t$, one can choose $(\phi_s(u_s),\psi_s^i(u_s),\ i = 1,\dots,K) = u_s$ and identify $U = \Phi$. This greatly simplifies the model.

4 EXISTENCE THEOREMS

Theorem 1: Assume:

(i) $|Y_1| < \bar{K}$ and Y_1 does not depend on the control law u.

(ii) $\int_0^1 \lambda_s^i ds < \bar{K}$ a.s.P_O

(iii) Φ closed and convex in $L^2_{loc}(W)\otimes L^2_{loc}(N^1,\dots,N^K)$ (with the obvious norm $\|\phi\|_2^W + \sum_{i=1}^{K} \|\psi^i\|_2^N)$ and $\mathcal{D}(\Phi) \subset L_2(\Omega, \mathcal{F}, P_0)$.

(iv) $\mathcal{F}_t = \mathcal{G}_t$ (complete information).

Then an optimal control u*(i.e. processes $\phi_s^* = \phi_s(u_s^*)$ and $\psi_s^i{}^* = \psi_s^i(u_s^*)$) exists.

Proof: (See [3, Thm. 5.1.1].) Since J(u) is a linear functional of $\varepsilon(\phi, \Psi)$ it suffices to show that $\mathcal{D}(\Phi)$ is convex and (weakly) closed. □

Theorem 2: Let $\psi_t^i \in [-1, \bar{M}]$, $\int_0^1 \lambda_s^i ds \le M$ and $|\phi_t|^2 \le \bar{K}(1+W_t^2)$ for all t, and suppose there exists a sufficiently large q (depending on K and M), such that

$$E_0(|Y_1|^q . I_{\{|Y_1|>c\}}) \to 0 \text{ as } c \to \infty, \tag{8}$$

where Y_1 is independent of u; let $\mathcal{D}(\Phi)$ be closed in $L_1(\Omega, \mathcal{F}, P_0)$. Then there exists an optimal control u* (i.e. $(\phi^*, \Psi^*) \in \Phi$ s.t. $J(\phi^*, \Psi^*) \le J(\phi, \Psi)$ for all $(\phi, \Psi) \in \Phi$).

Remark: The above theorem does not require complete information.

Proof: (See also [3, Thm. 5.1.2].) Beneš [1, Lemma 1] has shown that the bound on ϕ implies existence of $\alpha > 1$ s.t. $\sup_{\phi:(\phi,\psi)\in\Phi} E(\varepsilon(\phi)_1^\alpha) < \infty$. Similarly, the bound on ψ^i implies for any $\gamma > 0$:

$$E_0(\varepsilon(\psi^i, i=1,\ldots,K)_1^\gamma) = E_0\left\{ \prod_{i=1}^{K} \prod_{s<1} \left[1+\psi_s^i \, I_{\{N_s^i \ne N_{s^-}^i\}}\right]^\gamma \times \right.$$

$$\left. \times \exp \gamma \sum_{i=1}^{K} \int_0^1 \psi_s^i . \lambda_s^i ds \right\}$$

$$\le E_0\left[(1+M)^{\gamma . N}\right] e^{\gamma M(1+M)} < \infty,$$

where N is a Poisson random variable with intensity K.M. (Hence stochastically larger than $\sum_{i=1}^{K} N_1^i$). By the Hölder inequality with $1<\gamma<\alpha$

$$E_O(\varepsilon(\phi,\Psi)_1^{\gamma}) \leq E_O(\varepsilon(\phi)_1^{\alpha})^{\frac{\gamma}{\alpha}} \cdot E_O(\varepsilon(\psi^i, i=1,\ldots,K)_1^{\frac{\gamma\alpha}{\alpha-\gamma}})^{\frac{\alpha-\gamma}{\alpha}} < \infty \quad (9)$$

and $\mathcal{D}(\Phi)$ is uniformly integrable. By the Dunford-Pettis compactness criterion $\mathcal{D}(\Phi)$ is weakly (sequentially) compact and (Eberlein-Smulian theorem) every minimizing sequence (ϕ_j,ψ_j^i) has a subsequence s.t. (in terms of $\sigma(L_1,L_\infty)$-convergence)

$$\varepsilon(\phi_j,\psi_j^i)_1 \to \varepsilon(\phi,\psi^i)_1 \in \mathcal{D}(\Phi).$$

By the assumption on the cost $(\frac{1}{q} + \frac{1}{\gamma} = 1)$

$$E_O[\varepsilon(\phi,\psi^i)_1 \cdot Y_1 \cdot I_{\{|Y_1|>c\}}]$$

$$\leq E_O[\varepsilon(\phi,\psi^i)^{\gamma}]^{\gamma} E_O[|Y_1|^q I_{\{|Y_1|>c\}}]^{\frac{1}{q}} \quad (10)$$

$\longrightarrow 0$ (by (8) and (9))

as $c\to\infty$, and for c sufficiently large, the lefthand side will be smaller than $\frac{\varepsilon}{3}$ for any $\varepsilon>0$. Finally, by weak L_1-convergence for $j \geq j_O$,

$$|E_O[Y_1 I_{\{|Y_1|\leq c\}}(\varepsilon(\phi,\psi^i)_1 - \varepsilon(\phi_j,\psi_j^i)_1)]| < \frac{\varepsilon}{3}. \quad (11)$$

(10) and (11) show that for $j>j_O$ and any $\varepsilon>0$: $|J(\phi,\psi^i) - J(\phi_j,\psi_j^i)|<\varepsilon$. Hence $J(\phi,\psi^i) = J^*$ and (ϕ,ψ^i) is optimal.□

5 OPTIMALITY CRITERION

Davis and Varaiya [5] have obtained a necessary and sufficient condition for optimality, in a global or local (Bellman-Hamilton-Jacobi equation) form, for systems with Brownian motion noise. In [2,4,10] this has been extended to systems with jump disturbances, and one could easily write similar results for the model of this paper. We have attempted to obtain a condition of maximum principle type (following Elliott's [7] result for continuous systems) where the (optimal) dual variable is evaluated at a fixed optimal control. For the partial observation case, we have only succeeded in doing so for the following specific cost function ($c_s(u_s)>0$,

$k_s(u_s)\in L^2(W)$, $\bar{k}^i_s(u_s)\in L^1(N^i)$)

$$Y^u_t = \int_0^t c_s(u_s)\,ds + \int_0^t k_s(u_s)(dW_s-\phi_s(u_s)\,ds) + \sum_{i=1}^{K}\int_0^t \bar{k}^i_s(u_s)(dN^i_s-(1+\psi^i_s)\lambda^i_s\,ds) \tag{12}$$

which is an $(\mathcal{F}_t,P_u)$-semimartingale.

<u>Theorem 3:</u> Assume an optimal control $u^*\in\mathcal{U}$ exists corresponding to cost structure (12). Then there exist $\mathcal{F}_t$-predictable processes $g^*\in L^2_{loc}(W)$ and $h^{i*}\in L^1_{loc}(N^i)$ s.t. for any control law $u\in\mathcal{U}$

$$E_{u^*}\left[(g^*_t-k_t(u^*_t))\phi_t(u_t)+\sum_{i=1}^{K}(h^{*}_t-\bar{k}^i_t(u^*_t))\psi^i_t(u_t)\lambda^i_t+c_t(u_t)\,|\,\mathcal{G}_t\right]\geq 0$$

with equality holding if and only if u is optimal.

<u>Proof:</u> (See [3, Thm. 5.2.2].) By the principle of optimality

$$(\mathcal{U}(t,u) = \{v\in\mathcal{U}\,|\,v_s = u_s,\ s\geq t\})$$

$$U_t(u) = Y^u_t + P_u\text{-ess}\inf_{v\in\mathcal{U}(t,u)}\left[Y^v_1 - Y^v_t\,|\,\mathcal{G}_t\right] \tag{13}$$

is a (G_t, P_u)-submartingale for all u, a martingale if and only if u is optimal.

Introducing $\tilde{W}_t = E_{u*}\left[Y_1^u - Y_t^u \mid F_t\right]$, one then verifies that

$$E_{u*}\left[Y_t^u + W_t \mid G_t\right] \leq E_{u*}\left[E_u(Y_{t+h}^u + W_{t+h} \mid F_t) \mid G_t\right] \tag{14}$$

with equality if and only if u is optimal. Hence $Y_t^{u*} + \tilde{W}_t$ is an (F_t, P_{u*})-martingale

$$Y_t^{u*} + \tilde{W}_t = J* + \int_0^t g_s^*(dW_s - \phi^*_s ds)$$
$$+ \sum_{i=1}^{K} \int_0^t h_s^{i*} \, (dN_s^i - (1+\psi_s^{i*})\lambda_s^i ds). \tag{15}$$

Combining (13), (14) and (15) proves the theorem.□

The results in this paper can be extended to zero-sum 2-person games. One could for example imagine the case where one player controls the continuous process $(\phi_t(u_t))$, another player the jump rates $(\psi_t^i(v_t))$. This problem is treated in [11], which paper also gives some Hamiltonian type existence results for the control problem treated above.

6 ACKNOWLEDGEMENT

R. Boel gratefully acknowledges the hospitality of the SFB 72 of the Deutsche Forschungsgemeinschaft in Bonn during part of his work on this topic.

7 REFERENCES

1. Beneš, V.E. "Existence of optimal stochastic control laws", *SIAM J. Control*, **9**, 3, 446-472, (1971).

2. Boel, R. and Varaiya, P. "Optimal control of jump processes", *SIAM J. Control and Optimization*, **15**, 1, 92-119, (1977).

3. Boel, R. and Kohlmann, M. "Semimartingale models of stochastic optimal control, with applications to double martingales", to appear in *SIAM J. Control and Optimization*

4. Davis, M. and Elliott, R.J. "Optimal control of a jump process", *Z.Wahrscheinlichkeitstheorie verw. Gebiete,* **40**, 183-202, (1977).

5. Davis, M. and Varaiya, P. "Dynamic programming conditions for partially observable stochastic systems", *SIAM J. Control,* **11**, 226-261, (1973).

6. Elliott, R.J. "Double martingales", *Z.Wahrscheinlichkeitstheorie verw. Gebiete,* **34**, 17-28, (1976).

7. Elliott, R.J. "The optimal control of a stochastic system", *SIAM J. Control and Optimization,* **5**, 5, 756-778, (1977).

8. Gertner, I. and Rapaport, D. "Stochastic control of system with unobserved jump parameter process", *Information Sciences,* **13**, 269-282, (1977).

9. Jacod, J. and Mémin, J. "Charactéristiques locales et conditions de continuité absolue pour les semimartingales", *Z.Wahrscheinlichkeitstheorie verw. Gebiete,* **35**, 1, 1-37, (1976).

10. Kohlmann, M. "On control of jump processes, a martingale approach", Preprint 184, SFB 72, University of Bonn.

11. Kohlmann, M. "A game with Wiener noise and jump process disturbances", Submitted to *Stochastics*.

12. van Schuppen, J. and Wong, E. "Transformation of local martingales under a change of law", *Annals. of Probability,* **2**, 879-888, (1974).

CHAPTER 2

STOCHASTIC OPTIMISATION

2.1 OPTIMAL STOPPING-TIME PROBLEMS FOR REFLECTED DIFFUSION PROCESSES

A. Bensoussan

(University Paris IX and Laboria, France)

J. L. Lions

(Collège de France and Laboria, France)

1 INTRODUCTION

In this article we briefly review the results of the authors on the correspondence between variational inequalities with Neumann boundary conditions and optimal stopping time problems for reflected diffusion processes. We introduce several penalized approximations of the V.I. (variational inequality), we give their probabilistic interpretation and we obtain via analytic and probabilistic techniques estimates on the penalization error.

2 REVIEW ON REFLECTED DIFFUSION PROCESSES AND VARIATIONAL INEQUALITIES

2.1 Variational inequalities

Let σ be a bounded open subset of R^n. Let a_{ij}, a_i, a_0 be given on σ such that

$$a_{ij}, a_i, a_0 \text{ are measurable and bounded} \tag{2.1}$$

$$\sum_{ij} a_{ij}\xi_i\xi_j \geq \alpha|\xi|^2, \ \forall\xi \in R^n \tag{2.2}$$

$$a_0 \geq \beta > 0. \tag{2.3}$$

Let ψ be a given measurable function on σ, and let us set

$$\begin{cases} V = H^1(\sigma) \\ K = \{v | v \in V,\ v \leq \psi,\ \text{a.e.}\}. \end{cases} \tag{2.4}$$

We assume that

$$K \neq \emptyset. \tag{2.5}$$

Let us define for $u,v \in V$

$$a(u,v) = \int_\theta \Sigma\, a_{ij} \frac{\partial u}{\partial x_j} \frac{\partial v}{\partial x_i}\, dx + \int_\theta \sum_i a_i \frac{\partial u}{\partial x_i}\, v\, dx + \int_\theta a_0 uv\, dx. \tag{2.6}$$

For $f \in L^2(\sigma)$, one calls a V.I. (variational inequality) the following problem:

to find u such that

$$a(u,v-u) \geq (f,v-u) \ \forall v \in K,\ u \in K. \tag{2.7}$$

Assume the following coercivity assumption holds:

$$a(v,v) \geq \gamma \|v\|^2 \ \forall v \in V,\ \gamma > 0. \tag{2.8}$$

Then it is standard (cf. Lions-Stampacchia [2]) that there exists one and only one solution u of (2.7).

For what follows, it is important to state some regularity results. We assume that

$$\partial\sigma = \Gamma \text{ is Lipschitz} \tag{2.9}$$

$$f \in L^p(\sigma),\ \psi \in W^{2,p}(\sigma),\ p \geq 2. \tag{2.10}$$

$$\frac{\partial \psi}{\partial \nu_A}\Big|_\Gamma \geq 0 \tag{2.11}$$

where

$$\frac{\partial}{\partial \nu_A} = \Sigma\, a_{ij} \frac{\partial}{\partial x_j} \nu_i$$

with ν the outward unit normal on Γ. We also assume that

$$a_{ij} \in W^{1,\infty}(\sigma). \tag{2.12}$$

Under the assumptions (2.1), (2.2), (2.3), (2.8), (2.9), (2.10), (2.11), (2.12) the solution u of (2.7) satisfies the regularity property

$$u \in W^{2,p}(\sigma). \tag{2.13}$$

Moreover, one has

$$f \wedge A\psi \leq Au \leq f. \tag{2.14}$$

In this situation, u can be characterized by a set of unilateral conditions with complementary slackness relations, namely

$$\begin{cases} Au \leq f,\ u \leq \psi \quad \text{a.e.} \\ (Au-f)\ (u-\psi) = 0 \quad \text{a.e.} \\ \left.\dfrac{\partial u}{\partial \nu_A}\right|_\Gamma = 0. \end{cases} \tag{2.15}$$

From now on we will assume that

$$p > \frac{n}{2}\,. \tag{2.16}$$

Therefore the solution u of (2.13), (2.15) satisfies

$$u \in C^0(\bar{\sigma}). \tag{2.17}$$

2.2 Probabilistic interpretation

We now give the probabilistic interpretation of the function u. Setting

$$a(x) = \frac{1}{2}\,\sigma^2(x)$$

which uniquely defines a matrix $\sigma(x)$ such that $\sigma \in W^{1,\infty}$, we denote

$$g_i(x) = \sum_j \frac{\partial a_{ij}}{\partial x_j} - a_i(x) \quad (2.18)$$

$$\gamma_i(x) = \sum_j a_{ij}(x)\nu_j(x). \quad (2.19)$$

To give a probabilistic interpretation of $u(x)$, we use the submartingale problem studied by D. Stroock and S.R.S. Varadhan [3]. One needs an additional regularity assumption on the boundary, namely that

$$\sigma = \{x|\phi(x) < 0, \text{ where } \phi \in C^2(R^n),\ \phi,\ \frac{\partial\phi}{\partial x_i},\ \frac{\partial^2\phi}{\partial x_i\, x_j} \text{ bounded,}$$
$$|\frac{\partial\phi}{\partial x}| \geq \beta > 0 \text{ on } \Gamma\}. \quad (2.20)$$

We then define

$$\begin{cases} \Omega = C^0([0,\infty[;\bar{\sigma}),\ y(x;\omega) = \omega(s) \\ F^s = \sigma(y(\lambda),\ 0 \leq \lambda \leq s). \end{cases} \quad (2.21)$$

A probability measure P^x on Ω, $(x \in \bar{\sigma})$ is a solution of the submartingale problem starting from x, if for any smooth function $\phi(x,t)$ with compact support $R^n \times [0,\infty[$, and such that

$$\frac{\partial\phi}{\partial x}\cdot\gamma|_\Gamma \leq 0 \quad (2.22)$$

one has

$$\begin{cases} \phi(y(s),s) - \int_0^s \frac{\partial\phi}{\partial x}\cdot g(y(\lambda),\lambda)d\lambda - \frac{1}{2}\int_0^s \text{tr}\, \frac{\partial^2\phi}{\partial x^2}\, a(y(\lambda),\lambda)d\lambda - \\ - \int_0^s \frac{\partial\phi}{\partial s}(y(\lambda),\lambda)d\lambda \text{ is an } F^s \text{ sub-martingale} \end{cases} \quad (2.23)$$

$$P^x[y(0)=x] = 1. \quad (2.24)$$

Under the above assumptions, the submartingale problem has one and only one solution. In addition, there exists a scalar process $\xi(t)$ which is adapted and non decreasing, and a standard Wiener process $w(t)$ with values in R^n, such that

$$dy(t) = g(y(t))dt + \sigma(y(t))dw(t) - \chi_\Gamma(y(t))\gamma(y(t))d\xi(t) \quad (2.25)$$

$$\text{a.s.} \int_{t_1}^{t_2} \chi_\sigma(y(t))d\xi(t) = 0, \ \forall\ t_1 \leq t_2. \quad (2.26)$$

For any stopping-time θ of F^t, one sets

$$J_x(\theta) = E^x \Big[\int_0^\theta f(y(t))(\exp - \int_0^t a_0(y(s)ds)dt + \psi(y(\theta))\exp - \int_0^\theta a_0(y(t))dt \Big]. \quad (2.27)$$

Then one has the following:

<u>Theorem 2.1</u> : <u>Under the assumptions</u> (2.1), (2.2), (2.3), (2.8), (2.10), (2.11), (2.12), (2.16), (2.20), <u>one has</u>

$$u(x) = \operatorname*{Inf}_\theta J_x(\theta). \quad (2.28)$$

<u>If one sets</u>

$$\hat{\theta} = \inf\{t \geq 0 \mid u(y(t)) = \psi(y(t))\} \quad (2.29)$$

<u>then</u>

$$u(x) = J_x(\hat{\theta}). \square \quad (2.30)$$

<u>Remark 2.1</u> : In the case when $f - A\psi \leq C$ (constant), it can be shown that the assumption (2.8) is unnecessary. □

3 PENALIZED PROBLEMS

Unlike the Dirichlet case, two penalized problems can be associated with the V.I. (2.7). They are

$$a(u^\varepsilon,v) + \frac{1}{\varepsilon}((u^\varepsilon-\psi)^+,v) = (f,v) \quad \forall v \in V \tag{3.1}$$

$$a(u^\varepsilon,v) + \frac{1}{\varepsilon}((u^\varepsilon-\psi)^+,v) + \frac{1}{\varepsilon}((u^\varepsilon-\psi)^+,v)_\Gamma = (f,v) \quad \forall v \in V. \tag{3.2}$$

We use the same notation u for the solutions of (3.1), (3.2), although they are not at all identical. This will not be confusing since the two functions will not be used simultaneously.

Our objective is to give the probabilistic interpretation of (3.1), (3.2) and to derive estimates on the penalization error $u^\varepsilon-u$.

3.1 Estimates obtained by analytic techniques

One can state the following theorems.

Theorem 2.1 : Under the assumptions (2.1), (2.2), (2.3), (2.9) and

$$f \in L^2(\sigma),\ \psi \in H^2(\sigma) \tag{3.3}$$

one has the estimate

$$\|u^\varepsilon-u\| \leq C\,\varepsilon^{1/4} \tag{3.4}$$

where u^ε denotes the solution of (3.1) and where $\|\cdot\|$ denotes the norm in $H^1(\sigma)$. If in addition, one assumes (2.11) then one obtains a more accurate estimate, namely

$$\|u^\varepsilon-u\| \leq C\,\varepsilon^{1/2}.\ \square \tag{3.5}$$

Theorem 3.2 : Under the assumptions (2.1), (2.2), (2.3), (2.9) and (3.3), one has the estimate

$$\|u^\varepsilon-u\| \leq C\,\varepsilon^{1/2} \tag{3.6}$$

where u^ε denotes the solution of the penalized problem (3.2).□

Remark 3.1 : The assumption (2.11) is not required in the statement of Theorem 3.2; hence the penalization scheme (3.2) leads to a better approximation of the function u.□

3.2 Probabilistic interpretation of the penalized problems.

We call an admissible control a process v(t) adapted to $\mathcal{F}^t$, such that $0 \leq v(t) \leq 1$.

One defines a functional

$$J_x^\varepsilon(v) = E^x\Big[\int_0^\infty (f(y(t)) + \frac{1}{\varepsilon}\,\psi(y(t))v(t))(\exp - \int_0^t (a_0(y(s)) + \frac{v(s)}{\varepsilon})ds)dt\Big]. \tag{3.7}$$

We can state the following

Theorem 3.3 : We assume (2.1), (2.2), (2.3), (2.8), (2.12), (2.20) and

$$f \in L^p(\sigma),\ p > \frac{n}{2} \tag{3.8}$$

$$\psi \in C^0(\bar{\sigma}). \tag{3.9}$$

Then the solution u^ε of the penalized problem (3.1) is given explicitly by the formula

$$u^\varepsilon(s) = \operatorname*{Inf}_{v(\cdot)} J_x^\varepsilon(v). \tag{3.10}$$

Moreover, there exists an optimal control given by

$$\hat{v}^\varepsilon(t) = \begin{cases} 1 & \text{if } u^\varepsilon(y(t)) \geq \psi(t)) \\ 0 & \text{otherwise.}\square \end{cases} \tag{3.11}$$

We next introduce the functional

$$\begin{cases} J_x^\varepsilon(v)=E^x\Big[\int_0^\infty (f(y(t))+\frac{1}{\varepsilon}\,\psi(y(t))v(t))(\exp-\int_0^t (a_0+\frac{1}{\varepsilon}v)ds-\frac{1}{\varepsilon}\int_0^t vd\xi)dt \\ \qquad +\int_0^\infty \frac{1}{\varepsilon}\phi(y(t))v(t))(\exp-\int_0^t (a_0+\frac{1}{\varepsilon}v)ds-\frac{1}{\varepsilon}\int_0^t vd\xi)d\xi(t)\Big]. \end{cases} \tag{3.12}$$

Again we have used the same notation for the functionals (3.7) and (3.12), although they are different. However, since they will not be used simultaneously, there will be no confusion.

<u>Theorem 3.4</u> : <u>Under the assumptions of Theorem</u> 3.3, <u>the solution of the penalized problem</u> (3.2) u^ε <u>is given explicitly by the formula</u>

$$u^\varepsilon(x) = \operatorname*{Inf}_v J_x^\varepsilon(v). \tag{3.13}$$

Moreover, there exists an optimal control, still defined by (3.11).□

3.3 Obtaining a penalization error estimate by probabilistic techniques

Let us first consider the situation of Theorem 2.1. We have identified u(x) as (2.28). Now Theorem 3.3 applies, hence u^ε is characterized by (3.10). One can then make direct comparisons between the functionals $J_x^\varepsilon(v)$ and $J_x^\varepsilon(\theta)$. This yields an estimate of the difference $u^\varepsilon-u$. We can state the

<u>Theorem</u> 3.5 : <u>We make the assumptions of Theorem</u> 2.1. <u>Then one has</u>

$$\|u^\varepsilon-u\|_{C^0(\bar{\sigma})} \leq C\,\varepsilon^{\frac{a}{1+a}} \tag{3.14}$$

<u>where</u> a <u>is such that</u> $\frac{p}{1+a} > \frac{n}{2}$.□

We next state a result for the penalization scheme (3.2). We have the following

$$\psi \in W^{2,p}(\sigma),\ p > n. \tag{3.15}$$

If u^ε is the solution of (3.2), *the following estimate holds true*

$$\|u^\varepsilon - u\|_{C^0(\bar{\sigma})} \leq C\, \varepsilon^{\frac{a}{1+a}} \tag{3.16}$$

where a *is the same as in Theorem* 3.5.□

Remark 3.2 : We do not assume any sign property on $\left.\frac{\partial\psi}{\partial\nu_A}\right|_\Gamma$.□

Sketch of the proof of Theorem 3.6.

By virtue of the regularity of the function ψ, one has the relations

$$\begin{cases} J_x^\varepsilon(v) - \psi(x) = E^x\Big[\displaystyle\int_0^\infty (f - A\psi)(y(t))\,(\exp -(\int_0^t (a_0 + \frac{1}{\varepsilon}v)\,ds + \frac{1}{\varepsilon}\int_0^t v\,d\xi))\,dt \\ \qquad - \displaystyle\int_0^\infty \frac{\partial\psi}{\partial\nu_A}(y(t))\,(\exp - \int_0^t (a_0 + \frac{v}{\varepsilon})\,ds - \int_0^t \frac{v\,d\xi}{\varepsilon})\,d\xi(t)\Big]. \end{cases} \tag{3.17}$$

and

$$\begin{cases} J_x(\theta) - \psi(x) = E^x\Big[\displaystyle\int_0^\theta (f - A\psi)(y(t))\,(\exp - \int_0^t a_0\,ds)\,dt\ - \\ \qquad - \displaystyle\int_0^\theta \frac{\partial\psi}{\partial\nu_A}(y(t))\,(\exp - \int_0^t a_0\,ds)\,d\xi(t)\Big]. \end{cases} \tag{3.18}$$

Since we have not assumed (2.11), we cannot apply Theorem 2.1, and thus we do not know yet the relation (2.28). However defining

$$w(x) = \operatorname*{Inf}_\theta J_x(\theta)$$

one can first show that

$$u^{\varepsilon}(x) \geq w(x). \tag{3.19}$$

Next, let θ be any $\mathcal{F}^t$ stopping time and let us define

$$v_{\theta}(t) = \begin{cases} 1 \text{ if } t < \theta \\ 0 \text{ if } t \geq \theta. \end{cases}$$

An easy computation shows that

$$\begin{cases} J_x^{\varepsilon}(v_{\theta}) = J_x(\theta) + E^x \displaystyle\int_{\theta}^{+\infty} (f - A\psi)(y(t))\left(\exp - \int_0^t a_0 ds\right) \exp - \frac{t-\theta}{\varepsilon} \exp - \frac{\xi(t) - \xi(\theta)}{\varepsilon} dt \\ \qquad - E^x \displaystyle\int_{\theta}^{+\infty} \frac{\partial\psi}{\partial\nu_A}(y(t))\left(\exp - \int_0^t a_0 ds\right) \exp - \frac{t-\theta}{\varepsilon}\left(\exp - \frac{\xi(t) - \xi(\theta)}{\varepsilon}\right) d\xi(t). \end{cases} \tag{3.20}$$

But

$$E^x \int_{\theta}^{\infty} |f - A\psi| \left(\exp - \int_0^t a_0 ds\right) \exp - \frac{t-\theta}{\varepsilon}\, dt \leq C \| f - A\psi \|_{L^p} \varepsilon^{\frac{a}{1+a}}$$

$$E^x \int_{\theta}^{+\infty} \left|\frac{\partial\psi}{\partial\nu_A}(y(t))\right| \left(\exp - \int_0^t a_0 ds\right) \exp - \frac{t-\theta}{\varepsilon} \; \exp - \frac{\xi(t) - \xi(\theta)}{\varepsilon}\Big) d\xi(t) \leq$$

$$C\, E^x \int_{\theta}^{+\infty} \exp - \frac{\xi(t) - \xi(\theta)}{\varepsilon}\, d\xi(t) = -\varepsilon C\, E^x \int_{\theta}^{+\infty} d \, \exp - \frac{\xi(t) - \xi(\theta)}{\varepsilon} = C\, \varepsilon.$$

Hence we obtain (3.16), with w instead of u. But this estimate shows that $u^{\varepsilon} \to w$ in $C^0(\bar{\sigma})$, as $\varepsilon \to 0$. On the other hand, using for instance Theorem 3.2, we know that $u^{\varepsilon} \to u$ in $H^1(\sigma)$.

Therefore u=w, and the result follows.□

We finally give the interpretation of the V.I., when the obstacle ψ is only continuous. We have the

Theorem 3.7 : Under the assumptions of Theorem 3.3 the solution u(x) of (2.7) belongs to $C^O(\bar{\sigma})$. Moreover, the properties (2.28), (2.30) still hold true, and the solutions u^ε of both penalized problems converge towards u in $C^O(\bar{\sigma})$.

Proof:

The proof relies on the fact that if $\psi \in C^O(\bar{\sigma})$, then one can find a sequence $\psi^k \to \psi$ in $C^O(\bar{\sigma})$, such that $\psi^k \in W^{2,p}(\sigma)$, $\frac{\partial \psi^k}{\partial \nu_A} \geq 0$.

Moreover if ψ_1, ψ_2 are two obstacles, u_1, u_2 the solutions of the corresponding V.I. and u_1^ε, u_2^ε the solutions of both corresponding penalized problems, then one has the estimates

$$\|u_1-u_2\|_{C^O(\bar{\sigma})} \leq \|\psi_1-\psi_2\|_{C^O(\bar{\sigma})}$$

$$\|u_1^\varepsilon-u_2^\varepsilon\|_{C^O(\bar{\sigma})} \leq \|\psi_1-\psi_2\|_{C^O(\bar{\sigma})}$$

From these considerations, the desired result easily follows.□

Remark 3.3 : One can even obtain an estimate for the penalization error in $C^O(\bar{\sigma})$, but it is less precise than (3.16), cf. A. Bensoussan-J. L. Lions [1].□

Remark 2.4 : We also refer to the book [1] of the Authors for error estimates in parabolic evolution problems.□

4 REFERENCES

1. Bensoussan, A. and Lions, J. L., "Impulsive control and Quasi variational inequalities", book to be published.

2. Lions, J. L. and Stampacchia, G. "Variational inequalities", Comm. Pure Applied Math., (1967).

3. Stroock, D. and Varadhan, S. R. S. "Diffusion processes with boundary conditions", Comm. Pure Applied Math., **24**, 147-225, (1971).

2.2 ON THE OPTIMAL STOPPING SEMIGROUP FOR CONTROLLED MARKOV PROCESSES

M. Nisio

(Department of Mathematics, Kobe University, Kobe, Japan)

1 INTRODUCTION

In the recent papers ([3], [10], [11], [12], [15]) we proved that the evolution of the optimal value in stochastic control problems turns out to be a nonlinear semigroup. In [12] we treated semigroups of optimal stopping problems of controlled Markov processes. This article is a continuation of [12] and we show that, as $t \to \infty$, the limit of semigroup becomes the least excessive majorant.

First we recall the optimal stopping semigroup for Markov process. Let W be the path space, i.e. the set of all R^n-valued right continuous functions on $[0, \infty)$ with left limits. Let F_t be the right continuous modification of σ-algebra σ_t on W generated by $\{\omega(s), s \leq t\}$, namely $F_t = \bigcap_{s>t} \sigma_s$. F denotes the σ-algebra generated by $\{\omega(s), s < \infty\}$. Let $X(t,\omega)$ be the t-th coordinate of $\omega \in W$ and θ_s the shift operator by s, namely $X(t, \theta_s\omega) = X(t + s, \omega)$. Let $X = (X(t), W, P_x, x \in R^n)$ be an n-dimensional Markov process [7] whose transition semigroup P(t) acts on the Banach lattice C of all bounded and uniformly continuous functions on R^n, endowed with the usual supremum norm and usual order. By m we denote the set of all stopping times, i.e. $\tau \in m$ if and only if τ is $[0, \infty]$-valued function on W such that $(\tau \leq t) \in F_t$ for any $t \geq 0$. Put

$$m(t) = \{ \tau \wedge t;\ \tau \in m\}.^{(1)} \qquad (1.1)$$

---

(1) $a \wedge b = \min(a, b)$, $a \vee b = \max(a, b)$

In [12] we showed that the operator V(t) defined by (1.2)

$$V(t)\phi(x) = \sup_{\tau \in m(t)} E_x \phi(x(t)) \exp\left(-\int_0^\tau c(X(s))ds\right) + \int_0^\tau f(X(s)) \exp\left(-\int_0^s c(X(z))dz\right)ds \qquad \phi \in C \tag{1.2}$$

is a monotone contraction semigroup[1] on C, under certain conditions. Moreover its generator G is expressed by

$$G\phi = 0 \vee (A\phi - c\phi + f) \qquad \phi \in D(A) \tag{1.3}$$

where A is the generator of P(t).

We call V(t) the optimal stopping semigroup for (P(t),c,f). Its generator G is related to the so-called equality-inequality equation of optimal stopping. Furthermore if inf c(x) > 0, then $V(t)\phi$ is increasing to $\Phi \in C$, which is defined by the right side of (1.2) replacing m(t) by m. Φ is the least T(t)-excessive majorant of ϕ, where T(t) is the semigroup with generator $A\phi - c\phi + f$ [1], [2], [4], [6], [9], [13].

The objective of this note is to prove analogous results for controlled Markov processes, [8], [11], [12], [15]. Let Γ be a compact[2] subset of R^k, called a control region. Let $X^u = (X(t), W, P_x^u, x \in R^n)$ be a Markov process whose transition semigroup $P^u(t)$ acts on C. Let d be a Γ-valued function on $[0, \infty) \times W$ such that $d(s, \cdot)$ is σ_s-measurable and

$$d(t, \omega) = d(k2^{-N}, \omega) \text{ for } t \in [k2^{-N}, (k+1)2^{-N}). \tag{1.4}$$

By D_N we denote the set of all functions d expressed by (1.4). Put $D = \bigcup_{N=1}^{\infty} D_N$. An element of D is called a switching control. Corresponding to $d \in D_N$ we can define the unique probability measure Q_x^d on W such that [14]

(1) we mean a strongly continuous semigroup
(2) For simplicity we assume compactness, but it can be replaced by σ-compactness with little modification.

$$Q_x^d(X(t+s) \in \cdot / F_s)$$
$$= P_{X(s)}^{d(s)}(X(t) \in \cdot) \quad \text{a.e. for } k2^{-N} \leq s < s+t \leq (k+1)2^{-N} . \tag{1.5}$$

For continuous functions $c^u(x) = c(x, u)$ and $f^u(x) = f(x, u)$ on $R^n \times \Gamma$, we put

$$I(\phi, s, t, d) = \phi(X(t))\exp(-\int_s^t c(X(a), d(a))da)$$
$$+ \int_s^t f(X(a), d(a))\exp(-\int_s^a c(X(b), d(b))db)da \tag{1.6}$$

and

$$H(t, u)\phi(x) = H^u(t)\phi(x) = E_x^u \phi(X(t))\exp(-\int_0^t c(X(s), u)ds) \tag{1.7}$$

where E_x^u is the expectation with respect to P_x^u namely $H^u(t)$ is the semigroup with killing rate c^u.

Theorem 1 Suppose the following conditions (A1) - (A4) hold.

(A1) $D \equiv \bigcap_{u \in \Gamma} D(A^u)$ is dense in C and

$$\sup_u \| A^u \phi \| < \infty \qquad \phi \in D$$

where A^u is the generator of $P^u(t)$.

(A2) For a positive h

$$\sup_u \| g^u \| \leq h \text{ and } \sup_u |g^u(x) - g^u(y)| \leq h|x - y| \quad g = f, c$$

(A3) $c^u \geq 0 \qquad u \in \Gamma$

(A4) $H(t, u)\phi(x)$ is continuous in (x, u), and for $T > 0$ there exists a constant $\lambda = \lambda(T)$ such that

$$\sup_u |H(t, u)\phi(x) - H(t, u)\phi(y)| \leq |x - y|e^{\lambda T} \quad t \leq T \tag{1.8}$$

whenever ϕ is Lipschitz continuous with constant 1 and $\|\phi\| \leq 1$. Define $v(t)$ by

$$v(t)\phi(x) = \sup_{d \in D} \sup_{\tau \in m(t)} E_x^d \, I(\phi, 0, \tau, d) \tag{1.9}$$

where E_x^d means the expectation with respect to Q_x^d.

Then $v(t)$ is the envelope of the optimal stopping semigroup $V^u(t)$ for (P^u, c^u, f^u), $u \in \Gamma$, namely $v(t)$ is the monotone contraction semigroup on C such that

(i) $V^u(t)\phi \leq v(t)\phi, \quad u \in \Gamma$

(ii) If $\Lambda(t)$ is a semigroup with (i) then $v(t)\phi \leq \Lambda(t)\phi$.

Moreover its generator $\mathcal{G}$ is expressed by

$$\mathcal{G}\,\phi = 0 \vee \sup_u (A^u\phi - c^u\phi + f^u) \qquad \phi \in D \cap D(\mathcal{G}). \tag{1.10}$$

The operator $\mathcal{A}: \mathcal{A}\phi = \sup_u (A^u\phi - c^u\phi + f^u)$, is related to a so-called Bellman equation [5].

Theorem 2 Suppose $\inf_{x,u} c(x, u) > 0$. Then as $t \to \infty$, $v(t)\phi$ is increasing to $\tilde{\Phi} \in C$ defined by

$$\tilde{\Phi}(x) = \sup_{d \in D} \sup_{\tau \in m} E_x^d I(\phi, 0, \tau, d). \tag{1.11}$$

Moreover $\tilde{\Phi}$ is the least $\tilde{S}(t)$-excessive majorant of ϕ, where $\tilde{S}(t)$ is the envelope of $T^u(t)$, $u \in \Gamma$, [3], [10].

These theorems will be proved in §§2 and 3.

2 PROOF OF THEOREM 1

Let $\Sigma(K)$ be the set of all Lipschitz continuous functions ϕ with Lipschitz constant K and $\|\phi\| \leq K$. Put $\Sigma = \bigcup_{K>0} \Sigma(K)$

and $\Delta = 2^{-N}$. According to [12], we sketch the outline of construction of semigroup $S(t)$ of optimal stopping for controlled Markov processes. Let V^u be the optimal stopping semigroup for (P^u, c^u, f^u). Define $J = J(N)$ by

$$J\phi(x) = \sup_u V^u(\Delta)\phi(x), \quad \phi \in \Sigma.$$

Then J maps Σ into Σ. So we define an approximate $S^{(N)}(t)$ by

$$S^{(N)}(t)\phi = J^k(N)\phi \quad \text{for} \quad t = k2^{-N}, \quad \phi \in \Sigma.$$

$S^{(N)}(t)\phi$ is increasing to $S(t)\phi$, as $N \to \infty$. Our assumptions imply $S(t)\phi$ can be extended to $t \geq 0$ and $\phi \in C$.

For the proof of Theorem 1, we use a following different approximation to the same $S(t)$. Define $B = B(N)$ by

$$B\phi(x) = \phi \vee \sup_u T(\Delta, u)\phi(x) \quad \phi \in \Sigma. \tag{2.1}$$

Then (A2) and (A4) imply that B maps Σ into Σ. So $B^k\phi$ is defined successively. Setting $B^0\phi = \phi$, we have

Lemma 2.1

$$B^{k+1}\phi = \phi \vee \sup_u T(\Delta, u)B^k\phi \tag{2.2}$$

Proof For $k = 0$, (2.2) comes from the definition of B. Suppose (2.2) holds for k.

$$B^{k+1}\phi = B(B^k\phi) = B^k\phi \vee \sup_u T(\Delta, u)B^k\phi$$

$$= \phi \vee \sup_u T(\Delta, u)\ B^{k-1}\phi \vee \sup_u T(\Delta, u)B^k\phi$$

Since $B^{k-1}\phi \leq B^k\phi$ and $T(\Delta, u)$ is monotone, we have (2.2) for $k + 1$. Q E D.

Using the same method as in [12], we define

$$U(t, N)\phi = B^k(N)\phi \quad \text{for } t = k\Delta. \tag{2.3}$$

Then $U(t, N)\phi$ is increasing and $U(t)\phi = O_i\text{-lim}\ U(t, N)\phi$ [(1)] exists in Σ. Moreover $U(t)$ can be extended to C and any $t \geq 0$, and becomes a monotone contraction semigroup on C with property (i) of Theorem 1. Hence by the minimum property of S(t)

$$S(t)\phi \leq U(t)\phi \qquad \phi \in C. \tag{2.4}$$

Since S(t) satisfies (i), we see

$$B(N)\phi \leq S(\Delta)\phi.$$

Hence we have

$$U(t, N)\phi \leq S(t)\phi \quad \text{for } t = k\Delta.$$

This implies

$$U(t)\phi \leq S(t)\phi \text{ for binary } t \text{ and } \phi \in \Sigma.$$

This is the converse of inequality (2.4). Therefore

$$U(t) = S(t). \tag{2.5}$$

Put $\Delta = 2^{-N}$ and for $t = k\Delta$, $m_N(k) = \{\tau \in m(t);$ range of $\tau \subset \{j\Delta, j = 0, 1, 2 \ldots\}\}$. According to [14] we show the following lemma.

Lemma 2.2

$$B^k(N)\phi(x) = \sup_{d \in D_N} \sup_{\tau \in m_N(k)} E_x^d I(\phi, 0, \tau, d) \tag{2.6}$$

Proof First we prove (2.6) for k = 1. Let $d(0, \omega) = u$ for $X(0, \omega) = x$. Then

$$E_x^d I(\phi, 0, \tau, d) = E_x^u I(\phi, 0, \tau, u)$$

$$= E_x^u (I(\phi, 0, \Delta, u) \ ; \ \tau = \Delta) + \phi(x)\ P_x^u(\tau = 0).$$

(1) $\psi = O_i\text{-lim}\psi_N$ means "ψ_N is increasing to $\psi \in C$ at each point"

Since X^u is a strong Markov process by (A4),

$P_x^u(\tau = 0) = 1$ or 0.

Hence

$$E_x^d I(\phi, 0, \tau, d) = T(\Delta, u)\phi(x) \text{ or } \phi(x).$$

So we have (2.6) for $k = 1$.

Suppose (2.6) holds for k.

$$E_x^d I(\phi, 0, \tau, d) = E_x^d(I(\phi, 0, \tau, d) \ ; \ \tau \leq k\Delta) \\ + E_x^d(I(\phi, 0, (k+1)\Delta, d) \ ; \ \tau = (k+1)\Delta). \tag{2.7}$$

Since $(\tau = (k+1)\Delta) = (\tau \leq k\Delta)^c \in F_{k\Delta}$, we have

$$= E_x^d(I(\phi, 0, \tau \wedge k\Delta, d) \ ; \ \tau \leq k\Delta)$$

$$+ E_x^d(E_x^d(I(\phi, 0, (k+1)\Delta, d)/F_{k\Delta}) \ ; \ \tau = (k+1)\Delta).$$

Recalling the definition of Q_x^d, we see that

$$E_x^d(I(\phi, 0, (k+1)\Delta, d)/F_{k\Delta})$$

$$= \int_0^{k\Delta} f(X(s), d(s)) \exp(-\int_0^s c(X(a), d(a))da)ds \\ + \exp(-\int_0^{k\Delta} c(X(s), d(s))ds) \ T(\Delta, d(k\Delta))\phi(X(k\Delta)) \tag{2.8}$$

$$\leq I(B\phi, 0, k\Delta, d)$$

From (2.7) and (2.8) we can derive

$$E_x^d I(\phi, 0, \tau, d) \leq E_x^d I(B\phi, 0, \tau \wedge k\Delta, d) \tag{2.9}$$

Since $\tau \wedge k\Delta \in m_N(k\Delta)$, our assumption implies

$$E_x^d I(\phi, 0, \tau, d) \leq B^k B\phi(x) = B^{k+1}\phi(x). \tag{2.10}$$

Put

$$D(j, k) = \{x \in R^n; \ \phi(x) = B^{k-j}\phi(x)\}, \quad j = 0,\ldots k. \tag{2.11}$$

$$\tau_k(\omega) = \min\{j\Delta; \ X(j\Delta, \omega) \in D(j, k)\}. \tag{2.12}$$

By virtue of "$D(k, k) = R^n$", τ_k belongs to $m_N(k)$. On the other hand $T(\Delta, u)\phi(x)$ is continuous in (x, u) and Γ is compact. Hence the implicit function theorem ensures the existence of a Γ-valued Borel function $u_{k,k-1}(x)$ such that

$$T(\Delta, u_{k,k-1}(x))\phi(x) = \sup_u T(\Delta, u)\phi(x) \equiv J\phi(x). \tag{2.13}$$

Since $B^\ell\phi$ belongs to C, we can choose Γ-valued Borel functions $u_{k,j}$, $j = k-2,\ldots 0$, such that

$$T(\Delta, u_{k,k-1-j}(x))B^j\phi(x) = J\, B^j\phi(x), \ j = 1,\ldots k-1. \tag{2.14}$$

Define $d_k \in D_N$ by

$$d_k(t, \omega) = u_{k,j}(\omega(j\Delta)) \text{ for } j\Delta \leq t < (j+1)\Delta, \ j = 0,..k-1. \tag{2.15}$$

Now we will prove

$$E_x^{d_k} I(\phi, 0, \tau_k, d_k) = B^k\phi(x). \tag{2.16}$$

For $x \in D(0, k)$ we see $\phi(x) = B^k\phi(x)$ and $P_x^u(\tau_k = 0) = 1$. Hence (2.16) holds. For $x \notin D(0,k)$ we show (2.16) by induction.

For $k = 1$ we see $\phi(x) < B\phi(x)$ and $P_x^u(\tau_1 = \Delta) = 1$. Hence

$$E_x^{d_1} I(\phi, 0, \Delta, d_1) = T(\Delta, d_1)\phi(x) = J\phi(x) \tag{2.17}$$

From "$\phi(x) < B\phi(x)$" (2.16) holds. Suppose (2.16) holds for k. Putting $\theta = \theta_\Delta$ and $\theta\xi(\omega) = \xi(\theta\omega)$, we have

$$\begin{aligned} \theta\tau_k(\omega) &= \tau_k(\theta\omega) \\ &= \min\{j\Delta;\ \phi(X(j\Delta, \theta\omega)) = B^{k-j}\phi(X(j\Delta, \theta\omega))\} \\ &= \tau_{k+1}(\omega) - 1 \end{aligned} \tag{2.18}$$

$$\begin{aligned} \theta d_k(t, \omega) &= d_k(t, \theta\omega) = u_{k,j}(\theta\omega(j\Delta)) \\ &= u_{k+1,j+1}(\omega(j\Delta + \Delta)) = d_{k+1}(t + \Delta, \omega). \end{aligned} \tag{2.19}$$

From "$\phi(x) < B^{k+1}\phi(x)$" we see

$$B^{k+1}\phi(x) = JB^k\phi(x) = T(\Delta, d_{k+1})B^k\phi(x) \tag{2.20}$$

Using (2.18) and (2.19)

$$= E_x^{d_{k+1}} I(\phi, 0, \tau_{k+1}, d_{k+1}),$$

namely (2.16) holds for k+1. Combining (2.16) with (2.10), we complete the proof of Lemma 2.2.

For a binary t the right side of (2.6) is increasing to the right side of (1.9) as $N \to \infty$, by virtue of the right continuity of the paths. On the other hand the left side of (2.6) tends to $U(t)\phi$. Hence (2.5) implies "$v(t)\phi(x) = S(t)\phi(x)$ for binary t" Since $S(t)\phi$ is continuous in t and $v(t)\phi$ is increasing in t, we have "$v(t)\phi = S(t)\phi$ for any $t \geq 0$." This completes the proof of Theorem 1.

3 PROOF OF THEOREM 2

Put $f_0 = \sup_{xu} |f(x, u)|$ $c_0 = \inf_{xu} c(x, u)$ and $F(K, s) = Ke^{\lambda s} + f_0(e^{\lambda s} - 1)\lambda^{-1}$. From (A2) - (A4) we obtain

$$|v(t)\phi(x)| \leq \|\phi\| + f_0 c_0^{-1} \tag{3.1}$$

$$|B\phi(x) - B\phi(y)| \leq F(K, \Delta)|x - y| \text{ for } \phi \in \Sigma(k) \tag{3.2}$$

where $B = B(N)$ and $\Delta = 2^{-N}$. Since $|B\phi(x)| \leq K + f_0\Delta \leq F(K, \Delta)$, $B\phi$ belongs to $\Sigma(F(K, \Delta))$. Repeating this calculation, we have

$$B^j\phi \in \Sigma(F(K, j\Delta)). \tag{3.3}$$

Therefore we can derive

$$v(T)\phi \in \Sigma(F(K, T)) \quad \text{for } \phi \in \Sigma(k). \tag{3.4}$$

On the other hand, for $d \in D$ and $\tau \in m$

$$\begin{aligned} &|E_x^d I(\phi, 0, \tau, d) - E_x^d I(\phi, 0, \tau \wedge T, d)| \\ &\leq (f_0 + \|\phi\|)\exp(-c_0 T). \end{aligned} \tag{3.5}$$

Combining (3.5) with (3.4), for $\varepsilon > 0$ and $\phi \in \Sigma$, we can choose $\delta = \delta(\varepsilon, \phi) > 0$ such that

$$|v(t)\phi(x) - v(t)\phi(y)| < \varepsilon \quad \text{for } |x - y| < \delta \text{ and } t \geq 0. \tag{3.6}$$

Since Σ is dense in C and v is contraction, (3.6) holds for $\phi \in C$ with a positive $\tilde{\delta}(\varepsilon, \phi)$. Therefore recalling (3.1), we derive $\tilde{\Phi} = O_i\text{-}\lim_{t\to\infty} v(t)\phi$.

By the definitions of $\tilde{S}(t)$ and $S(t)$

$$\tilde{S}(t)\phi \leq S(t)\phi. \tag{3.7}$$

Since $\tilde{\Phi}$ is $S(t)$-invariant, we have $\tilde{S}(t)\tilde{\Phi} \leq S(t)\tilde{\Phi} = \tilde{\Phi}$.

From "$\phi \le S(t)\phi \le \tilde{\Phi}$" we conclude that $\tilde{\Phi}$ is $\tilde{S}(t)$-excessive majorant of ϕ.

Let Ψ be $\tilde{S}(t)$-excessive majorant of ϕ. Then

$$B\phi = \phi \vee \sup_u T(\Delta, u)\phi \le \phi \vee \tilde{S}(\Delta)\phi$$
$$\le \Psi \vee \tilde{S}(\Delta)\Psi = \Psi.$$

Hence $B^2\phi \le B\Psi \le \Psi \vee \tilde{S}(\Delta)\Psi \le \Psi$. Repeating this calculation, we have $B^k\phi \le \Psi$. This implies "$S(t)\phi \le \Psi$ for $t \ge 0$". As $t \to \infty$ we have "$\tilde{\Phi} \le \Psi$". Q E D.

4 REFERENCES

1. Bensoussan, A. and Lions, J. L. "Problèmes de temps d'arrêt optimal et inequations variationnelles paraboliques", *Appl. Analy.*, **3**, 267-294, (1973).

2. Cowan, R. and Zabczyk, J. "A new version of the best choice problem", preprint.

3. Davis, M. H. A. "Nonlinear semigroups in the control of partially-observable stochastic system", Measure Th. and Appl. Stoch. Analy., (Oberwolfach 1977), Lect. Notes in Math. Springer-Verlag, (to appear).

4. Dynkin, E. B. and Yushkevich, A. A. "Markov processes, Theorems and Problems", Plenum Press, (English Transl.), (1969).

5. Fleming, W. H. and Rishel, R. V. "Deterministic and Stochastic Optimal Control", *Appl. Math.*, **1**, Springer-Verlag, (1975).

6. Grigelionis, B. I. and Shiryaev, A. N. "On Stefan problem and optimal stopping rules for Markov processes", *Th. Prob. Appl.*, **11**, 541-558, (1966).

7. Itô, K. "Stochastic processes", Lect. Notes, 16, Aarhus Univ., (1969).

8. Krylov, N. V. "Control of a solution of stochastic integral equations", *Th. Prob. Appl.*, **17**, 114-131, (1972).

9. Nagai, H. "On an optimal stopping problem and a variational inequality", *Jour. Math. Soc. Japan*, **30**, 303-312, (1978).

10. Nisio, M. "On stochastic optimal controls and envelope of Markovian semi-groups", Proc. Int. Symp. Stoch. Diff. Equat., Kyoto, 297-325, (1976).

11. Nisio, M. "On nonlinear semigroup associated with stochastic optimal control and its excessive majorant", Proc. Sem. Proba. Th., Banach Center, (to appear), (1976).

12. Nisio, M. "On nonlinear semigroup associated with optimal stopping for Markov processes", *Appl. Math. Opt.*, **4**, 143-169, (1978).

13. Shiryaev, A. N. "Statistical Sequential Analysis", Nauk, (Russian), (1976).

14. Zabczyk, J. "Optimal control by means of switchings", *Studia Math.*, **43**, 161-171, (1973).

15. Zabczyk, J. "Semigroup method in stochastic control theory", Centre Rech. Math., Univ. Montréal, (1978).

2.3 ON THE OPTIMAL SEARCH FOR A TARGET WHOSE MOTION IS A DIFFUSION PROCESS

U. Pursiheimo and M. Ruohonen

(Institute for Applied Mathematics, University of Turku, Turku, Finland)

1 INTRODUCTION

We consider here a search for a target whose motion is a diffusion process in n-dimensional Euclidean space R^n. Similar kind of considerations have been treated by Hellman [3], [4], Perko [6] and Saretsalo [10]. All these works involve rather strong continuity assumptions for the search density function, which we here denote by $\alpha(x,t)$. As a consequence of these assumptions the results of Koopman [5] and Arkin [1] concerning the optimal search in the case of a stationary target cannot be obtained from above results, when the diffusion process is slowed down such that the target becomes stationary. For further references on the theory of search the reader is referred to Stone [11].

In this work we approach the problem by noticing that the diffusion process can be represented with the model of the generalized conditionally deterministic motion in [7]. By using this model we get a necessary and sufficient condition for the optimality of the search density function α. The objective function to be minimized is the probability of non-detection of the target during a fixed search time. This problem can be interpreted also as the maximization of the probability of stopping a diffusion process with stopping rate $\alpha(x(t),t)$ with appropriate constraints (see Theorem 2 below).

Further, we present an algorithm for determining the optimal search density in a discrete approximation of the process. The discretization is performed both in time and in space. The algorithm is a modification of an algorithm presented by Brown in [2]. Our algorithm contains as an essential feature the use of maximal error as termination criterion. The maximal error is obtained by the use of duality as shown in [8].

The starting point for the algorithm is the necessary and sufficient condition for the search density to be optimal in the diffusion process case. Although originally developed for diffusion processes, the algorithm itself applies to Markov processes in general. It has been applied to the one-dimensional Ornstein-Uhlenbeck diffusion model for which a natural discretization is available.

2 MOTION OF THE TARGET

We shall consider here the following type of diffusion process ω in R^n, which describes the motion of the target during the search. The transition probability

$$P(t,x,s,A) = P\{\omega(s) \in A \text{ and the target is not found} | \omega(t)=x\}$$

when $t < s$ and $A \in B^n$ (= the Borel sets of R^n), satisfies the following conditions for all $\varepsilon > 0$

$$\lim_{h\to 0+} \int_{|x-y|\geq\varepsilon} P(t,x,t+h,dy) = 0 \tag{1}$$

$$\lim_{h\to 0+} \frac{1}{h}\int_{|x-y|\leq\varepsilon} P(t,x,t+h,dy)(y^i-x^i) = b^i(x,t) \tag{2}$$

$$\lim_{h\to 0+} \frac{1}{h}\int_{|x-y|\leq\varepsilon} P(t,x,t+h,dy)(y^j-x^j)(y^i-x^i) = a^{ij}(x,t) \tag{3}$$

$$\lim_{h\to 0+} \frac{1}{h}\left[1 - \int_{|x-y|\leq\varepsilon} P(t,x,t+h,dy)\right] = \alpha(x,t) \tag{4}$$

where we have used notations $x = (x^1,\ldots,x^n)$, $y = (y^1,\ldots,y^n)$ and $|x|$ as the usual Euclidean distance in R^n. We further suppose that functions a^{ij} and b^i are continuous and have continuous first and second derivatives. The function α is supposed to be a measurable function on $R^n\times[0,T]$ satisfying additionally the following seminorm conditions

$$\int_{R^n} \operatorname*{ess\,sup}_{t\in[0,T]} |\alpha(x,t)|dx < \infty$$

$$\int_0^T \operatorname*{ess\,sup}_{x\in R^n} |\alpha(x,t)|dt < \infty$$

and further that $\alpha(x,t) \geq 0$ for a.e. $(x,t) \in R^n \times [0,T]$.

In [10] a partial differential equation of parabolic type for the transition rate function $p(t,x,s,y)$ in the case when α is continuous and bounded is presented. The same result appears also in [4], but with still stronger continuity assumptions. We shall now prove a similar result, but now α satisfies only conditions stated above.

<u>Theorem 1</u>. The transition rate function $p(t,x,s,y)$ corresponding to the transition probability in (1) - (4) satisfies in a small neighbourhood of almost every $(y,s) \in R^n \times [0,T]$ the following equation

$$\frac{\partial}{\partial s} p(t,x,s,y) = \frac{1}{2} \sum_{i=1}^{n} \sum_{j=1}^{n} \frac{\partial^2}{\partial y^i \partial y^j}(a^{ij}(y,s)p(t,x,s,y))$$
$$- \sum_{i=1}^{n} \frac{\partial}{\partial y^i} (b^i(y,s)p(t,x,s,y)) - \alpha(y,s)p(t,x,s,y) \tag{5}$$

with initial condition $p(t,x,t,y) = \delta(x-y)$.

<u>Proof</u>: Let $f(x)$ be a finite and bounded function with continuous first and second order partial derivatives. For such a function we have by using the Chapman-Kolmogorov equation and a Taylor expansion for f

$$\int_{R^n} P(t,x,s+h,dy)f(y) = \int_{R^n}\int_{R^n} P(t,x,s,dz)P(s,z,s+h,dy)f(y)$$

$$= \int_{R^n}\int_{U_\varepsilon(z)} \Big[f(z) + \sum_{i=1}^{n} \frac{\partial f(z)}{\partial z^i} (y^i - z^i) +$$

$$+ \frac{1}{2} \sum_{i=1}^{n} \sum_{j=1}^{n} \frac{\partial^2 f(z)}{\partial z^i \partial z^j} (y^i-z^i)(y^j-z^j) + \gamma|y-z|^2\Big]\times$$

$$\times P(t,x,s,dz)P(s,z,s+h,dy) +$$

$$+ \int_{R^n}\int_{V_\varepsilon(z)} P(t,x,s,dz)P(s,z,s+h,dy)f(y),$$

where $U_\varepsilon(z) = \{y \mid |y-z| \leq \varepsilon\}$, $V_\varepsilon(z) = \{y \mid |y-z|y > \varepsilon\}$, $|\gamma| = |\gamma(z,y)| < n^2\delta/2$, when $|y-z| \leq \varepsilon$; see [12]. We integrate the first integral in four parts:

$$\int_{R^n}\int_{U_\varepsilon(z)} f(z)P(t,x,s,dz)P(s,z,s+h,dy)$$

$$= \int_{R^n} f(z)P(t,x,s,dz) \int_{U_\varepsilon(z)} P(s,z,s+h,dy)$$

$$= \int_{R^n} f(z)P(t,x,s,dz)\left[1 - \alpha(z,s)h + o(h)\right]$$

according to the condition (4). Secondly

$$\int_{R^n}\int_{U_\varepsilon(z)} \sum_{i=1}^{n} \frac{\partial f(z)}{\partial z^i} (y^i - z^i)P(t,x,s,dz)P(s,z,s+h,dy)$$

$$= \int_{R^n} \sum_{i=1}^{n} \frac{\partial f(z)}{\partial z^i} \left[b^i(z,s)h + o(h)\right]P(t,x,s,dz)$$

$$= \int_{R^n} \sum_{i=1}^{n} \frac{\partial f(z)}{\partial z^i} \left[b^i(z,s)p(t,x,s,z)\right]dz\cdot h + o(h)$$

$$= -\int_{R^n} \sum_{i=1}^{n} \frac{\partial}{\partial z^i} \left[b^i(z,s)p(t,x,s,z)\right]f(z)dz\cdot h + o(h),$$

where (2) and integration by parts has been used. Similarly

$$\int_{R^n}\int_{U_\varepsilon(z)} \frac{1}{2} \sum_{i=1}^{n} \sum_{j=1}^{n} \frac{\partial^2 f(z)}{\partial z^i \partial z^j} (y^i - z^i)(y^j - z^j) P(t,x,s,dz) P(s,z,s+h,dy)$$

$$= \int_{R^n} \frac{1}{2} \sum_{i=1}^{n} \sum_{j=1}^{n} \frac{\partial^2 f(z)}{\partial z^i \partial z^j} [a^{ij}(z,s) p(t,x,s,z)] dz \cdot h + o(h)$$

$$= \int_{R^n} \frac{1}{2} \sum_{i=1}^{n} \sum_{j=1}^{n} \frac{\partial^2}{\partial z^i \partial z^j} [a^{ij}(z,s) p(t,x,s,z)] f(z) dz \cdot h + o(h).$$

Finally, the absolute value of the residual term of the first integral is

$$\left| \int_{R^n}\int_{U_\varepsilon(z)} \gamma |y-z|^2 P(t,x,s,dz) P(s,z,s+h,dy) \right|$$

$$\leq \int_{R^n}\int_{U_\varepsilon(z)} |\gamma| \, |y-z|^2 P(t,x,s,dz) P(s,z,s+h,dy)$$

$$\leq \frac{n^2}{2}\delta \int_{R^n} \left[\sum_{i=1}^{n} a^{ii}(z,s) h + o(h) \right] P(t,x,s,dz)$$

$$= \frac{n^2}{2}\delta \left[\sum_{i=1}^{n} \int_{R^n} a^{ii}(z,s) P(t,x,s,dz) h + o(h) \right]$$

The second integral may be approximated as follows

$$\int_{R^n}\int_{V_\varepsilon(z)} f(y) P(s,z,s+h,dy) P(t,x,s,dz)$$

$$\leq ||f|| \int_{R^n}\int_{V_\varepsilon(z)} P(s,z,s+h,dy) P(t,x,s,dz)$$

$$= ||f|| \int_{R^n} o(h) P(t,x,s,dz) = ||f|| \, o(h),$$

where $||f||$ denotes the maximum value of f. Collecting these results together we get in a small neighbourhood of almost all $s \in [0,T]$

$$\int_{R^n} P(t,x,s+h,dy)f(y) = \int_{R^n} P(t,x,s,dz)f(z) -$$

$$- \int_{R^n} \alpha(z,s)P(t,x,s,dz)h -$$

$$- \int_{R^n} \sum_{i=1}^{n} \frac{\partial}{\partial z^i} \left[b^i(z,s)p(t,x,s,z)\right] f(z)dz \cdot h + \tag{6}$$

$$+ \int_{R^n} \frac{1}{2} \sum_{i=1}^{n} \sum_{j=1}^{n} \frac{\partial^2}{\partial z^i \partial z^j} \left[a^{ij}(z,s)p(t,x,s,z)\right] f(z)dz \cdot h + r,$$

where

$$|r| \leq \frac{n^2}{2}\delta\left[\int_{R^n} \sum_{i=1}^{n} a^{ii}(z,s)P(t,x,s,dz)h + o(h)\right] + o(h).$$

Now

$$\left|\frac{r}{h}\right| \leq \frac{n^2\delta}{2}\left[\int_{R^n} \sum_{i=1}^{n} a^{ii}(z,s)P(t,x,s,dz) + \frac{o(h)}{h} \right] + \frac{o(h)}{h} .$$

If we let $h \to 0$ then $o(h)/h \to 0$ and thus $o(h)/h < \delta$, if $h < h_0$. Hence, the residual term can be made arbitrarily small by choosing ε and h_0 conveniently. For this reason (6) gives the following result when $h \to 0$.

$$\int_{R^n} \left[\frac{\partial p(t,x,s,y)}{\partial s} + \alpha(y,s)p(t,x,s,y) + \right.$$

$$+ \sum_{i=1}^{n} \frac{\partial}{\partial y^i} (b^i(y,s)p(t,x,s,y)) -$$

$$\left. - \frac{1}{2} \sum_{i=1}^{n} \sum_{j=1}^{n} \frac{\partial^2}{\partial y^i \partial y^j} (a^{ij}(y,s)p(t,x,s,y))\right] f(y)dy = 0.$$

From this it follows that the expression in brackets must be zero for almost all y in R^n.

3 MOTION AS A GENERALIZED CONDITIONAL DETERMINISTIC MODEL

If we let $\alpha(x,t)$ denote the search density function, then condition (4) states that the probability that the target is found during a small time interval $(t,t+h)$, supposing that the target is at point x, is equal to

$$\alpha(x,t)h + o(h)$$

for almost all $(x,t) \in R^n \times [0,T]$. This means that if some amount u of search density accumulates during time $[0,T]$ on the target at point x then the target is not found during this time $[0,T]$ with probability $\exp(-u)$.

Now if we have fixed search density α and we let the target move along some route $\omega(t)$ $(0 \leq t \leq T)$, then amount

$$u = \int_0^T \alpha(\omega(t),t)dt$$

of search density accumulates on the target and hence, the target is not found during this time $[0,T]$ when it moves along the route $\omega(t)$ with probability

$$\exp(-\int_0^T \alpha(\omega(t),t)dt) \tag{7}$$

Let (Ω,S,P) be the probability space consisting of all possible routes $\omega \in \Omega$. All the routes ω are some realizations of the diffusion process corresponding to equations (1) - (4) with $\alpha \equiv 0$, i.e., when no search is carried out. By using (7) we directly get the probability that the target is not found during time $[0,T]$ as

$$\int_\Omega \exp(-\int_0^T \alpha(\omega(t),t)dt)dP(\omega). \tag{8}$$

In order to get this in the form of the generalized conditionally deterministic motion we define routes $\beta(x,\omega,t)$ depending on the initial point x, parameter ω, and time t as follows

$\beta(x,\omega,t) = x + \omega(t) - \omega(0)$,

which obviously satisfies the condition $\beta(x,\omega,0) = x$. The Jacobian determinant corresponding to the mapping of the initial point x to the location point $y = \beta(x,\omega,t)$ at the moment t for fixed ω is equal to 1. The inverse function of this mapping is also obvious: if

$$y = x + \omega(t) - \omega(0)$$

then

$$x = y - \omega(t) + \omega(0).$$

In this way however, many such routes β that are not included in the original Ω come into consideration. For this reason, to get only such routes β, for which really $\beta(x,\omega,0) = \omega(0)$, i.e., for which $x = \omega(0)$, we define a probability measure P^* in $\mathcal{B}^n \otimes \mathcal{S}$, where $\mathcal{B}^n$ is the set of all Borel sets in R^n, as follows: for $F \in \mathcal{B}^n \otimes \mathcal{S}$ let

$$P^*\{(x,\omega) \in F\} = \iint_F f(x)\,dx\,dP(\omega|x)$$

where f(x) is the probability density function of the initial point x of the process and $P(\omega|x)$ is the conditional probability of ω on $\mathcal{S}$ when the initial point of the route is supposed to be x.

Hence, the probability of non-detection (8) can be written as

$$P(T|\alpha) = \int_{R^n}\int_{\Omega} f(x)\exp\left(-\int_0^T \alpha(\omega(t),t)\,dt\right)dx\,dP(\omega|x).$$

This expression is now in the form where we can use results directly from [7] and get the following theorem in the case when the search effort $\int\alpha(x,t)\,dx$ at each moment is limited.

Theorem 2. A necessary and sufficient condition for α^* to be the solution of

$$P(T|\alpha^*) = \min_{\alpha} \{P(T|\alpha) \mid \int_{R^n} \alpha(x,t)\,dx \leqq M\}$$

is that there exists function $\lambda(t) \geq 0$ such that for almost all $(x,t) \in R^n \times [0,T]$ we have

$$p(x,t)q(x,t,T) \begin{cases} = \lambda(t), & \text{if } \alpha^*(x,t) > 0 \\ \leq \lambda(t), & \text{if } \alpha^*(x,t) = 0 \end{cases} \tag{9}$$

where p satisfies partial differential equation

$$\frac{\partial}{\partial t} p(x,t) = \frac{1}{2} \sum_{i=1}^{n} \sum_{j=1}^{n} \frac{\partial^2}{\partial x^i \partial x^j} \left[a^{ij}(x,t)p(x,t)\right] - \sum_{i=1}^{n} \frac{\partial}{\partial x^i} \left[b^i(x,t)p(x,t)\right] - \alpha(x,t)p(x,t) \tag{10}$$

with initial condition

$p(x,0) = f(x)$,

and q the equation

$$\frac{\partial}{\partial t} q(x,t,T) = -\frac{1}{2} \sum_{i=1}^{n} \sum_{j=1}^{n} a^{ij}(x,t) \frac{\partial^2}{\partial x^i \partial x^j} q(x,t,T) - \sum_{i=1} b^i(x,t) \frac{\partial}{\partial x^i} q(x,t,T) + \alpha(x,t)q(x,t,T) \tag{11}$$

with boundary condition

$q(x,T,T) = 1$.

These equations must be satisfied and the partial derivatives in question must be continuous in a small neighbourhood of almost all $(x,t) \in R^n \times [0,T]$.

<u>Remark</u>. The function $p(x,t)$ is the generalized probability density function of the target at moment t including the probability that the target is not found during time $[0,t]$. Hence,

$$P(T|\alpha) = \int_{R^n} p(x,T)\,dx.$$

Function q(x,t,T) is the probability that the target has not been found during time [t,T] when it has been at point x at moment t. Hence, we have also

$$P(T|\alpha) = \int_{R^n} f(x)q(x,0,T)dx.$$

<u>Proof</u>. The expression D in the necessary and sufficient condition for the optimality in [7] gives the following expression (with changed sign)

$$\int_{\Omega} f(x-\omega(t)+\omega(0))\exp(-\int_0^T \alpha(x-\omega(t)+\omega(\tau),\tau)d\tau)dP(\omega|x-\omega(t)+\omega(0))$$

which can be written, too, in the form

$$\int_{\Omega\cap\{\omega(t)=x\}} \exp(-\int_0^T \alpha(\omega(\tau),\tau)d\tau)dP(\omega), \qquad (12)$$

which is the generalized a posteriori probability density function of the location x of the target at moment t including the probability that the target has not been found during time [0,T]. If we define functions p(x,t) and q(x,t,T) as in the remark above, then (12) can be stated simply in the form

$$p(x,t)q(x,t,T),$$

which gives the part (9) of the theorem.

We still have to prove that the partial differential equations (10) and (11) must hold for p and q. To obtain (10) we only have to notice that

$$p(x,t) = \int_{R^n} f(y)p(0,y,t,x)dy$$

and use this on both sides of (5). The initial condition follows from the initial condition of (5).

To obtain (11) we first note that

$$q(x,t,T) = \int_{R^n} q(y,t+h,T)P(t,x,t+h,dy).$$

As in the proof of theorem 1 we have now

$$q(x,t,T) = \int_{U_\varepsilon(x)} \Big[q(x,t+h,T) + \sum_{i=1}^{n} (y^i - x^i) \frac{\partial q(x,t+h,T)}{\partial x^i} +$$

$$+ \frac{1}{2} \sum_{i=1}^{n} \sum_{j=1}^{n} (y^i - x^i)(y^j - x^j) \frac{\partial^2 q(x,t+h,T)}{\partial x^i \partial x^j} +$$

$$+ \gamma |y-x|^2 \Big] P(t,x,t+h,dy)$$

$$+ \int_{V_\varepsilon(x)} q(x,t+h,T) P(t,x,t+h,dy),$$

where $|\gamma| \leq n^2\delta/2$, when $|y-x| < \varepsilon$. Apart from the residual term the first integral above is

$$q(x,t+h,T)\left[1 - \alpha(x,t)h + o(h)\right] + \sum_{i=1}^{n} b^i(x,t) \frac{\partial q(x,t+h,T)}{\partial x^i} +$$

$$+ \frac{1}{2} \sum_{i=1}^{n} \sum_{j=1}^{n} a^{ij}(x,t) \frac{\partial^2 q(x,t+h,T)}{\partial x^i \partial x^j} + o(h).$$

The residual term and the second integral are treated as in the proof of theorem 1. This gives (11) if q with its derivatives is a continuous function of t.

4 THE OPTIMIZATION ALGORITHM

In the previous section we have considered necessary and sufficient conditions for the optimal α^*. In this section we consider methods to obtain an α satisfying these conditions. Now the interpretations of functions $p(x,t)$ and $q(x,t,T)$ and the optimality condition (9) give the basic ideas (originally due to Brown [2], here somewhat developed further) to an optimization algorithm for finding α^*. In order to get a computer algorithm we discretize the model with respect to time and space and find the discrete analogues for functions p and q. Let the time interval $[0,T]$ be divided into M sub-intervals of length Δt and let the region of definition of our Markov process be divided into N sub-regions of volume Δx.

We use the following notation for the discretized process:

$X(j)$ = the sub-region, where the target is at the interval j

$q(i,k:j) = P\{X(j+1)=k \mid X(j)=i\}$

$P(i,j) = P\{X(j)=i$ and the target has not been found at the intervals $1,\ldots,j-1\}$

$Q(i,j) = P\{$the target is not found at the intervals $j+1,\ldots,M \mid X(j)=i\}$

$\alpha(i,j)$ = the search effort in the region i at the interval j

$$\sum_{i=1}^{n} \alpha(i,j) = m(j)$$

where $m(j)$ is the available amount of search effort at the interval j.

With this notation

$$P(T|\alpha) = \sum_{i=1}^{N} P(i,1)Q(i,1)\exp(-\alpha(i,1)\Delta t/\Delta x).$$

The probability $P(T|\alpha)$ may also be expressed as

$$P(T|\alpha) = \sum_{i=1}^{N} P(i,j)Q(i,j)\exp(-\alpha(i,j)\Delta t/\Delta x).$$

The product $P(i,j)Q(i,j)$ is a same kind of a term that appears in the necessary and sufficient condition for optimality in theorem 2.

Suppose we have some allocation α. Then if we change α in such a way that in the interval j we substitute the values $\alpha(.,j)$ with feasible values that are optimal with respect to the generalized distribution $P(i,j)Q(i,j)$, then with this new allocation $P(T|\alpha)$ is decreased. The algorithm of Brown is based on this fact. We have modified this algorithm in order to speed up the convergence. Brown determines the optimal allocation for each stage always starting from the first time interval 1 and going to the last time interval M. Then the $P(i,j)$ and $Q(i,j)$ are updated from M to 1 without optimization. Our modification uses a zig-zag algorithm in which the optimization is performed at each time interval from 1 to M and also backwards from M to 1. We have also used a different termination rule. Instead of comparing two successive values of $P(T|\alpha)$ as Brown we have made use of duality and determined the maximal error if we use the contemporary α instead of the optimal α^*.

The algorithm uses the following updating formulae

$$P(i,j) = \sum_{k=1}^{N} q(k,i:j-1)P(k,j-1)\exp(-\alpha(k,j-1)\Delta t/\Delta x) \qquad (13)$$

$$Q(i,j) = \sum_{k=1}^{N} q(i,k:j)Q(k,j+1)\exp(-\alpha(k,j+1)\Delta t/\Delta x). \qquad (14)$$

The algorithm is as follows:

STEP 1. Set

$P(i,j) = 0,\ j=2,\ldots,M,\ i=1,\ldots,N$

$P(i,1) = g(i),\ i=1,\ldots,N$ (initial distribution)

$Q(i,j) = 1\ \forall\ i,j$

$\alpha(i,j) = 0\ \forall\ i,j$

$j=1$.

STEP 2. For j, calculate the generalized distribution

$D(i,j) = P(i,j)Q(i,j),\ i=1,\ldots,N$

and determine the corresponding optimal $\alpha(i,j)$, $i=1,\ldots,N$.

STEP 3. Set $j=j+1$ and calculate $P(i,j)$ according to formula (13).

STEP 4. Check whether $j=M$. If yes, then go to STEP 5, otherwise go to STEP 2.

STEP 5. Set $j=j-1$ and calculate $Q(i,j)$ according to formula (14).

STEP 6. For j, determine the generalized distribution

$D(i,j) = P(i,j)Q(i,j),\ i=1,\ldots,N$

and the corresponding optimal $\alpha(i,j)$ $i=1,\ldots,N$

STEP 7. Check whether $j=1$. If not, go to STEP 5. If yes go to STEP 8.

STEP 8. Calculate

$$\sum_{i=1}^{N} P(i,1)Q(i,1)\exp(-\alpha(i,1)\Delta t/\Delta x).$$

STEP 9. Check the termination rule; then go to STEP 2 or stop.

One suitable termination rule is such that the algorithm performs a predetermined number of stages, where one stage consists of the STEPS 2-8. After this the maximal error is calculated and the user decides whether to go on or to stop. Of course this decision could be added to the algorithm itself, but it is usually more convenient to follow the convergence by man-machine interaction in order to avoid excessive computations in case of very slow convergence. The maximal error is

$$E(\alpha) = \sum_{i=1}^{N} \sum_{j=1}^{M} (\nu(j) - D(i,j,\alpha))\alpha(i,j),$$

where

$$D(i,j,\alpha) = P(i,j)Q(i,j)\exp(-\alpha(i,j)\Delta t/\Delta x)\Delta t/\Delta x$$

$$\nu(j) = \max_{i} D(i,j,\alpha),$$

see [8]. For this it is necessary to calculate P(i,j) again step by step according to formula (13) while the values of Q(i,j) are already available.

The algorithm has been applied to one-dimensional Ornstein-Uhlenbeck diffusion model for which a natural discretization is available. The results are reported in [9]. In this application slow convergence of the algorithm was noticed. This is due to the fact that usually the optimum is very flat and two consecutive values of $P(T|\alpha)$ may be near each other and yet be far from the optimum. For this reason the use of maximal error as a criterion for the termination of the algorithm is essential. With this error $E(\alpha)$ we can determine upper and lower bounds for the optimal $P(T|\alpha^*)$,

$$P(T|\alpha) - E(\alpha) \leqq P(T|\alpha^*) \leqq P(T|\alpha),$$

and for the optimal value $\alpha = \alpha^*$ the three values coincide.

5. REFERENCES

1. Arkin, V. I. "Uniformly optimal strategies in search problems", *Theor. Prob. Appl.*, **9**, 674-680, (1964).

2. Brown, S. S. "Optimal search for a moving target in discrete time and space", submitted for publication.

3. Hellman, O. "On the effect of search upon the probability distribution of a target whose motion is a diffusion process", *Ann. Math. Stat.*, **41**, 1717-1724, (1970).

4. Hellman, O. "On the optimal search for a randomly moving target, *SIAM J. Appl. Math.*, **22**, 545-552, (1972).

5. Koopman, B. O. "The theory of search, Part III, the optimum distribution of searching effort", *Oper. Res.*, **5**, 613-626, (1957).

6. Perko, A. "Some problems of search and detection", Rep. Inst. Appl. Math. Univ. Turku, 21, (1971).

7. Pursiheimo, U. "On the optimal search for a target whose motion is conditionally deterministic with stochastic initial conditions on location and parameters", *SIAM J. Appl. Math.*, **32**, 105-114, (1977).

8. Pursiheimo, U. "Conjugate duality in optimization of search for a target with generalized conditionally deterministic motion", to appear in Journal of Optimization Theory and Applications.

9. Pursiheimo, U. and Ruohonen, M. "An optimization algorithm for search for a Markov moving target", Rept. Inst. Appl. Math. Univ. Turku, 87, (1978).

10. Saretsalo, L. "On stochastic models of search for stationary and moving objects", Publ. Inst. Appl. Math. Univ. Turku, 2, (1971).

11. Stone, L. D. "The theory of optimal search", Academic Press, New York, (1975).

12. Wentzell, A.D. "Theorie Zufälliger Prozesse", Berlin, Akademie-Verlag, (1979).

2.4 EQUIVALENT MARKOVIAN DECISION PROBLEMS

W. R. S. Sutherland

(Mathematics Department, Dalhousie University, Halifax, N.S., Canada, B3H 4H8)

1 INTRODUCTION

Optimal economic growth theory deals with the question of how an economy chooses between consumption and investment of its stock of capital. For an introduction to the extensive literature on this question see [2] or [8]. The simplest continuous-time model, due to Cass [3] and Koopmans [5], has been studied using methods of optimal control theory. We will consider a discrete-time version, due to Gale [4], as extended to the case of uncertain returns on investment by Brock and Mirmàn [1]. This discrete-time model has been analysed by using methods of dynamic or mathematical programming. The purpose of this paper is to suggest how linear programming may be used to study this question of investment versus consumption.

Specifically, assume that there is a single type of capital. Let $y_t \geqq 0$ be the stock of capital at time $t = 0,1,2,...$ and $c_t \geqq 0$, $x_t \geqq 0$ be the stocks which are, respectively, consumed and invested during time period t. These must satisfy the budget constraint, $x_t + c_t \leqq y_t$. An investment x_t in period t produces a stock $y_{t+1} = f_\omega(x_t)$ in period t+1 where $Y = F(x) = f_\omega(x)$ is a random variable with a known distribution for $\omega \epsilon \Omega$. A typical random production function is illustrated in Fig. 1.

The satisfaction of consumption c is denoted by a utility function u(c) which is assumed to be discounted by a factor $0 < \delta < 1$. A non-negative sequence (x_t, c_t) defines a feasible plan, from a given initial stock $y_0 > 0$, if the budget

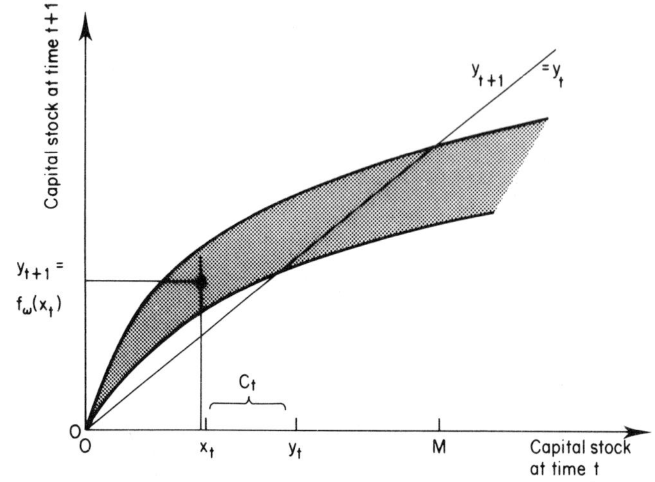

Fig. 1 The Consumption/Investment Possibilities

constraints are satisfied. A feasible plan is optimal if it maximizes the expected value of the sum of discounted utilities. Letting E denote expected value, this economic growth model can be stated as a nonlinear programming problem:

Given $y_0(\omega_0)$ find random variables

$$x_t(\omega_0,\ldots,\omega_t),\ c_t(\omega_0,\ldots,\omega_t),\ y_t(\omega_0,\ldots,\omega_t)$$

which maximize $E(\Sigma_{t=0}^{\infty}\ \delta^t u(c_t))$ subject to

$$x_t + c_t \leqq y_t,\ y_{t+1} = f_{\omega_{t+1}}(x_t),$$

$$x_t \geqq 0,\ c_t \geqq 0,\ y_t \geqq 0 \text{ almost surely for } t = 0,1,2,\ldots\ .$$

Under the usual assumptions[1] on the utility and production functions, Brock and Mirman [1] (see also [7]) have shown that there exists an optimal plan given by $c_t = \pi(y_t)$, $x_t = y_t - \pi(y_t)$ where the policy function π is a continuous, increasing

1 For each $\omega\varepsilon\Omega$, $f_\omega(x)$ is an increasing concave differentiable function with $f_\omega(0) = 0$, $f'_\omega(0) = +\infty$, $f'_\omega(+\infty) = 0$, and $u(c)$ is an increasing concave differentiable function with $u(0) = 0$, $u'(0) = +\infty$.

function of the observed stock y_t. Letting $V(y_0)$ denote the expected value of this optimal plan, then V satisfies the functional equation

$$V(y_0) = \underset{0 \le c_0 \le y_0}{\text{maximum}}\{u(c_0) + \delta E(V(f_\omega(y_0 - c_0)))\}$$

which is the basis of the dynamic programming approach. The optimal plan is characterized by the following necessary and sufficient conditions:

$$\text{(i)} \quad u'(c_t) = \delta \int_\Omega f'_\omega(x_t) u'(c_{t+1}) d\omega, \text{ all } t,$$

$$\text{(ii)} \quad E(\delta^t u'(c_t) x_t) \to 0 \text{ as } t \to \infty.$$

These conditions are analogous to the Euler equation and a transversality condition for the continuous-time version of the model. In the discrete-time model these conditions are obtained by applying convex duality theory to a finite planning model ($t = 0,1,\ldots,T$) and letting $T \to \infty$. As well as proving existence of optimal plans, this dynamic programming approach yields a number of qualitative properties of optimal plans: most notably-the turnpike property - the convergence of the investment process to a steady-state distribution for all initial stocks.

The purpose of this paper is to propose an alternate approach, based on linear programming, to the existence and computation of the optimal plans. This approach is outlined in Section 2 and culminates in a conjecture about the solution of a continuous linear programming problem. Section 3 supports this conjecture by showing that the approach is successful for a discretized version of the economic growth model. Finally Section 4 applies this discretized approach to a numerical example for which the optimal policy function π is known.

2 AN EQUIVALENT MODEL APPROACH

This section proposes an alternative to the usual dynamic programming approach. It is based on a concept of equivalent economic growth models -that is, models which have the same optimal policy function π-and is applicable in principle to more general Markovian decision problems. First we need to be more specific about the formula for the expected value of a feasible plan.

Let P denote the family of functions $\pi: I \rightarrow I$ such that $\pi(y) \leqq y$, where $I = \{y: 0 \leq y \leq M\}$ for some sufficiently large constant M. A sequence of functions $\pi = \{\pi_0, \pi_1, \ldots\}$ where $\pi_t \varepsilon P$ is called a <u>policy</u>. If $\pi_t = \pi_0$ for all t, then this is a <u>stationary</u> policy. Every policy defines a feasible plan by means of the stochastic difference equations

$$c_t = \pi_t(y_t)$$

$$x_t = y_t - \pi_t(y_t)$$

$$y_{t+1} = f_{\omega_{t+1}}(y_t - \pi_t(y_t))$$

for $t = 0,1,\ldots$. The one-step transition probabilities for the Markov process $\{y_0, y_1, y_2, \ldots\}$ associated with a policy π are assumed to be represented by the density functions $g^{\pi_t}(y_t, y_{t+1})$. Let $g^{\pi_{-1}}(y_{-1}, y_0)$ be the degenerate density for the single point $\{y_0\}$. The satisfaction associated with this policy is assumed to be represented by a function $r: P \times I \rightarrow R$ where $r^{\pi}(y) = u(\pi(y))$. Thus, if the economy follows the policy π from given initial stock y_0, then the expected <u>value</u> is given by

$$V(r,\pi)(y_0) =$$

$$\sum_{t=0}^{\infty} \delta^t \int_I \cdots \int_I \left[g^{\pi_0}(y_0, y_1) g^{\pi_1}(y_1, y_2) \cdots g^{\pi_{t-1}}(y_{t-1}, y_t)\right] r^{\pi_t}(y_t)\, dy_1 dy_2 \ldots dy$$

which is assumed to exist for all policies π. A policy π is <u>optimal</u> from initial stock y_0 if

$$V(r,\pi)(y_0) \geqq V(r,\theta)(y_0)$$

holds for all policies θ.

The alternative approach is based on two special concepts: trivial models and equivalent models. These ideas were previously developed in [9] for the discrete Markov programming problem. Since we will be considering various economic growth models which have the same random production function, we henceforth refer to a model by simply its utility or reward function (r).

Definition: A model (r) is trivial if

(i) $r^z(y) \leqq 0$ holds for all $0 \leq z \leq y$ where $y \varepsilon I$.

(ii) for each $y \varepsilon I$ there exists some $z = \pi(y)$, $0 \leq z \leq y$, such that $r^{\pi(y)}(y) = 0$.

Lemma 1. If (r) is a trivial model then π of (ii) is an optimal policy.

□ Let π be the function given in condition (ii), then as $\pi \varepsilon P$ it defines a feasible plan such that $V(r,\pi)(y_0) = 0$ for each $y_0 \varepsilon I$. Let θ be any policy, then by condition (i), $V(r,\theta)(y_0) \leqq 0$ proving the optimality of the stationary policy π for all initial states. □

Definition: Two models (r) and (s) are equivalent if the function $V(r,\pi)(y_0) - V(s,\pi)(y_0)$ is independent of the choice of policy π.

Lemma 2: Equivalent models have the same optimal policies.

□ Since $V(r,\pi)(y_0) - V(s,\pi)(y_0) = V(r,\theta)(y_0) - V(s,\theta)(y_0)$ holds for all policies π and θ, then $V(r,\pi)(y_0) - V(r,\theta)(y_0) = V(s,\pi)(y_0) - V(s,\theta)(y_0)$ so that π is an optimal policy for (r) if and only if π is an optimal policy for (s). □

The basis of the alternative approach is to show that the original model is equivalent to some trivial model (which then yields the opitmal policy function π). We next prove that equivalence between two models is characterized by a functional equation. To obtain this we need to assume that there is a special choice of consumption, denoted by $\pi(y) = *$, which terminates the economy with probability one and which yields a final reward or scrap value $r^*(y)$. Policies now consist of functions in P^*. That is, $\pi_t : I \to I \cup \{*\}$ such that $0 \leq \pi_t(y) \leq y$ or $\pi_t(y) = *$. Also $g^*(y_0,y_1)$ is the degenerate density for the single point $\{*\}$ in $I \cup \{*\}$, hence $g^*(y_0,y_1) = 0$ for all $y_1 \varepsilon I$.

Lemma 3: Two models (r) and (s) are equivalent if and only if the functional equation

$$r^{\pi}(y) - r^*(y) + \delta \int_I g^{\pi}(y,z) r^*(z) dz$$

$$= s^{\pi}(y) - s^*(y) + \delta \int_I g^{\pi}(y,z) s^*(z) dz$$

holds for all $\pi \varepsilon P$, all $y \varepsilon I$.

□ Suppose that the functional equation is satisfied, then for any policy π we have

$$V(r,\pi)(y_0) - V(s,\pi)(y_0)$$

$$= \sum_{t=0}^{\infty} \delta^t \int_I .. \int_I [g^{\pi_0}(y_0,y_1)..g^{\pi_{t-1}}(y_{t-1},y_t)](r^{\pi_t}(y_t) - s^{\pi_t}(y_t)) dy_1 .. dy_t$$

$$= \sum_{t=0}^{\infty} \delta^t \int_I \ldots \int_I [g^{\pi_0}(y_0,y_1)\ldots g^{\pi_{t-1}}(y_{t-1},y_t)] [(r^*(y_t) - s^*(y_t))$$

$$-\delta \int_I g^{\pi_t}(y_t,y_{t+1})(r^*(y_{t+1}) - s^*(y_{t+1}))dy_{t+1}] dy_1 \ldots dy_t$$

and rearranging the order of integration this becomes a "telescoping" infinite series given by

$$\lim_{T\to\infty} [(r^*(y_0) - s^*(y_0)) - \delta^{T+1} \int_I \ldots \int_I [g^{\pi_0}(y_0,y_1)\ldots g^{\pi_T}(y_T,y_{T+1})]$$

$$(r^*(y_{T+1}) - s^*(y_{T+1}))dy_1 \ldots dy_{T+1}].$$

But as the integral is convergent and $0 < \delta < 1$, we then have $V(r,\pi)(y_0) - V(s,\pi)(y_0) = r^*(y_0) - s^*(y_0)$
which is independent of the choice of policy π. Conversely, suppose that (r) and (s) are equivalent, then setting

$\pi = \{\pi,*,*,...\}$ and $\theta = \{*,*,*,...\}$ for some $\pi\varepsilon P$, the equation

$$V(r,\pi)(y_0) - V(r,\theta)(y_0) = V(s,\pi)(y_0) - V(s,\theta)(y_0) \qquad \square$$

reduces to the required functional equation.

Note that the functional equation is linear in the functions r and s. If we take the original economic model and define

$$b^{\pi}(y) = r^{\pi}(y) - r^*(y) + \delta \int_I g^{\pi}(y,z)r^*(z)dz$$

and let $w^z(y) \geqq 0$ be a "weighting" function, then any feasible solution of the following continuous <u>linear programming problem</u> will define an equivalent model which satisfies condition (i) for triviality.

Find $s^z(y) \leqq 0$ for $z\varepsilon P^*$, $y\varepsilon I$ which maximizes

$$\int_{P^*}\int_I w^z(y)s^z(y)dydz \quad \text{subject to}$$

$$s^z(y) - s^*(y) + \delta\int_I g^z(y,x)s^*(x)dx = b^z(y)$$

for all $z\varepsilon P$, $y\varepsilon I$.

At present it is just a conjecture that this is a well-formulated linear programming problem that has some optimal solution $s^{\pi}(y)$. Moreover the success of this approach depends on there being an optimal solution such that the model (s) is trivial and $\pi(y) \neq *$ for all $y\varepsilon I$. The motivation for such a conjecture is that this is the situation for the discretised-stock version of the economic growth mode.

3 THE DISCRETIZED MODEL

The equivalent model approach is now applied to a discrete-stock version r of the economic growth model. In order to discretize this model, the interval $I = [0,M]$ is replaced by the finite set $N = \{1,2,...,n\}$ where i denotes the stock $y_i = (2i-1)M/(2n)$, the midpoint of the subinterval

$[(i-1)M/n, iM/n]$ for $i = 1,\ldots,n$. The consumption versus investment choice is described by $\alpha \varepsilon A_i = \{0,1,\ldots,i-1\}$ where α denotes consumption $c = \alpha M/n$ units of stock y_i yielding a reward of $r_i^\alpha = u(\alpha M/n)$. The one-step transition probabilities are given by

$$p_{ij}^\alpha = \int_{(j-1)M/n}^{jM/n} g^{\alpha M/n}((2i-1)M/2n, z)\,dz$$

where $1 \leqq i \leqq n$, $1 \leqq j \leqq n$ and $0 \leqq \alpha \leqq i-1$. The special choice $\alpha = *$ has $p_{ij}^* = 0$ for all i,j with arbitrary rewards r_i^*.

A policy π is now a sequence of functions $\pi_t : N \to \{*,0,1,\ldots,n-1\}$ where $\pi_t(i) \in A_i^* = A_i \cup \{*\}$. The expected value of a policy π is given by the n-vector

$$V(r,\pi) = \sum_{t=0}^{\infty} \delta^t (P^{\pi_0} \ldots P^{\pi_{t-1}}) r^{\pi_t}$$

where P^{π_t} is the n×n matrix with entries $p_{ij}^{\pi_t(i)}$ and r^{π_t} is the n-vector with entries $r_i^{\pi_t(i)}$. The definitions and lemmas of Section 2 are unchanged (see [9]) except that the functional equation of lemma 3 is replaced by the system of linear equations

$$r_i^\alpha - r_i^* + \delta\, \Sigma_{j=1}^n p_{ij}^\alpha r_j^* = s_i^\alpha - s_i^* + \delta\, \Sigma_{j=1}^n p_{ij}^\alpha s_j^*$$

where $1 \leqq i \leqq n$, $0 \leqq \alpha \leqq i-1$. Letting b_i^α equal the left-hand side of these equations, the discretized version s of the linear programming problem is:

Find $x_i^\alpha \geqq 0$ for $0 \leqq \alpha \leqq i-1$ or $\alpha = *$ and

$1 \leqq i \leqq n$ which minimize

$$w\, \Sigma_{i=1}^n x_i^* + \Sigma_{i=1}^n \Sigma_{\alpha=0}^{i-1} x_i^\alpha$$

subject to

$$x_i^\alpha - x_i^* + \delta\, \Sigma_{j=1}^n p_{ij}^\alpha x_j^* = -b_i^\alpha$$

for $0 \leqq \alpha \leqq i - 1$, $1 \leqq i \leqq n$.

The weight w in the objective is any scalar $w \geqq 0$ such that $w + w_i > 0$ holds for $1 \leqq i \leqq n$ where

$$w_i = i - \delta\, \Sigma_{k=1}^n\, \Sigma_{\alpha=0}^{k-1}\, p_{ki}^\alpha .$$

The dual of this linear programming problem can be written in the following form:

Find $y_i^\alpha \geqq 0$ for $0 \leqq \alpha \leqq i-1$ or $\alpha = *$ and $1 \leqq i \leqq n$ which maximize $\Sigma_{i=1}^n\, \Sigma_{\alpha=0}^{i-1}\, y_i^\alpha\, b_i^\alpha$

subject to

$$y_i^* + \Sigma_{\alpha=0}^{i-1}\, y_i^\alpha - \delta\, \Sigma_{k-1}^n\, \Sigma_{\alpha=0}^{k-1}\, y_k^\alpha\, p_{ki}^\alpha = w + w_i$$

for $1 \leqq i \leqq n$.

Since the dual linear programming problem has the initial basic feasible solution $y_i^\alpha = 0$, $y_i^* = w + w_i$, then it is easy to apply the Simplex method.

<u>Theorem</u>. The primal linear programming problem has an optimal solution X. Moreover the model (s), where $s_i^\alpha = -x_i^\alpha$, is a trivial model equivalent to the original model (r).

The proof of this theorem (see [9]) is a direct consequence of the strong duality theorem of linear programming. Furthermore, the trivial policy π satisfies $\pi(i) \neq *$ for $1 \leqq i \leqq n$ provided that the "scrap values" r_i^* are set sufficiently low.

4 AN EXAMPLE

The linear programming problem of the previous section is applied to an example given by Mirman and Zilcha [7].

Merton [6] has also given the "turnpike" distribution for several examples.

The utility function is $u(c) = \ell n(c)$ and the production function is $f_\omega(x) = a_\omega x^\omega$ where $a_\omega \geq 0$ and $0 < \alpha \leq \omega \leq \beta < 1$ holds for some α,β. Mirman and Zilcha show that the policy $c_t = \pi(y_t) = (1-\delta E(\omega))y_t$ satisfies the sufficient conditions (see Section 1) for optimality.

Fig. 2 compares this analytic solution against the solution obtained from the discretized linear programming problem of Section 3. The specific production function is $f_\omega(x) = \frac{3}{2}(1-\omega)x^\omega$ where ω is uniformly distributed on $\frac{1}{3} \leq \omega \leq \frac{2}{3}$. The discount factor is $\delta = 0.95$, the interval $I = [0,1]$ and the grid size $n = 25$. The optimal policy function is $c = \pi(y) = 0.525y$. The linear programming problem has 25 constraints, 350 variables and required about 18 seconds of time on a CDC 6400 using IMSL code ZX3LP.

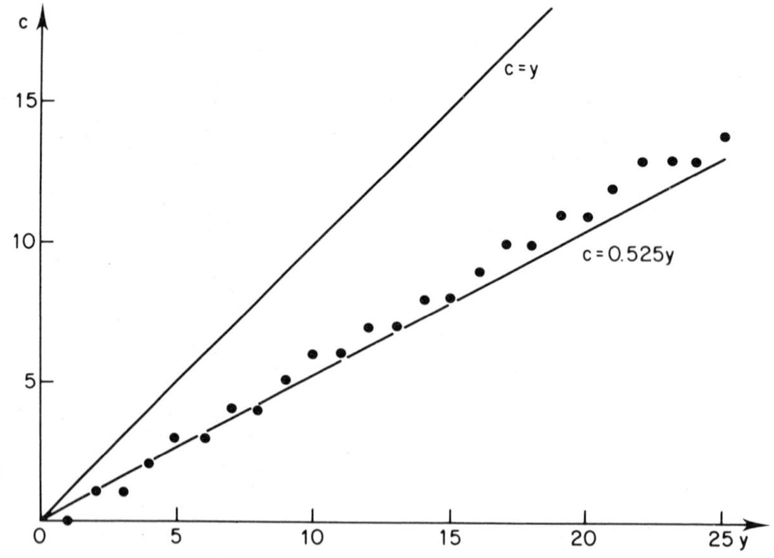

Fig. 2 Comparison of LP and Actual Optimal Policies

5 ACKNOWLEGEMENT

This research has been supported by the National Research Council of Canada under grant A7751.

6 REFERENCES

1. Brock, W. A. and Mirman, L. J. "Optimal economic growth and uncertainty: The discounted case", *Journal of Economic Theory*, **4**, 479-513, (1972).

2. Burmeister, E. and Dobell, A. R. "Mathematical Theories of Economic Growth", MacMillan, New York, (1970).

3. Cass, D. "Optimal growth in an aggregative model of capital accumulation", *Review of Economic Studies*, **32**, 233-240, (1965).

4. Gale, D. "A mathematical theory of optimal economic development", *Bulletin of the American Mathematical Society*, **74**, 207-223, (1968).

5. Koopmans, T. C. "On the concept of optimal economic growth", *Pontificau Academiae Scientarium Scripta Varia*, **28**, 225-300, (1965).

6. Merton, R. C. "An asymptotic theory of growth under uncertainty", *Review of Economic Studies*, **42**, 375-393, (1975).

7. Mirman, L. J. and Zilcha, I. "On optimal growth under uncertainty", *Journal of Economic Theory*, **11**, 329-339, (1975).

8. Shell, K. "Essays on the Theory of Optimal Economic Growth", M.I.T. Press, Cambridge, Mass., (1967).

9. Sutherland, W. R. S. "Optimality in transient Markov chains and linear programming", *Mathematical Programming*, (1980).

CHAPTER 3

STOCHASTIC PROCESSES

3.1 EXIT PROBABILITIES FOR DIFFUSIONS DEPENDING ON SMALL PARAMETERS

W. H. Fleming*

(Department of Mathematics, Brown University, Providence, Rhode Island, 02912, USA)

1 INTRODUCTION

This paper is concerned with Markov diffusion processes which obey stochastic differential equations depending on a small positive parameter ε. The parameter enters as a coefficient in the noise term of the stochastic differential equation. The problem is to estimate the probability that the solution $\xi^{\varepsilon}(t)$ exits from a given region D during a given time interval $s \leq t \leq T$. We first consider a nondegenerate case when noise enters in a nonsingular way (§ 2). Then the Ventcel-Freidlin estimates [3], [5] give the desired estimate for the exit probability. In § 2 we outline a new method for deriving this estimate, based on stochastic control ideas. Then in § 3 we consider a case in which noise enters only certain components of the stochastic differential equations, and quote an estimate for the exit probability. This estimate was proved by O. Hernandez-Lerma [4].

Only statements of results and general indications of methods of proof are given. Detailed proofs appear in [1] and [4].

2 NONDEGENERATE CASE

Consider a Markov diffusion process ξ^{ε} on n-dimensional R^n which obeys the stochastic differential equations

$$d\xi^{\varepsilon} = b[t,\xi^{\varepsilon}(t)]dt + \sqrt{\varepsilon}\,\sigma[t,\xi^{\varepsilon}(t)]dw, \quad s \leq t \leq T, \tag{1}$$

*Supported in part by NSF Grant MCS76-07261 and U.S.A.F. Grant AFOSR76-3063B

with initial data $\xi^{\varepsilon}(s) = x$ and with w an n-dimensional Brownian motion. It is assumed that b, σ and the inverse matrix σ^{-1} are bounded, Lipschitz functions on R^{n+1}. Let $D \subset R^n$ be open, bounded, with C^2 boundary ∂D. Let τ^{ε} denote the exit time of $\xi^{\varepsilon}(t)$ from D, starting from $x \in D$. Let $q^{\varepsilon} = P(\tau^{\varepsilon} \leq T)$ be the exit probability. The Ventcel-Freidlin estimates imply that

$$-\varepsilon \log q^{\varepsilon} \to I^{O} \tag{2}$$

as $\varepsilon \to 0$, where I^{O} is defined from the following problem of calculus of variations type [3, p. 346]. Let

$$L(s,x,v) = \frac{1}{2}(b(s,x) - v)'a(s,x)^{-1}(b(s,x) - v), \tag{3}$$

where $a = \sigma\sigma'$. Then

$$I^{O} = \min_{\phi} \int_{s}^{\theta} L[t, \phi(t), \dot{\phi}(t)]dt, \tag{4}$$

where θ is the exit time of $\phi(t)$ from D and the inf is taken among all $\phi \in C^1([s,T], R^n)$ such that $\theta \leq T$, $\phi(s) = x$.

In [1, §7] this result is obtained by a method based on the following idea. We introduce the logarithmic substitution $I^{\varepsilon} = -\varepsilon \log q^{\varepsilon}$. Then (2) states that $I^{\varepsilon} \to I^{O}$ as $\varepsilon \to 0$. As a function of the initial data (s,x), the exit probability q^{ε} obeys the backward partial differential equation associated with (1). The function $I^{\varepsilon}(s,x)$ obeys a nonlinear parabolic partial differential equation, which is just the dynamic programming equation for the following stochastic control problem.

Let $\eta(t)$, $v(t)$ denote respectively the state and the control at time t ($\eta(t)$ and $v(t)$ take values in R^n). The dynamics are

$$d\eta = v(t)dt + \sqrt{\varepsilon}\, \sigma\, [t, \eta(t)]dw, \quad s \leq t \leq T, \tag{5}$$

with $\eta(s) = x$. Let Φ be a Lipschitz function on R^{n+1}, such that $\Phi(s,x) = 0$ for $x \in \partial D$ and $\Phi(T,x) > 0$ for $x \in D$. Let

$$J^{\varepsilon} = \min_{v} E \int_{s}^{\alpha} L[t,\eta(t),v(t)]dt + \Phi[\alpha,\eta(\alpha)] \tag{6}$$

where $\alpha = \min[T$, exit time from D of $\eta(t)]$. The minimum is taken among all bounded, nonanticipative control processes v. In outline our proof that $I^{\varepsilon} \to I^{0}$ as $\varepsilon \to 0$ is as follows; for details see [1, §7]. By a rather easy argument one can first show that $\limsup_{\varepsilon\to 0} I^{\varepsilon} \leq I^{0}$. This is done by taking as comparison controls in (5) $v(t) = \dot{\phi}(t)$, with ϕ any $C^{(1)}$ (deterministic) function with exit time $\theta < T$, $\phi(s) = x$. The opposite estimate $I^{0} \leq \liminf_{\varepsilon\to 0} I^{\varepsilon}$ uses a bit of stochastic control theory, specifically the so-called Verification Theorem [2, Sec. 6.4]. We show that $\liminf_{\varepsilon\to 0} J^{\varepsilon} \geq J^{0}$, where J^{0} is the solution of the calculus of variations problem obtained by setting $\varepsilon = 0$ in (5) and omitting expectations in (6). We next take $\Phi = \Phi^{M} = M\Psi$, and let $M \to \infty$. In essence, Φ^{M} is a penalty function; as $M \to \infty$ the penalty for not reaching ∂D becomes infinite.

3 DIFFERENTIAL EQUATIONS WITH MARKOV PARAMETERS

Let us now suppose that ξ^{ε} obeys the ordinary differential equations

$$d\xi^{\varepsilon} = F[t,\xi^{\varepsilon}(t),\beta^{\varepsilon}(t)], \tag{7}$$

where $\beta^{\varepsilon}(t)$ represents a vector of random parameters. Let us suppose that the parameter process is nearly deterministic. In fact, we suppose that β^{ε} is a diffusion obeying the same kind of stochastic differential equations as in (1):

$$d\beta^{\varepsilon} = b[t,\beta^{\varepsilon}(t)]dt + \sqrt{\varepsilon}\,\sigma[t,\beta^{\varepsilon}(t)]dw. \tag{8}$$

Here $\beta^{\varepsilon}(t), w(t)$ have values in some R^{m}. We again let $q^{\varepsilon} = P(\tau^{\varepsilon} \leq T)$, with τ^{ε} the exit time of $\xi^{\varepsilon}(t)$ from a given bounded region D; and let $I^{\varepsilon} = -\varepsilon \log q^{\varepsilon}, \varepsilon > 0$. The appropriate minimum problem to define I^{0} is no longer (4). Instead, let us consider the following (deterministic) minimum problem. Let $(\eta^0(t),\gamma(t))$ denote C^1 functions, with $\eta^0(t) \in R^n$, $\gamma(t) \in R^n$, such that

$$\begin{aligned} d\eta^0 &= F[t,\eta^0(t),\gamma(t)]dt, \quad s \leq t \leq T, \\ \eta^0(s) &= \xi^\varepsilon(s) = x, \quad \gamma(s) = \beta^\varepsilon(s) = y. \end{aligned} \tag{9}$$

Let θ denote the exit time of $\eta^0(t)$ from D; and let

$$I^0 = \min_\gamma \int_s^\theta L[t,\gamma(t),\dot{\gamma}(t)]dt, \tag{10}$$

where the minimum is taken among those γ for which $\theta \leq T$. It is assumed that this set of functions is not empty. In [4] it is shown that $I^\varepsilon \to I^0$ as $\varepsilon \to 0$ under the following additional assumptions. The functions b,σ in (8) satisfy the same assumptions as in §2. The backward operator of the Markov diffusion $(\xi^\varepsilon,\beta^\varepsilon)$ is hypoelliptic (note that this operator is degenerate parabolic since noise enters only (8) and not (7).) Finally, let $\nu(x)$ denote the exterior unit normal to ∂D at $x \in \partial D$. Let Γ^+, Γ^-, Γ^0 denote the sets of (t,x,y) where $F(t,x,y)\cdot\nu(x)$ is respectively positive, negative, and 0. It is easily seen that $P(\tau^\varepsilon,\xi^\varepsilon(\tau^\varepsilon),\beta^\varepsilon(\tau^\varepsilon)) \in \Gamma^+ \cup \Gamma^0) = 1$. It is assumed that $P((t,\xi^\varepsilon(t),\beta^\varepsilon(t)) \in \Gamma^0$ for some $t \in [s,T]) = 0$.

In the proof in [4] that $I^\varepsilon \to I^0$ as $\varepsilon \to 0$ it is first shown that the exit probability $q^\varepsilon(s,x,y)$ is a solution of the backward equation, smooth in $(-\infty,T) \times D \times R^m$ and continuous at boundary points in Γ^+. The proof then proceeds along lines rather similar to the nondegenerate case indicated in §2.

It appears that the same method can be applied to systems where in (8) b also depends on $\xi^\varepsilon(t)$. This includes, for instance, the case of a low intensity white noise driven nth order linear differential equation, after rewriting as a first order system.

4 REFERENCES

1. Fleming, W. H., "Exit probabilities and optimal stochastic control", *Applied Math. and Optimization,* **4**, (1978).

2. Fleming, W. H. and Rishel, R. W., "Deterministic and Stochastic Optimal Control", Springer-Verlag, (1975).

3. Friedman, A. "Stochastic Differential Equations and Applications", vol. II, Academic Press, (1976).

4. Hernandez-Lerma, O "Stability of differential equations with Markov parameters and the exit problem", Ph.D. Thesis, Brown University, (1978).

5. Ventcel, A. D. and Freidlin, M. I. "On small random perturbations of dynamical systems", *Russian Math. Surveys,* **25**, 1-56, (1970), [*Uspehi Mat. Nauk,* **25**, 3-55, (1970)].

3.2 A NEW APPROACH TO THE THEORY OF NONLINEAR STOCHASTIC SYSTEMS

N. A. A. Virani and R. W. H. Sargent

(Imperial College, London, UK)

1 INTRODUCTION

A natural way of formulating a mathematical model for the dynamic behaviour of a physical system is in terms of the evolution of the state $\{x_t\}$ of the system under the influence of controls $\{u_t\}$ and disturbances $\{\xi_t\}$ as the solution of a set of differential equations:

$$\dot{x}_t = f(t,x_t,u_t,\xi_t), \; t\in T, \qquad (1.1)$$

with measurements $\{y_t\}$ given by

$$y_t = g(t,x_t,u_t,\xi_t), \qquad (1.2)$$

where $T = [t_0, t_f]$ is the time interval of interest and $\{x_t\}$, $\{u_t\}$, $\{y_t\}$, $\{\xi_t\}$ are vector stochastic processes taking values in R^n, R^m, R^p, R^q respectively.

It is in fact more convenient to describe the measurements in terms of an observation process $\{z_t\}$ defined by

$$\left.\begin{aligned} z_t &= \int_{t_0}^{t} y_s \, ds, \\ \text{so that } y_t &= \dot{z}_t = g(t,x_t,u_t,\xi_t), \; z_{t_0} = 0 \end{aligned}\right\} \qquad (1.3)$$

The information I_t available to the controller at time t is a specified subset of the a priori information I_0 at time t_0, and the observation process $\{z_s, t_0 \leqslant s \leqslant t\}$; the controller determines the controls $\{u_t\}$ as a function of this available information:

$$u_t = h(t, I_t) \tag{1.4}$$

and the controls may be restricted to take values in a specified set $U \subseteq R^m$. The optimal controller design problem is then to determine the function $h(\cdot,\cdot)$ on the interval T which minimizes a suitable index of the performance of the system. The latter can be expressed in the form

$$J = E\ \{\theta(x_{t_f}) \mid I_0\}, \tag{1.5}$$

for some function θ: $R^n \to R$, if necessary by appropriate re-definition of the state.

The classical approach to dealing with this stochastic system is to model the disturbances as the solution of an Ito stochastic differential equation:

$$d\xi_t = \mu(t,\xi_t)dt + \nu(t,\xi_t)d\beta_t, \tag{1.6}$$

where $\{\beta_t\}$ is a vector Brownian motion. Equation (1.6) together with equations (1.1) and (1.3) then form a special case of the Ito stochastic differential equation:

$$d\chi_t = \phi(t,\chi_t,u_t)dt + \sigma(t,\chi_t)d\beta_t, \tag{1.7}$$

where $\chi_t^T = \left[x_t^T, \xi_t^T, z_t^T\right]$ is an extended "state". Thus (1.7) can be taken as a fairly general model for a controlled stochastic process.

Kushner [1] has derived a general nonlinear filter for system (1.7) which gives the evolution of the appropriate conditional probability density of χ_t, and his formal work has since been rigorously verified [2,3]. This general filter is infinite-dimensional, and in practical applications it is necessary to use approximations, such as the extended Kalman filter. Nevertheless, the basic theory is essentially complete, and the remaining problems are concerned with finding adequate practical approximations for specific cases.

There are however several unsatisfactory features about this classical approach. It is well known [4] that the solution $\{\chi_t\}$ to (1.7) is a Markovian process with continuous sample functions, but these, as for Brownian motion, have unbounded variation, as has been shown by Davis [5]. This does not accord with intuitive notions of physical behaviour, as discussed by Doob [4, p. 398] and indeed Wong and Zakai [6] have shown that extra correction terms appear when an Ito equation like (1.7) is obtained as a limiting form of a "physical" process. Further, the general filter and its approximations, which involve the "innovations process", are again Ito stochastic differential equations; in applications however they are formally applied as ordinary differential equations. Although this causes no difficulty, since in practice the observations do not exhibit unbounded variation, the theoretical inconsistency in the approach remains.

A more practical disadvantage is that it appears impossible to derive bounds for the errors in approximate filters or in the approximation to the optimal control laws based on them; it does not even appear possible to show that these errors become small as the nonlinearities or the disturbances become small in some well-defined sense.

All these difficulties can be avoided if we model $\{\xi_t\}$ as a stochastic process with almost all sample functions of bounded variation. Again following Doob, we retain the notion that a physically realizable process should have continuous sample paths, and in fact strengthen these two conditions to require almost all sample functions of such a process to be absolutely continuous. Stochastic systems described by equations (1.1) and (1.2) or (1.3) are thus brought within the scope of the classical theory of Lebesgue integration.

In this paper we consider the properties of such "physically realizable" systems, derive a general nonlinear filter for them in finite-difference form, and discuss the question of convergence to a continuous filter.

Again the general filter is infinite-dimensional, and in a later paper we shall present various approximations to it and their use in obtaining approximations to the optimal feedback control law (1.4). It will emerge that firm bounds on the errors involved can be obtained, thus generalizing earlier work along these lines [7,8] from the case of perfect state observation to the partially observable case. All these topics are treated in more detail by Virani [9].

2 PHYSICALLY REALIZABLE STOCHASTIC SYSTEMS

In this section we define a physically realizable stochastic process and consider the conditions under which equations (1.1), (1.2) and (1.3) define such processes.

We assume an underlying probability space (Ω, F, P), where F is a σ-algebra of subsets of the basic space Ω and P is a probability measure. Similarly $\mathcal{J}$ is a σ-algebra of Lebesgue measurable sets in $T = [t_0, t_f] \subset R$.

<u>Definition</u>

A stochastic process $\{\xi_t, t \in T\}$ defined on (Ω, F, P) and taking values in $S \subseteq R^q$ is said to be physically realizable if it is separable and measurable with respect of $F \times \mathcal{J}$, and almost all of its sample functions are absolutely continuous on T with

$$E\{\|\xi_{t_0}\|\} < \infty, \quad \int_T E\{\|\dot{\xi}_t\|\}dt < \infty.$$

Conversely, suppose we are given a stochastic process $\{\xi_t, t \in T\}$ with almost all sample functions absolutely continuous. Then this process is continuous in probability and it follows [10] that a separable and measurable modification of the process always exists. Thus we can always find a physically realizable process $\{\tilde{\xi}_t, t \in T\}$ defined on the same probability space such that $P(\xi_t = \tilde{\xi}_t) = 1$ for each $t \in T$. Hence to assume that $(\xi_t, t \in T)$ has been replaced by $(\tilde{\xi}_t, t \in T)$ will never have any significant consequences.

It immediately follows from the definition that almost all sample functions of a physically realizable process are Lebesgue integrable on any finite time interval, for we have

$$E\{\|\xi_t\|\} \leqslant E\{\|\xi_{t_0}\|\} + \int_T E\{\|\dot{\xi}_t\|\}dt < \infty$$

and

$$\int_T E\{\|\xi_t\|\}dt \leqslant (t_f - t_0)(E\{\|\xi_{t_0}\|\} + \int_T E\{\|\dot{\xi}_t\|\}dt) < \infty,$$

and the result follows from Fubini's Theorem.

Definition

We shall call the system described by (1.1) a *physically realizable system* if the following conditions hold:

(i) $\{\xi_t, t\in T\}$ is a physically realizable stochastic process taking values $S\subseteq R^q$, a.s.

(ii) Controls $\{u_t\}$ are

either given Lebesgue measurable functions $u: T\to U\subseteq R^m$,
or a physically realizable stochastic process $\{u_t, t\in T\}$ taking values in $U\subseteq R^m$, a.s.

(iii) The initial state x_0 is a random variable $x_0:\Omega\to R^n$ such that $x_0\in X_0$ a.s., where X_0 is a given bounded convex set.

(iv) The function $f: T\times R^n\times U\times S\to R^n$ is Lebesgue measurable in t and x for fixed u, ξ, and continuous in u, ξ for fixed t, x.

(v) There exists a function F(t) integrable on T, and a function $\psi(s)$ positive and continuous for $s\geq 0$, but not integrable on $[0,\infty)$, such that

$\|f(t,x,u,\xi)\| \leq F(t)/\psi(\|x\|)$

for almost all t,x,u,ξ.

(vi) There exist $\varepsilon > 0$ and a function L(t) integrable on T such that

$|(x - x^*)^T(f(t,x,u,\xi) - f(t,x^*,u,\xi))| \leq L(t).\|x - x^*\|^2$

for almost all t,u,ξ and x, x* such that $\|x - x^*\|<\varepsilon$.

With the additional condition that $f(t,x,u,\xi)$ be continuous in x, conditions (iv) to (vi) are the classical conditions [11] for the existence of a unique absolutely continuous function $x(t)\in R^n$, $t\in T$, with $x(t_0) = x_0$ satisfying

$$\dot{x}(t) = f(t,x(t), u(t), \xi(t)) \quad \text{a.e.,} \tag{2.1}$$

where x_0 is a given initial condition and u(t), $\xi(t)$ are given Lebesgue measurable functions on T. Moreover the family of all such solutions for fixed $x_0\in X_0$ and all permissible functions u(t), $\xi(t)$ is uniformly bounded on T.

Without the extra continuity condition on $f(t,x,u,\xi)$, we may retain these properties by defining the solution of (2.1) in the sense of Filippov [12]. This will enable us to treat controlled physically realizable systems without requiring any continuity conditions on the function $h(t,I_t)$ of (1.4).

Thus under conditions (i) - (vi), equation (1.1) defines a physically realizable stochastic process $\{x_t\}$, and (1.3) similarly defines a physically realizable stochastic process $\{z_t\}$.

We note also that if $g(t,x,u,\xi)$ is Lipschitz continuous in all its arguments, then almost all sample functions of the measurement process $\{y_t\}$ are absolutely continuous, and hence this process too is physically realizable.

3 THE GENERAL FILTER

In this section we consider a signal process $\{x_t\}$ taking values in $V\subseteq R^n$ a.s. and an observation process $\{z_t\}$ taking values in $W\subseteq R^p$ a.s., both physically realizable stochastic processes defined on (Ω,F,P), and obtain expressions describing the evolution of the conditional density $p(x_t|z_t,\ t_0 \leq s \leq t)$ and the corresponding conditional moments.

We define $T_n = \{t_0, t_1, t_2, \ldots t_{n-1}, t_n \equiv t_f\} \subset T$ and, assuming the relevant densities exist, use the notation:

$$X_t = \{x_s | s \leq t;\ s,\ t \in T_n\}, \quad Z_t = \{z_s | s \leq t;\ s,\ t \in T_n\},$$

$$p_t \equiv p(x_t|Z_t) \equiv p(t,v|Z_t)\ , \quad p(z_t|\cdot) \equiv p(t,w|\cdot),\ t \in T_n$$

$$E^t\{\cdot\} \equiv E\ \{\cdot|Z_t\}.$$

For $t,\ t+h \in T_n$:

$$\delta x_t = x_{t+h} - x_t,\ \delta z_t = z_{t+h} - z_t,\ \delta p_t = p_{t+h} - p_t,$$

$$\delta\zeta_t = \delta z_t - E^t\{\delta z_t\}$$

We shall assume that almost everywhere on Ω and for all $v \in V$, $w \in W$, t, $t+h \in T_n$, and all subsets $T_n \subset T$, the density

$$\rho(t,h,v,w) \equiv p(t+h,v|Z_t, z_{t+h} = w)$$

exists, is absolutely continuous in h, and has a derivative $\rho_w(t,h,v,w)$ with respect to w which is uniformly continuous on W.

Theorem 1.

For the processes $\{x_t\},\{z_t\}$ $t \in T_n$, the conditional density p_t satisfies the equation

$$\delta p_t = E^t\{\delta p_t\} + \psi_w(t,h,v,E^t\{z_{t+h}\}).p(t+h,v|Z_t).\delta\zeta_t + o(\|\delta\zeta_t\|) \text{ a.s.} \tag{3.1}$$

$$\text{where } \psi(t,h,v,w) = \frac{E\{p(t+h,w|X_{t_f}, Z_t)|x_{t+h} = v, Z_t\}}{E\{p(t+h,w|X_{t_f}, Z_t)|Z_t\}} \tag{3.2}$$

Proof

We have $E^t\{z_t\} = z_t$, and hence

$$z_{t+h} - E^t\{z_{t+h}\} = \delta z_t - E^t\{\delta z_t\} = \delta\zeta_t.$$

It follows from the Mean Value Theorem that

$$p_{t+h} = \rho(t,h,v,z_{t+h}) = \rho(t,h,v,E^t\{z_{t+h}\}) + \rho_w(t,h,v,E^t\{z_{t+h}\}).\delta\zeta_t$$

$$+ o(\|\delta\zeta_t\|), \text{ a.s.}$$

Whence, since $E^t\{\delta\zeta_t\} = 0$, we have

$$E^t\{p_{t+h}\} = \rho(t,h,v,E^t\{z_{t+h}\}) + o(\|\delta\zeta_t\|), \text{ a.s.}$$

and hence, using also $E^t\{p_t\} = p_t$, we have

$$\delta p_t = E^t\{\delta p_t\} + \rho_w(t,h,v,E^t\{z_{t+h}\}).\delta\zeta_t + o(\|\delta\zeta_t\|) \text{ a.s.} \quad (3.3)$$

Now using Bayes' Rule we have

$$\rho(t,h,v,w) = p(t+h, v|Z_t, z_{t+h} = w)$$

$$= \frac{p(t+h, w|x_{t+h} = v, Z_t).p(t+h,v|Z_t)}{p(t+h,w|Z_t)}$$

$$= \frac{E\{p(t+h,w|X_{t_f}, Z_t)|x_{t+h} = v,Z_t\}}{E\{p(t+h,w|X_{t_f}, Z_t)|Z_t\}} p(t+h,v|Z_t)$$

Thus from (3.2) we have

$$\rho(t,h,v,w) = \psi(t,h,v,w).\ p(t+h,v|Z_t), \quad (3.4)$$

and substitution of this into (3.3) yields (3.1)

Q.E.D.

A corresponding filter equation for conditional expectations is easily obtained from (3.1), since for any integrable function $\theta: R^n \to R$, provided that the relevant conditional expectations exist, multiplication of (3.1) by $\theta(v)$ and integration over v yields

$$\delta E^t\{\theta(x_t)\} = E^t\{\delta\theta(x_t)\} + E^t\{\theta(x_{t+h}).\psi_w(t,h,x_{t+h}, E^t\{z_{t+h}\})\}\delta\zeta_t + o(\|\delta\zeta_t\|) \text{ a.s.} \quad (3.5)$$

Equation (3.5) is the kind of result to be expected, for the first term $E^t\{\delta\theta(x_t)\} = E^t\{\theta(x_{t+h}) - \theta(x_t)\}$ is the predicted dynamic change in the expected value based on information at time t, while the second term is the correction due to the innovation process, $\delta\zeta$, the new information from the measurements.

In Section 5 we shall show that $\|\delta\zeta_t\| = o(h)$, which shows that the error in ignoring the remainder term in (3.1), (3.3) and (3.5) can be made arbitrarily small by choosing a sufficiently small value of h. Indeed the same is true of the cumulative error in applying these equations recursively, with $h_i = t_{i+1} - t_i$, $i=0,1,2,\ldots.(k-1)\leq n$, for if $\bar{h} = \sup(h_i | 0 \leq i \leq n)$

we obtain

$$p_{t_k} - p_{t_0} = \sum_{i=1}^{k-1} E^{t_i}\{\delta p_{t_i}\} + \sum_{i=0}^{k-1} \rho_w(t_i,h_i,v,E^{t_i}\{z_{t_{i+1}}\})\delta\zeta_{t_i} + \tag{3.6}$$

$+ o(\bar{h})$, a.s.,

and

$$E^{t_k}\{\theta(x_{t_k})\} - E^{t_0}\{\theta(x_{t_0})\} =$$

$$\sum_{i=0}^{k-1} E^{t_i}\{\delta\theta(x_{t_i})\} + \sum_{i=0}^{k-1} E^{t_i}\{\theta(x_{t_{i+1}})\psi_w(t_i,h_i,x_{t_{i+1}},E^{t_i}\{z_{t_{i+1}}\})\}.\delta\zeta_{t_i} \tag{3.7}$$

$+ o(\bar{h})$, a.s.

Similarly we may define the innovations sequence:

$$\zeta_{t_0} = 0,\ \zeta_{t_k} = \sum_{i=0}^{k-1} \delta\zeta_{t_i} = z_{t_k} - z_{t_0} - \sum_{i=0}^{k-1} E^{t_i}\{\delta z_{t_i}\},\ \text{a.s.} \tag{3.8}$$

Since $E^{t_k}\{z_{t_k}\} = z_{t_k}$ it is easy to show that $\{\zeta_{t_k}\}$ is a discrete martingale on the family of σ-fields F_t generated by Z_t. We can similarly show that

$$m_{t_k} = E^{t_k}\{\theta(x_{t_k})\} - E^{t_0}\{\Theta(x_{t_0}) - \sum_{i=0}^{k-1} E^{t_i}\{\delta\theta(x_{t_i})\} \tag{3.9}$$

is a discrete martingale on F_t, so that (ignoring the remainder), equation (3.7) represents a Doob decomposition into a contribution adapted to $F_{t_{k-1}}$ and a martingale. Again since $E^t\{p_t\} = p_t$ a similar result holds for (3.6).

Finally we note that for any $t,t+h \in T$ we can choose a subset $T_n \subset T$ such that $t,t+h \in T_n$ and (cf. Doob [4,p.21]

$$E^t\{\cdot\} \equiv E\{\cdot|Z_t\} = E\{\cdot|(z_s|s \in [t_0,t] \subseteq T)\}. \tag{3.10}$$

Thus equations (3.1), (3.3) and (3.5) are valid for either of the interpretations of $E^t\{\cdot\}$ given in (3.10).

The continuous limit of equations (3.6), (3.7) and (3.8) will be examined in Section 5.

4 AN EXPLICIT REPRESENTATION

To use the results of Section 3 we need to evaluate $\psi_w(t,h,x_{t+h},E^t\{z_{t+h}\})$ and from (3.2) it can be seen that this requires an expression for the density $p(z_{t+h}|X_{t_f},Z_t)$.

Now for a system described by (1.1) and (1.3), the probability law of $\{\xi_t\}$ determines through (1.3) the finite-dimensional family of densities of $p(Z_{t_f}|X_{t_f})$, and hence the required density. However in applications it is rare that the probability law of $\{\xi_t\}$ is completely specified, and one is usually content with a characterization based on the first few moments. In these circumstances it is more convenient to assume an explicit form for $p(z_{t+h}|X_{t_f},Z_t)$, choosing its parameters to give a wide sense approximation for the disturbance process.

Of course the explicit form chosen must be compatible with the definition of $\{x_t\}$ and $\{z_t\}$ as physically realizable stochastic processes. Since the Gaussian density describes an unbounded random vector it cannot serve as a model. However a simple modification suffices:

<u>Definition</u>

The bounded Gaussian density is defined by:

$$P_y(w) = I_S(w)\,\frac{\exp\{-\tfrac{1}{2}(w-m)^T Q^{-1}(w-m)\}}{\int_S \exp\{-\tfrac{1}{2}(w-m)^T Q^{-1}(w-m)\}\,dw} \tag{4.1}$$

where $w\in R^q$ is the generic value of y, $S\in R^q$ is a pre-compact open ball of radius c>0 with centre at m, and $I_S(w)$ is an indicator function:

$$I_S(w) = 1 \text{ if } w \in S \tag{4.2}$$

$$I_S(w) = 0 \text{ otherwise}$$

It is easy to show that

$$E\{y\} = m;\ \|y-m\| < c, \text{ a.s.} \tag{4.3}$$

It should be noted that Q is not the covariance matrix of y, but nevertheless plays a role similar to that of the covariance matrix for a Gaussian density. For example $\|Q\| = 0$ implies that the density is a Dirac delta-function, while $\|Q\| = \infty$ implies an improper cylindrical density; more generally Q determines the concentration of the density about the mean value.

Theorem 2

Suppose that the density $p(t+h,w|X_{t_f},Z_t)$ is given by (4.1) with

$$Q = hR, \quad c = K(t)h, \quad m = E\{z_{t+h}|X_{t_f},\ Z_t\}, \tag{4.4}$$

where R is a fixed positive definite matrix and K(t) is bounded on T. Then

$$\psi_w(t,h,v,E^t\{z_{t+h}\}) = \frac{1}{h}\left[E\{\delta\ z_t|x_{t+h} = v,Z_t\} - E^t\{\delta z_t\}\right]^T R^{-1} + \varepsilon \tag{4.5}$$

where $\|\varepsilon\| = O(h)$.

Proof

Substitution of (4.1) and (4.4) into (3.2) yields (for $v,w \in S$)

$$\psi(t,h,v,w) = \frac{E\{e^{-a}|x_{t+h} = v,\ Z_t\}}{E\{e^{-a}|Z_t\}} = E\{\frac{e^{-a}}{E^t\{e^{-a}\}}|x_{t+h} = v,\ Z_t\}$$

where $a = \frac{1}{2h}(w - m)^T R^{-1}(w-m)$. Differentiation with respect to w then yields

$$\psi_w(t,h,v,w) = E\left\{\frac{e^{-a}E^t\{a_w e^{-a}\} - a_w e^{-a}E^t\{e^{-a}\}}{(E^t\{e^{-a}\})^2} \middle| x_{t+h} = v,\ Z_t\right\} \tag{4.6}$$

Now from (4.3) and (4.4) we have

$$\left.\begin{aligned} &\|w - m\| \leqslant K(t)h, \quad \text{a.s.,} \\ &\text{and hence } |a| \leqslant \tfrac{1}{2}\,(K(t))^2\|R^{-1}\|h, \quad \text{a.s.,} \\ &\text{and } \|a_w\| \leqslant K(t)\,\|R^{-1}\|, \text{ a.s.} \end{aligned}\right\} \tag{4.7}$$

Expanding e^{-a} in (4.6) and using (4.7) we obtain

$$\psi_w(t,h,v,w) = E\{E^t\{a_w\} - a_w | x_{t+h} = v,\ Z_t\} + \varepsilon \tag{4.8}$$

where $\|\varepsilon\| = O(h)$.

Now substituting $w = E^t\{z_{t+h}\}$ we have

$$a_w = \frac{1}{h}(w-m)^T R^{-1} = \frac{1}{h}\left[E^t\{z_{t+h}\} - E\{z_{t+h}|X_{t_f},\ Z_t\}\right]^T R^{-1} \tag{4.9}$$

it immediately follows that $E^t\{a_w\} = 0$, and since we also have

$$\{z_t\} = E^t\{z_t\} = E\{z_t|X_{t_f},\ Z_t\}, \tag{4.10}$$

substitution of (4.9) and (4.10) into (4.8) yields the required result.

Q.E.D.

Substitution of this result into the filter equation (3.5) yields

$$\delta E^t\{\theta(x_t)\} = E^t\{\delta\theta(x_t)\} + \frac{1}{h}\,E^t\{\theta(x_{t+h}).\left[E\{\delta z_t|x_{t+h},\ Z_t\} - E^t\{\delta z_t\}\right]^T\}\,R^{-1}\,\delta\zeta_t + \varepsilon', \quad \text{a.s.,} \tag{4.11}$$

where again ε' and the resulting cumulative error can be made arbitrarily small by choosing h sufficiently small.

This filter is still infinite-dimensional, and it is necessary to approximate the expectations of the nonlinear functions involved. These can be evaluated with the help of the system equations (1.1) - (1.3), and details for some small-noise models are given by Virani [9]. Since discretization is in any case necessary for numerical computation, equation (4.11) is already in convenient form for practical use.

5 THE CONTINUOUS LIMIT

It remains to show that $\|\delta\zeta_t\| = o(h)$, as used in Section 3, and in this section we shall examine the limiting form of the filter equations as the time-steps tend to zero.

We consider a physically realizable stochastic process $\{x_t, z_t, t \in T\}$ defined on (Ω, F, P), and denote by F_t the increasing family of σ-fields generated by $Z_t = \{z_s | s \in [t_0, t] \subseteq T\}$ such that F_t includes all subsets of P-measure zero, writing $E^t\{\cdot\} \equiv E\{\cdot | F_t\}$. We shall need the following results:

Lemma 1

The Lebesgue integral $\int_{t_0}^{t} E^s \{\dot{x}_s\}\, ds$, $t \in T$, is well defined for almost all sample functions.

Proof

Since $\{x_t, t \in T\}$ is physically realizable, the derivative $\dot{x}_t$ exists a.e. on $\Omega \times T$ and

$$\int_T E\{\|\dot{x}_s\|\} ds < \infty.$$

Thus $$\int_T E\{E^s\{\|\dot{x}_s\|\}\} ds = \int_T E\{\|\dot{x}_s\|\} ds < \infty.$$

But $E^s\{\|\dot{x}_s\|\}$ is a non-negative measurable function on $\Omega \times T$, so from Tonelli's Theorem it follows that $E^s\{\|\dot{x}_s\|\}$ is integrable on $\Omega \times T$. The existence of the required integral follows from Fubini's Theorem. Q.E.D.

Lemma 2

$$E^{\tau}\{x_t - x_\tau\} = \int_\tau^t E^{\tau}\{\dot{x}_s\}ds, \quad \tau \leq t;\ \tau,\ t \in T, \text{ a.s.} \tag{5.1}$$

The proof is similar to that of Lemma 1.

Theorem 3

The process $\{m_t\ ,\ t\in T\}$ is defined by

$$m_t = E^t\{x_t\} - E^{t_0}\{x_{t_0}\} - \int_{t_0}^t E^s\{\dot{x}_s\}ds, \quad t\in T, \text{ a.s.} \tag{5.2}$$

Then $\{m_t, F_t\}$ is a martingale.

Proof

From Lemma 1 the process $\{m_t, t\in T\}$ is well defined and $E\{\|m_t\|\} < \infty$, $t\in T$. From (5.2) it follows immediately that $E^t\{m_t\} = m_t$, $t\in T$, so $\{m_t\}$ is adapted to F_t.

Also from (5.2), for $\tau \leq t$ we have

$$m_t - m_\tau = E^t\{x_t\} - E^{\tau}\{x_\tau\} - \int_\tau^t E^s\{\dot{x}_s\}ds, \quad \text{a.s.}$$

Whence $E^{\tau}\{m_t - m_\tau\} = E^{\tau}\{x_t - x_\tau\} - E^{\tau}\ \{\int_\tau^t \dot{x}_s ds\} = 0$, a.s.

Thus $E^{\tau}\{m_t\} = E^{\tau}\{m_\tau\} = m_\tau$, a.s. and $\{m_t, F_t\}$ is a martingale.

Q.E.D.

Now if $\Theta: V \to R$ is Lipschitz continuous, then $\{\Theta(x_t)\}$ has absolutely continuous sample paths, and we assume in addition that the appropriate finiteness conditions are satisfied so that $\{\Theta(x_t)\}$ is a physically realizable stochastic process. Then it is clear from Theorem 3 that the continuous limit of (3.8) is of the form

$$E^t\{\Theta(x_t)\} = E^{t_0}\{\Theta(x_{t_0})\} + \int_{t_0}^t E^s\{\Theta(x_s)\dot{x}_s\}ds + m_t, \quad t\in T, \text{ a.s.} \tag{5.3}$$

where $\{m_t, F_t\}$ is a martingale. However we must still show that (3.8) converges to this limit as $h \to 0$, which requires examination of the behaviour of $\delta\zeta_t$.

Now from Lemma 2 we have

$$\delta\zeta_t = \delta z_t - E^t\{\delta z_t\} = \int_t^{t+h} (\dot{z}_s - E^t\{\dot{z}_s\})ds, \; [t,t+h] \subset T, \text{ a.s.},$$

and hence
$$\lim_{h\to 0^+} \frac{\delta\zeta_t}{h} = \dot{\zeta}_t^+ = \dot{z}_t - E^t\{\dot{z}_t\}, \text{ a.s. on } \Omega\times T, \qquad (5.4)$$

showing that the right-derivative $\dot{\zeta}_t^+$ is well defined a.e. on $\Omega\times T$. We note also that, although $\delta\zeta_t$ here is defined on T_n, it is possible to choose T_n so that $E^t\{\cdot\}$ can be interpreted here as $E\{\cdot|F_t\}$.

Now from Lemma 1 it follows that the right-hand side of (5.4) is integrable on T a.s., and we have

$$\zeta_t^* = \int_{t_0}^{t} \dot{\zeta}^+ ds = z_t - z_{t_0} - \int_{t_0}^{t} E^s\{\dot{z}_s\}ds, \; t\in T, \text{ a.s.} \qquad (5.5)$$

Clearly almost all sample functions of $\{\zeta_t^*\}$ are absolutely continuous, and since $E^t\{z_t\} = z_t$ it follows from Theorem 3 that $\{\zeta_t^*, F_t\}$ is a martingale. Now it is well known (see for example Davis [5]) that the continuous sample paths of a martingale are either of unbounded variation or constant, so we have $\zeta_t^* = \zeta_{t_0}^* = 0$, $t\in T$, a.s., and hence from (5.5) $\dot{\zeta}^+ = 0$ a.e. on $\Omega\times T$. Together with (5.4) this yields the desired result that $\delta\zeta_t = o(h)$.

Now if we define $F_{t^-} \subset F_t$ as the increasing family of σ-fields generated by $Z_{t^-} = \{z_s | s\in[t_0,t)\subset T\}$, again including all subsets of P-measure zero, and write $E^{t-}\{\cdot\} \equiv E\{\cdot|F_{t-}\}$, then proofs analogous to those of Lemma 1 and Theorem 3 show that the Lebesgue integral $\int_{t_0}^{t} E^{s-}\{\dot{x}_s\}.ds$, $t\in T$, is well defined a.s., and that if we define

$$m'_t = E^t\{x_t\} - E^{t_0}\{x_{t_0}\} - \int_{t_0}^{t} E^{s-}\{\dot{x}_s\}ds, \ t \in T, \text{ a.s.,} \tag{5.6}$$

then $\{m'_t, F_t\}$ is also a martingale.

We also have $\dot{\zeta}_t^-$ well defined and

$$\lim_{h \to 0} \frac{\delta\zeta_{t-h}}{h} = \dot{\zeta}_t^- = \dot{z}_t - E^{t-}\{\dot{z}_t\}, \text{ a.e. on } \Omega\times T, \tag{5.4a}$$

As before, we can use (5.6) to show that $\dot{\zeta}_t^- = 0$ a.e. on $\Omega\times T$, and hence the derivative $\dot{\zeta}_t = 0$ exists a.e. on $\Omega\times T$.

From (5.4) and (5.4a) we therefore obtain another basic property of physically realizable stochastic processes:

$$\dot{z}_t = E^t\{\dot{z}_t\} = E^{t-}\{\dot{z}_t\}, \text{ a.e. on } \Omega\times T. \tag{5.7}$$

From (5.6) we also have the alternative expression for $E^t\{\theta(x_t)\}$:

$$E^t\{\theta(x_t)\} = E^{t_0}\{\theta(x_{t_0})\} - \int_{t_0}^{t} E^{s-}\{\theta(x_s).\dot{x}_s\}ds + m'_t, \ t \in T, \text{ a.s.} \tag{5.3a}$$

and hence using (5.3):

$$m_t - m'_t = \int_{t_0}^{t} (E^s\{\theta(x_s).\dot{x}_s\} - E^{s-}\{\theta(x_s).\dot{x}_s\})ds, \ t \in T, \text{ a.s.}$$

But this shows that $\{m_t - m'_t\}$ has absolutely continuous sample paths a.s., and since $\{m_t - m'_t, F_t\}$ is a martingale we deduce that $m_t = m'_t$ a.e. on $\Omega\times T$, and hence

$$E^s\{\theta(x_s).\dot{x}_s\} = E^{s-}\{\theta(x_s).\dot{x}_s\}, \text{ a.e. on } \Omega\times T \tag{5.8}$$

Finally it is perhaps worth noting that similar continuous limits for p_t cannot be obtained in the same way since, although

$E^t\{p_t\} = p_t$, we cannot assert the existence of $\dot{p}_t$ a.e. on $\Omega\times T$ in order to apply Theorem 3.

6 CONCLUSIONS

We have shown that a coherent theory for nonlinear stochastic processes can be constructed on the basis of an intuitively reasonable "physically realizable" stochastic process with absolutely continuous sample functions.

A general nonlinear filter has been derived in finite-difference form, suitable for use in numerical computations, and an example of an explicit representation of this filter is given, based on a "bounded Gaussian" distribution as a probability law.

The continuous limit of the filter equations has been examined, and it is shown that the error in using the finite-difference form can be made arbitrarily small by use of sufficiently small time-steps.

7 ACKNOWLEDGEMENTS

The authors wish to express their thanks to Dr. M. H. A. Davis and Dr. J. M. Clarke for helpful discussions on the topics presented in this paper.

8 REFERENCES

1. Kushner, H. J. "On the Differential Equations Satisfied by Conditional Probability Densities of Markov Processes, with Applications", *J. SIAM Control*, **A2**, 1, 106, (1962).

2. Kallianpur, G. and Striebel, C., "Estimation of Stochastic Systems: Arbitrary system process with additive white noise observation errors", *Ann. Math. Statist.*, **39**, 785-801, (1968).

3. Fujisaki, M., Kallianpur, G. and Kunita, H. "Stochastic Differential Equations for the Nonlinear Filtering Problem", *Osaka J. Math.*, **9**, 19-40, (1972).

4. Doob, J. L. "Stochastic Processes", J. Wiley and Sons Inc., New York, (1953).

5. Davis, M. H. A. "Martingale Integrals and Stochastic Calculus", in J. K. Skwrizynski ed., "Communication Systems

and Random Process Theory", Sijthoff and Noordhoff, Alphen aan den Rijn, 687-704, (1978).

6. Wong, E. and Zakai, M. "On the Convergence of Ordinary Integrals to Stochastic Integrals", *Ann. Math. Statist.*, **36**, 1560, (1965).

7. Joffe, B. L. and Sargent, R. W. H. "A Naive Approach to the Optimal Control of Nonlinear Stochastic Systems", in D. J. Bell (Ed.) "Recent Mathematical Developments in Control", Academic Press, London, (1973).

8. Perkins, J. D. and Sargent, R. W. H. "Nonlinear Optimal Stochastic Control - Some Approximations when the Noise is Small", Proceedings 7th IFIP Conference on Optimization Techniques, Lecture Notes in Computer Science, **41**, Part 2, Springer-Verlag, Berlin, 820, (1976).

9. Virani, N. A. A. "A New Approach to the Analysis of Nonlinear Stochastic Systems", Ph.D. Thesis, University of London, (1980).

10. Wong, E. "Stochastic Processes in Information and Dynamical Systems", McGraw Hill, New York, (1971).

11. McShane, E. J. "Integration", Princeton University Press, Princeton, (1944).

12. Filippov, A. F. "Differential Equations with Discontinuous Right-hand Side", *Amer. Math. Soc. Translations,* Ser. 2, **42**, 199, (1964).

3.3 GENERIC MODELS FOR STOCHASTIC SYSTEMS

W. L. Root

(Aerospace Engineering Department, The University of Michigan, USA)

1 INTRODUCTION

In the theory touched on in this paper a system is regarded as a mapping from a space of admissible inputs to a space of outputs. If the outputs have a natural interpretation as stochastic quantities, the system will be said to be stochastic.

The material to follow is from work largely motivated by problems in system identification, although system identification as such is not discussed here and what is done is fairly abstract. We elaborate a little on this. By a system identification we mean the construction of a suitable (for whatever purpose intended) mathematical model for an unknown "real-life system" from an experimental record of input-output data, using what prior knowledge of the system there may be. Usually in practice the prior knowledge is assumed to be sufficient to allow one to specify a "pre-model", i.e., a mathematical model complete except for a finite set of uncertain parameters or coefficients, e.g., in the form of a differential or difference equation with undetermined coefficients. However, there may be a question as to whether a particular pre-model is really adequate; there may be competing pre-models, or it may be difficult to guess an appropriate pre-model. In light of this it seems worthwhile to move to one higher level of abstraction and look at a setting in which pre-models can be compared or even constructed.

This line of thought has led to consideration of abstract families of mappings from input to output and thence to the rather obvious definition of compound system given below. It also has led to consideration of parameterized approximate representations of such families (see, e.g. [2] and [4]), but these representations are not treated here. The systems of interest are taken to be those for which inputs and outputs are time functions (possibly to be considered random) with outputs causally related to inputs. Thus there must first be established some function-space structure for input and output spaces, and then a structure in which to consider the families of causal mappings. The mappings are not linear in general and their domains will not be linear spaces. However we find it convenient to use metric subspaces of normed linear function spaces as input spaces. Certain conditions are imposed on these spaces, dictated both by what seems reasonable in a model and by mathematical convenience: most notably, invariance under time-shift and total boundedness in the appropriate metric. The causal mappings are not required to be invariant under time shifts; i.e., we deal with time-varying systems. Time-varying systems are perhaps awkward theoretically but they certainly occur in practice. Two situations may be noted where it appears possible to do some kind of analysis: (1) not much is known a priori about the time-variation except that it is slow (see [2]); (2) there is some assumed hyperstructure (possibly stochastic, possibly not) that determines the time variation, or at least its local properties. It is this second situation that is dealt with here.

Consider a single time-varying system operating for an indefinite period. The lack of time-invariance suggests focusing attention on trajectories $t \to F_t$ where F_t is a map characterizing the behaviour of the system at or in the neighbourhood of the time t, and a consideration of such trajectories is in fact the main topic of the paper. The maps F_t are typicall defined so that an image of an F_t (an "observation") is a section of the whole output function on a nonzero time interval; the time index t is then the right-hand endpoint of this observation interval.

Conditions are given for continuity of trajectories. The case that the future of a trajectory is generated by a semigroup of transformations (the so-called predictable case) is emphasized, and in particular the elementary effects of changing observation interval length are noted. These considerations apply to abstract systems, deterministic or stochastic.

A stochastic structure is set up which allows associating stochastic systems with a given (presumably, but not necessarily deterministic) "compound" system. Under certain conditions the stochastic system has a continuous trajectory if the given compound system has continuous trajectories.

A simple illustrative example is stated.

Along with some other background material, one non-standard concept, that of a fitted family of normed linear spaces, is discussed briefly in the next section. The fitted families are introduced to provide a flexible abstract setting for characterizing system input and output spaces (see [6]).

Very little in the way of proof is included here; for although the proofs are not difficult, they still would require space that is not available. Results for which no reference is cited are proved in [6], a report in preparation.

2 PRELIMINARIES

2.1 Systems

Let U and X be metric spaces and Y a real Banach space. Let $F:U \to Y$ and $f:X \times U \to Y$ be continuous maps. $X \times U$ is to be regarded as a metric space with distance function given by $d((x,u),(x',u')) = d(x,x') + d(u,u')$ for all $x,x' \in X$ and $u,u' \in U$. We define a <u>system</u> to be $s = (Y,F,U)$ and a <u>compound system</u> to be $S = (Y,f,X,U)$. S is <u>bounded in U(X)</u> if each $f(x,\cdot)$ $(f(\cdot,u))$ is bounded. Of course, a compound system is a system with $X \times U$ replacing U. The reason for introducing the more elaborate definition is that U and X are to play different roles; U is to be thought of as a space of possible inputs, X as a space of parameters determining the condition of the system. Usually the point of view is that the input $u \in U$ is known to and possibly controllable by the experimenter while the parameter $x \in X$ is unknown, at least <u>a priori</u>. The space X is needed in the abstract description of time-varying systems with which we are concerned; it is also needed in system identification problems. When there is no chance of confusion, the word system will be used for both system and compound system as defined above.

For any metric space U and any real Banach space Y, $\mathcal{F}(U,Y)$ denotes the vector space of bounded continuous maps $g:U \to Y$ made into a Banach space with norm $\|g\| = \sup_{u \in U} \|g(u)\|$. It will sometimes be helpful to distinguish various norms by use of subscripts; for example the above could be written,

$$\|g\|_{\mathcal{F}(U,Y)} = \sup_{u \in U} \|g(u)\|_Y .$$

It is easily shown (see [4]) that $S_I = (Y, f_1, \mathcal{F}(U,Y), U)$ with f_1 defined by $f_1(F,u) = F(u)$, $F \in \mathcal{F}(U,Y)$, is a compound system, of course bounded in U. If a system $S = (Y,f,X,U)$ is bounded in U, let $\psi : X \to \mathcal{F}(U,Y)$ be given by $\psi(x) = F$ where F in turn is defined by $F(u) = f(x,u)$, $u \in U$. $\psi(X)$ is a metric subspace of $\mathcal{F}(U,Y)$ and $S_1 = (Y, f_1, \psi(X), U)$ is, by an obvious definition, a compound subsystem of S_I. S_1 is called the <u>natural representation</u> of S and ψ the <u>natural mapping</u>. ψ is continuous iff $f(\cdot,u)$ defines equicontinuous functions of x for each $u \in U$; a sufficient condition is that U be compact. One notes S_1 has some algebraic structure that S does not necessarily have. In particular, f_1 is <u>prelinear</u> in its first variable, i.e.,

$$f_1(\alpha F + \beta F', u) = \alpha f_1(F,u) + \beta f_1(F', u)$$

whenever all three terms are defined. Thus f_1 can be extended linearly to $V[\psi(X)]$, the linear span of $\psi(X)$ in $\mathcal{F}(U,Y)$.

X and U may be interchanged in the above comments, but the structure that results when this is done is not needed here.

Suppose that for each* $t \in R$, or each $t \in R_+$, there is given $F_t \in \mathcal{F}(U,Y)$. Then either of the ordered sets of systems $((Y, F_t, U), t \in R)$ or $((Y, F_t, U), t \in R_+)$ is called a <u>trajectory of systems</u> (or just <u>trajectory</u>, and denoted simply by (F_t)). We shall also consider families of trajectories $\{(F_t(\theta))\}$, where $\theta \in \Theta$, an index set. According to this definition the systems comprising the trajectories in a family all belong to some subsystem of S_I; in other words they are in natural representation form. A trajectory (F_t) is <u>continuous</u> if $t \to F_t$ is a continuous map from R or R_+ into $\mathcal{F}(U,Y)$. A trajectory is <u>predictable</u> if $F_t = F_s$ implies $F_{t+v} = F_{s+v}$ for all $v \geq 0$; a family of trajectories is <u>predictable</u> if, whenever $F_t(\theta) = F_s(\theta')$ for any t, s, θ, θ', then $F_{t+v}(\theta) = F_{s+v}(\theta')$ for all $v \geq 0$. If a family is predictable so is any subfamily.

*R is the real line; $R_+ = \{t \in R : t \geq 0\}$.

The predictability condition is equivalent to: there exists a one-parameter semigroup of transformations $\tau(v)$, $v \geq 0$, such that $F_{t+v}(\theta) = \tau(v)\, F_t(\theta)$. For each $v, \tau(v): \mathcal{H} \to \mathcal{H}$, where $\mathcal{H} = \bigcup_{t,\theta} F_t(\theta)$ is regarded as a metric subspace of $\mathcal{F}(U,Y)$. Note that even with a single given trajectory (F_t) the introduction of τ implies that we are effectively considering the family consisting of all translates of F_t.

A (compound) stochastic system is taken to be a (compound) system for which the output quantities have the additional interpretation of being stochastic variables. Actually, we consider only the following special case. Let Y be a real, separable Banach space and $(\Omega, \mathcal{B}, \mu)$ a separable probability space. Take Z to be the L_p-space $L_2(\Omega, \mathcal{B}, \mu, Y)$ (in the notation of [1]). When the σ-algebra $\mathcal{B}$ is understood we write $Z = L_2(\Omega, \mu, Y)$. Then the system $\Sigma = (Z,f,X,U)$, where $f: X \times U \to Z$ is continuous, is a stochastic compound system. We note that this definition includes cases for which Ω is the parameter space X' of an associated deterministic system $S=(Y,f',X',U)$, where Y and U are the same as in the stochastic system Σ, and $z(w)=y=f(w,u)$. This kind of thing is developed later.

2.2 *Fitted families of normed linear spaces*

It is desirable not to be restricted to specific function spaces for inputs and outputs; for this reason the notion of fitted families of normed linear spaces was introduced in [5]. The definition is repeated here, since in everything that follows, except the example, the normed linear spaces can be arbitrary long as they belong to fitted families and satisfy whatever particular hypotheses are imposed. It will be seen that the conditions imposed by the definition are very weak. There are a multitude of examples (see [5]).

Let C be a linear space of functions from R into a separable real Banach space such that any translate of a function in C is also in C. Let $N = \{\|\cdot\|_{s,t}, -\infty < s \leq t < \infty\}$ be a family of seminorms on C satisfying:

1. If $g_1, g_2 \in C$ and $g_1(\tau) = g_2(\tau)$, $s \leq \tau \leq t$, then $\|g_1 - g_2\|_{s,t} = 0$.
2. $\|L_\tau g\|_{s-\tau,t-\tau} = \|g\|_{s,t}$, $g \in C$, $-\infty < \tau < \infty$, where L_τ is translation to the left by τ.

3. If $r \leq s \leq t$, $\|g\|_{s,t} \leq \|g\|_{r,t}$, $g \in C$.

4. If $r \leq s \leq t$, $\|g\|_{r,t} \leq \|g\|_{r,s} + \|g\|_{s,t}$, $g \in C$.

5. For some $\alpha > 0$ (α may be $+\infty$) and $K \geq 0$, and for all $g \in C$, $\|g\|_{r,s} \leq K \|g\|_{r,t}$, $r \leq s \leq t$, whenever $t - r < \alpha$.

The pair (C,N) is called a fitted family of seminorms on C. The normed linear space formed from equivalence classes of functions in C with norm $\|\cdot\|_{s,t}$ is denoted $A_{s,t}$ (or sometimes $B_{s,t}$ when two families appear in the same context). The set $\{A_{s,t}\}$ is the fitted family of normed linear spaces given by (C,N).

With (C,N) given, let $C_0 = \{g \in C: \lim_{s\to-\infty} \|g\|_{s,t}$ exists, $t \in R\}$. For $g \in C_0$, define $\|g\|_t = \|g\|_{-\infty,t} = \lim_{s\to-\infty} \|g\|_{s,t}$. Obviously $\|\cdot\|_t$ is a seminorm on C_0, and C_0 is linear and shift invariant. With $\bar{N} = \{\|\cdot\|_{s,t}, -\infty \leq s \leq t<\infty$, but $-\infty < t\}$ the properties (1),..., (4) hold for $(C_0,\bar{N})$ even when the left-most point is replaced by $-\infty$; (5) holds for $r = -\infty$ if $\alpha = \infty$, otherwise the property is irrelevant for $r = -\infty$. Hence $(C_0,\bar{N})$ is a fitted family in an extended sense. The normed linear space formed from equivalence classes of functions of C_0 with norm $\|\cdot\|_t$ is denoted A_t. A_t is linearly homeomorphic to A_0 under a shift operation.

(C,N) determines $(C_0,\bar{N})$; it also determines $C_{00} = \{g \in C_0:\sup_{t\in R}\|g\|_t < \infty\}$. C_{00} is linear and shift invariant. For $g \in C_{00}$ we put $\|g\| = \sup_{t\in R}\|g\|_t = \sup_{(s,t),s\leq t} \|g\|_{s,t}$. The normed linear space formed from equivalent classes of functions in C_{00} with norm $\|\cdot\|$ is denoted A and is called the bounding space for the family $\{A_{s,t}\}$.

One result concerning fitted families will be stated since it is auxiliary to the material on trajectory continuity in the next Section. First, a definition is needed.

Given a fitted family $\{A_{s,t}\}$, for any $c > 0$ and any t let

$$G(c,t) = \{g \in C_0 : \|g\|_\tau \leq c \quad \forall \tau \leq t\}.$$

$\{A_{s,t}\}$ is <u>tapered</u> if for any $\varepsilon > 0$, $c > 0$ and $t \in R$ there is a positive number $\delta = \delta(\varepsilon,c,t)$ such that $\|g\|_t \leq \|g\|_{t-\delta,t} + \varepsilon$ for all $g \in G(c,t)$. Since $G(c,s) = L_{t-s}[G(c,t)]$ and since $s < t$ implies $G(c,t) \subset G(c,s)$, it follows that if $\{A_{s,t}\}$ is tapered, δ does not in fact depend on t, and further, $\|g\|_s \leq \|g\|_{s-\delta,s} + \varepsilon$ for all $g \in G(c,t)$ and any $s \leq t$.

It is convenient to say that $A_{s,t}$ is shift-continuous if $\|L_h g - g\|_{s,t} \to 0$ as $h \to 0$, $g \in C$, and that $M \in C_0$ is shift-continuous in A_t if $\|L_h g - g\|_t \to 0$ as $h \to 0$, $g \in M$.

<u>Proposition 1</u> (Prop. 5 of [4]). For fixed a, b, $-\infty < a < b < \infty$, let $M_0 \subset C$ be a set of functions that vanish outside the interval $[a,b]$. Further, let it be required that M_0 is totally bounded in every $A_{s,t}$ with $a \leq s \leq t \leq b$. Then if $\{A_{s,t}\}$ is tapered and the $A_{s,t}$ are shift-continuous, the shift-invariant set $M = \{L_\tau g : g \in M_0, \tau \in R\}$ is totally bounded and shift-continuous in each A_t.□

3 TRAJECTORIES OF TIME-VARYING SYSTEMS

3.1 *Trajectories of truncated systems*

In this section we consider systems for which the inputs and outputs are (perhaps random) functions of time with the outputs causally related to the inputs. Let $\{A_{s,t}\}$ and $\{B_{s,t}\}$ be fitted families of normed linear spaces with bounding spaces A and B, respectively. Denote norms in $A_{s,t}$, A_t and A by $\|\cdot\|_{s,t}$, $\|\cdot\|_t$ and $\|\cdot\|$, respectively, and in $B_{s,t}$, B_t and B by $|||\cdot|||_{s,t}$, $|||\cdot|||_t$, $|||\cdot|||$, respectively. The input space U is to be a subset (metric subspace) of A with the property

(i) $L_t U = U$.

U_t is the metric subspace of A_t determined by U; the elements of U_t are equivalence classes of elements of U (which in turn are equivalence classes of elements of C). The output space Y = B. F is a map from U into Y which is required to have the properties

(a) if $u_1, u_2 \in U$ satisfy $\|u_1 - u_2\|_t = 0$, $t \in R$, then $|||F(u_1) - F(u_2)|||_t = 0$; thus F induces a causal map $\tilde{F}_t$ from U_t into B_t.

(b′) $\tilde{F}_t$ is continuous and bounded.

Thus $(B_t, \tilde{F}_t, U_t)$ is a system if B_t is a Banach space, which it can always be taken to be. B_t is linearly isometric to B_0 and U_t is pre-linearly isometric to U_0 under the translation L_t. Consequently, corresponding to $\tilde{F}_t$ is a map $F_t \in \mathcal{F}(U_0, B_0)$ given by

$$F_t(u) = L_t\tilde{F}_t(R_t u), \quad u \in U_0$$

where $R_t = L_{-t}$. Thus $((B_0, F_t, U_0), t \in R)$ is a trajectory, and indeed a trajectory of systems truncated from the bounding system (Y, F, U). Note that (Y, F, U) may not properly be a system since it has not been required that F be continuous.

It follows at once from conditions (a) and (b′) that F also induces a bounded continuous map $\tilde{F}_{t-a,t}: U_t \to B_{t-a,t}$ for any $a \geq 0$. The number a represents the length of the observation interval. Thus, with $F_{(a)t} = L_t\tilde{F}_{t-a,t}R_t$, $((B_{-a,0}, F_{(a)t}, U_0), t \in R)$ is also a trajectory of systems truncated from (Y, F, U). B_0 or $B_{-a,0}$, as the case may be, is called the observation space and may be denoted Y_0 when it is fixed in context.

Example

Let $\phi: R \to R_+$ satisfy the conditions: $\phi(t) = 0$ for $t < 0$; $\phi(0) = 1$; ϕ monotone nonincreasing on

$[0,\infty);\phi(t) \to 0$ as $t \to \infty$, and $\int_0^\infty \phi(t)\,dt = a,\ 0 < a < \infty.$

Let C be the set of real-valued Lebesgue measurable functions g on R for which $\|g\|_{s,t}$ exists for all $s \le t$, where $\|\cdot\|_{s,t}$ is defined by

$$\|g\|^2_{s,t} = \int_s^t |g(v)|^2 \phi(t - v)\,dv, \quad -\infty < s \le t < \infty.$$

$A_{s,t}$ is the Hilbert space of equivalence classes of functions in C with respect to $\|\cdot\|_{s,t}$. $\{A_{s,t}\}$ is a fitted family that is both tapered and shift-continuous. A_t is also a Hilbert space. It follows from Proposition 1 that there are nontrivial shift-invariant sets U such that the U_t they determine are totally bounded and shift-continuous metric subspaces of A_t. We choose such a U as input space. For the family $\{B_{s,t}\}$ we take $B_{s,t} = L_2[r,t]$, $r = \max(s,t-a)$, the Lebesgue space of square-integrable real functions on $[r,t]$. With the observation interval fixed at $a > 0$, the output spaces are $Y_t = B_t = L_2[t-a,t]$. Recall the norm for Y_t is denoted $|||\cdot|||_t$.

We consider truncated input-output mappings $\tilde{F}_t$ of the form

$$[\tilde{F}_t(u)](\sigma) = \int_0^\infty h(s;v)u(s - v)\,dv, \quad t - a \le s \le t,$$

$u \in U_t$, where h is to satisfy a condition that may be stated as follows. Put $h_t(\sigma;v) = h(t + \sigma;v)$, $t \in R$, $-a \le \sigma \le 0$, then

$$\int_{-a}^0 \int_0^\infty |h_t(\sigma;v)|^2 \frac{dv\,d\sigma}{\phi(b + \sigma)} \le K_1^2 = \text{constant}, \quad t \in R.$$

For convenience, the L_2 space for which the square of the norm is given by the expression above is denoted $L_{2\phi}$, and the norm $\|\cdot\|_\phi$. The mappings $F_t : U_0 \to Y_0$ corresponding to the $\tilde{F}_t$ are given by

$$\left[F_t(u)\right](\sigma) = \int_0^\infty h_t(\sigma;v)u(\sigma - v)dv, \quad -a \leq \sigma \leq 0.$$

It can be readily verified that the F_t (and $\tilde{F}_t$) are uniformly continuous. Since U_0 is bounded,

$$\sup_{U_0} \|u\|_0 = K_0 = \text{constant, and}$$

$$\|F_t\| = \sup_{U_0} |||F_t(u)|||_t \leq K_0 \|h_t\|_\phi .$$

Let $X = \{h_t,\ t \in R\} \subset L_{2\phi}$, and define $f{:}X \times U_0 \to Y_0$ by $\left[f(h_t,u)\right](\sigma) = \left[F_t(u)\right](\sigma)$. f is continuous and $S = (Y_0, f, X, U_0)$ is compound system. Its natural mapping ψ is continuous.

It is also true that the bounding system map $F{:}U \to B$ is well-defined and that it and the input and output spaces satisfy the conditions of Proposition 3 below so that the trajectory (F_t) is continuous.

This example is continued later. We remark: (1) The choice of a linear operator is for convenience; the same sort of thing can be done with polynomial integral operators (see related material in [4]) but it is superficially more complicated. (2) We still get a compound system if X contains the h_t given by any set of functions h satisfying the condition.

3.2 Semigroup generation of trajectories

With the families $\{A_{s,t}\}$, $\{B_{s,t}\}$ given, and $U \subset A$ satisfying condition (i) above, let $\hat{H}$ be a set of maps $F{:}U \to B$, each of which satisfies conditions (a) and (b$'$). Then for any $a > 0$ there is a family of trajectories $\{((B_{-a,0}, F_{(a)t}, U_0), t \in R)\}$ where each $F_{(a)t}$ is determined by an $F \in \hat{H}$ (the index θ for $F(\theta)$ has been suppressed). Suppose further that for some $a = b > 0$, the family of trajectories is predictable.

There is then a semigroup of transformations $\tau_b(v)$, $v \geq 0$, with domain $\mathcal{H}_{(b)} = \bigcup_{t \in R, F \in \hat{\mathcal{H}}^F} (b)t$ regarded as a metric subspace of $\mathcal{F}(U_0, B_{-b,0})$.

How does changing observation interval length affect the properties of the trajectories? A partial answer is given in Proposition 2 and later in Proposition 4.

<u>Proposition 2.</u> If $\{B_{s,t}\}$ allows $\alpha \geq c \geq b$, then there is a semigroup $\tau_c(v):\mathcal{H}_{(c)} \to \mathcal{H}_{(c)}$ such that $F_{(c)t+s} = \tau_c(s)F_{(c)t}$, $s \geq 0$. Furthermore if $\tau_b(v)$ is a continuous transformation for some v, $0 < v \leq b$, then the transformations $\tau_c(v)$ are continuous for all $v > 0$.□

3.3 Continuity of trajectories

The following sufficient condition for trajectory continuity is proved in [5].

<u>Proposition 3.</u> Let $\{A_{s,t}\}$ and $U \subset A$ be such that (i) above is satisfied, and also:

(ii) U_0 is totally bounded in A_0,

(iii) $\|L_h u - u\|_0 \to 0$ as $h \to 0$ for all $u \in U$.

Let $\{B_{s,t}\}$ be such that:

(iv) $\|L_h y - y\|_0 \to 0$ as $h \to 0$ for all $y \in B$.

Finally let $F:U \to B$ satisfy (a) and also have the property:

(b) $\tilde{F}_t$ is uniformly continuous on U_t for each t.

Then the trajectory (F_t) is continuous.□

The relevance of Proposition 1 is now evident. It guarantees that under reasonable conditions there are useful classes of input functions which will satisfy the conditions (i), (ii), (iii) on U providing the family $\{A_{s,t}\}$ is tapered. The tapering condition itself ensures that the input-space topology does not allow a continuous input-output map to have "too much memory". Thus it is really a restriction on systems, as is clear in the example above. An example is given in [5] of a system with "too much memory" and resulting discontinuous trajectory.

Continuity of trajectories is preserved under certain circumstances if the length of the observation interval is changed. Obviously if $(F_{(b)t})$ and $(F_{(a)t})$ are trajectories derived from a bounding system (B, F, U) and $0 \leq a \leq b \leq \infty$, then continuity of $(F_{(b)t})$ implies continuity of $(F_{(a)t})$. We also have:

Proposition 4. With the same hypotheses as for Proposition 2, it follows that continuity of $(F_{(b)t})$ implies continuity of $(F_{(c)t})$.□

4 A TYPE OF STOCHASTIC SYSTEM TRAJECTORY

4.1 Associated stochastic systems

We start by indicating a procedure for constructing compound stochastic systems which are associated in a natural way with a given compound system bounded in U. It is convenient to consider a compound system $S_1 = (Y, f_1, \mathcal{H}, U)$ in natural representation form. We assume U to be compact and Y to be separable. It then follows that $\mathcal{F}(U, Y)$ is separable, and $\mathcal{H} = \psi(X)$ considered as a metric subspace of the Banach space $\mathcal{F}(U, Y)$ is separable.

Let μ be a probability measure on $\mathcal{B}$ (the σ-algebra of Borel sets of $\mathcal{H}$) which satisfies

$$\int_{\mathcal{H}} \|F\|^2_{\mathcal{F}(U, Y)} d\mu(F) = k^2 < \infty. \qquad (*)$$

Let ρ be the map from U to Y-valued functions on $\mathcal{H}$ given by $\rho(u) = f_1(\cdot,u)$. Since

$$\int_{\mathcal{H}} \| [\rho(u)](F)\|^2_Y d\mu(F) = \int_{\mathcal{H}} \|F(u)\|^2_Y d\mu(F) \leq k^2,$$

$\rho(u) \in L_2(\mathcal{H}, \mu, Y)$. It may be readily verified that $\rho: U \to L_2(\mathcal{H}, \mu, Y)$ is continuous, so $(L_2(\mathcal{H}, \mu, Y), \rho, U)$ is a stochastic system.

Now suppose that $\{\tau(\beta)\}$ is a set of transformations on $\mathcal{H}$ such that each pair $(\mu, \tau(\beta))$ has the properties, familiar from ergodic theory:

(α) $\tau(\beta):\mathcal{H} \to \mathcal{H}$ is Borel measurable

(β) if $E \in \mathcal{B}$ and $\mu(E) = 0$, then $\mu(\tau^{-1}(\beta)(E)) = 0$

(γ) $$\sup_{E \in \mathcal{B}, \mu(E) \neq 0} \frac{\mu[\tau^{-1}(\beta)(E)]}{\mu(E)} = M(\beta) < \infty.$$

We write μ_β for the measure defined by $\mu_\beta(E) = \mu(\tau^{-1}(\beta)(E))$.

Define the transformations $T(\beta)$ on $L_2(\mathcal{H}, \mu, Y)$ by $[T(\beta)g](F) = g([\tau(\beta)](F))$, $g \in L_2(\mathcal{H}, \mu, Y)$. $T(\beta)$ is a continuous linear operator on $L_2(\mathcal{H}, \mu, Y)$ with linear operator norm $|T(\beta)| = [M(\beta)]^{\frac{1}{2}}$ (see [1], VIII 5.7). Let $T_r(\beta)$ denote the restriction of $T(\beta)$ to the (compact) image of U, $A = \rho(U)$. Then

$$\|T_r(\beta)\|_{\mathcal{F}(A,\, L_2(\mathcal{H}, \mu, Y))} \leq [M(\beta)]^{\frac{1}{2}} k.$$

With $\mathcal{J} \overset{d}{=} \{T_r(\beta)\}$ regarded as a metric subspace of $\mathcal{F}(A, L_2(\mathcal{H}, \mu, Y))$, define $f:\mathcal{J} \times U \to L_2(\mathcal{H}, \mu, Y)$ by $f(T_r, u) = T_r(\rho(u))$. f is uniformly continuous on $\mathcal{J} \times U$, so we have:

Proposition 5. $\Sigma = (L_2(\mathcal{H}, \mu, Y), f, \mathcal{J}, U)$ is a compound stochastic system bounded in U. Its natural representation is $\Sigma_1 = (L_2(\mathcal{H}, \mu, Y), f_1, \tilde{\mathcal{J}}, U)$ where the elements of $\tilde{\mathcal{J}} = \psi(\mathcal{J})$ are the mappings $T_r(\beta) \circ \rho$.□

4.2 *Families of trajectories and associated stochastic trajectories*

When the index set for the set of mappings $\tau(\beta)$ used above is R_+ or R, the compound system Σ_1 described in Proposition 5 describes a trajectory of stochastic systems, i.e., $(T_r(t) \circ \rho)$

is a trajectory. A natural case to consider is that in which the compound system S_1 is comprised of the systems of a set of trajectories generated by a semigroup or group $\tau(v)$. The same semigroup (group) will generate a stochastic trajectory with respect to a probability measure μ provided the conditions (α), (β), (γ) are satisfied. The stochastic trajectory has the interpretation of being a randomization of the original set of trajectories. Does continuity of the original trajectories imply continuity of the stochastic trajectory? Apparently not in general, but we have the following result with $(\tau(v))$ a group.

Proposition 6. Let $\{((Y_0, F_t(\theta), U_0), t \in R, \theta \in \Theta\}$ be a set of trajectories of systems for which the following conditions are satisfied. Y_0 is separable; U_0 is compact; $F_t(\theta) \in \mathcal{F}(U_0, Y_0)$, and each trajectory is continuous. Furthermore $F_t(\theta)$ is generated by a one-parameter group of transformations $\tau(t), t \in R$. Each $\tau(t)$ is a Borel measurable 1:1 map from $\mathcal{H} = \bigcup_{t,\theta} F_t$ onto itself, where $\mathcal{H}$ is a metric subspace of $\mathcal{F}(U_0, Y_0)$; $\tau^{-1}(t) = \tau(-t)$.

Let μ be a probability measure on $\mathcal{B}$, the σ-algebra of Borel sets of $\mathcal{H}$, such that

$$\int_{\mathcal{H}} \|F_0\|^2_{\mathcal{F}(U_0, Y_0)} d\mu(F_0) < \infty,$$

$$\mu(E) = 0 \Rightarrow \mu(\tau(t)E) = 0, \; E \in \mathcal{B}, \; t \in R,$$

$$\sup_{E \in \mathcal{B}} \frac{(\tau(t)E)}{\mu(E)} = m(t) < \infty, \; t \in R.$$

Then, with $\rho: U_0 \to L_2(\mathcal{H}, \mu, Y_0)$ given by $[\rho(u)](F_0) = F_0(u)$, and $\Phi_t = T(t) \circ \rho$, the trajectory $((L_2(\mathcal{H}, \mu, Y_0), \Phi_t, U_0), t \in R)$ is continuous and satisfies $\Phi_{t+s} = T(t) \circ \Phi_s$. □

Example (continuation)

Everything is to be as it was above, except that the kernel h is specialized to be of the form $h_t(\sigma;v) = w_t(\sigma)g(v) = w(t+\sigma)g(v)$. Then $[F_t(u)](\sigma)$

$= w(t+\sigma)\int_0^\infty g(v)u(\sigma - v)dv$, $-a \le \sigma \le 0$. We require that the real function w satisfy $\int_{-a}^0 |w_t(\sigma)|^2 d\sigma \le K_2^2$ = constant, $t \in R$.

We also require that $\int_0^\infty \frac{|g(v)|^2}{\phi(b+v)} dv < \infty$, and for at least one $u \in U_0$, $\int_0^\infty g(v)u(\sigma - v)dv \neq 0$ a.e. in $[-a,0]$. Then $\|h_t\|_\phi \le$ constant, $t \in R$, and the natural map ψ, $\psi(h_t) = F_t$, is 1:1. Put $H = \psi(X)$; H is a metric subspace of $F(U_0, Y_0)$.

If further we require that for no pair (s,t) $s \neq t$, is $w_t(\sigma) = w_s(\sigma)$ a.e. in $[-a,0]$, then the correspondence between t and w_t as an element of $L_2[-a,0]$ is 1:1. Thus ξ, the composition of mappings indicated by $t \to w_t \to h_t \to F_t$, is 1:1. $F_t = \xi(t)$, $t \in R$, and $H = \xi(R)$. It is easy to see that $\xi: R \to H$ is continuous, and hence Borel measurable. Since H, as a metric subspace of $F(U_0, Y_0)$, is separable, ξ^{-1} is Borel measurable by a well-known theorem of Kuratowski ([3], p. 15). Let $\theta_r s = r + s$, $r, s \in R$, and put $\tau_r = \xi \circ \theta_r \circ \xi^{-1}$. Then (τ_r) is a group of Borel measurable 1:1 transformations on H. The trajectory (F_t) is now predictable as well as continuous.

To obtain a stochastic system we impose a Borel probability measure P on R. This induces a probability measure μ on the Borel sets of H according to $\mu(E) = P(\xi^{-1}(E))$. A simple calculation gives

$$\int_H \|F\|^2 d\mu(F) \le K_0^2 \int_0^\infty \frac{g^2(v)}{\phi(b+v)} dv \int_{-\infty}^\infty \int_{-b}^0 |w_t(\sigma)|^2 d\sigma\, dP(t)$$

which gives us a condition on P in order that the left side be finite. We assume this satisfied, then with $\rho : U_0 \to L_2(H, Y_0; \mu)$ defined as above, $(L_2(H, Y_0; \mu), \rho, U_0)$ is a stochastic system.

Certain choices of P will allow the hypotheses for Propositions 5 and 6 to be satisfied also; for example, let P be given by the density $p(t) = \frac{1}{2} e^{-|t|}$. Then

$$\frac{\mu[\sigma_r(E)]}{\mu(E)} \le e^{|r|}, \quad r \in R,$$

which with the earlier conditions is sufficient for both these theorems. In particular there is guaranteed a continuous stochastic trajectory.

5 REFERENCES

1. Dunford, N. and Schwartz, J. T. "Linear Operators Part I", Interscience, New York, (1958).

2. Fiske, P. H. and Root, W. L. "Identifiability of Slowly Varying Systems", *Information and Control,* **32**, 3, 201-230, (1976).

3. Parthasarathy, K. R. "Probability Measures on Metric Spaces", Academic Press, New York, (1967).

4. Root, W. L. "On the Modelling of Systems for Identification Part I, ε-representations of Classes of Systems", *SIAM J. Control,* **13**, 4, 945-974, (1975).

5. Root, W. L. "Considerations regarding input and output Spaces for Time-Varying Dynamical Systems", *App. Math. and Optimization,* **4**, 365-384, (1978).

6. Root, W. L. "Representations of Time-Varying Systems", Report in preparation, Univ. of Michigan.

3.4 MATHEMATICAL MODELS FOR RANDOM VIBRATION PROBLEMS

R. F. Curtain

(Mathematics Institute, University of Groningen, Netherlands)

1 INTRODUCTION

This paper is concerned with random vibration problems which occur in mechanically flexible systems such as space-craft, bridges, and other mechanical structures. Stochastic stability for such systems is of considerable interest, most of the known results being for finite dimensional approximations [8], [5], [4], [3], although some analysis of the distributed model exists [6], [7], [8]. Control problems are also of interest, both deterministic [1] and stochastic. The purpose of this paper is to investigate the applicability of the known results on the semigroup theory infinite dimensional stochastic systems described in [21], [22] and [14] to a large class of random vibration problems. While it has long been known how to formulate second order wave type equations as a semigroup, most of the results in the semigroup theory have been primarily directed towards applications to parabolic distributed systems and delay equations. This has been especially the case for problems involving boundary control and point sensing, which usually presuppose an analytic semigroup.

The following examples are typical of these of interest in applications ([1] - [8]).

Example 1 [6]. The lateral displacement of a string stretched between fixed ends and subject to a random transverse load $(f_0 + f(t))\dfrac{\partial v}{\partial x}$ with damping force $\alpha \dfrac{\partial v}{\partial t}$.

$$\frac{\partial^2 v}{\partial t^2} + \alpha\frac{\partial v}{\partial t} - \frac{\partial^2 v}{\partial x^2} + (f_0 + f(t))\frac{\partial v}{\partial x} = 0$$

$$v(0,t) = 0 = v(1,t)$$

Example 2 [6], [8]. Vibration of a panel in supersonic flow subjected to random end loads.

$$\frac{\partial^2 v}{\partial t^2} + \alpha \frac{\partial v}{\partial t} + m \frac{\partial v}{\partial x} + \left(f_0 + f(t)\right)\frac{\partial^2 v}{\partial x^2} + \frac{\partial^4 v}{\partial x^4} = 0$$

$$v(0,t) = 0 = v(1,t) = \frac{\partial^2 v}{\partial x^2}(0,t) = \frac{\partial^2 v}{\partial x^2}(1,t)$$

With m = 0 this is just a simply supported beam subjected to random axial loads. [5], [8].

Example 3 [2]. Flutter and random vibrations of a plate in supersonic flow.

$$\frac{\partial^2 v}{\partial t^2} + \alpha \frac{\partial v}{\partial t} + m \frac{\partial v}{\partial x} + \frac{\partial^4 v}{\partial x^4} = p(x,t)$$

$$v(0,t) = 0 = v(1,t) = \frac{\partial v}{\partial x}(0,t) = \frac{\partial v}{\partial x}(1,t)$$

Here p(x,t) is the external force due to pressure fluctuations.

With m = 0 and boundary conditions

$v(0,t) = 0 = v(1,t) = \frac{\partial^2 v}{\partial x^2}(0,t) = \frac{\partial^2 v}{\partial x^2}(1,t)$, this also models

the response of bridges to moving random loads p(x,t). [3].

Example 4 [6]. A cantilevered column subjected to a random follower force at its free end with viscous damping.

$$\frac{\partial^2 v}{\partial t^2} + \alpha \frac{\partial v}{\partial t} + f(t) \frac{\partial^2 v}{\partial x^2} + \frac{\partial^4 v}{\partial x^4} = 0$$

$$v(0,t) = 0 = \frac{\partial v}{\partial x}(0,t) = \frac{\partial^2 v}{\partial x^2}(1,t) = \frac{\partial^3 v}{\partial x^3}(1,t).$$

In section 2 we shall formulate a large class of second order stochastic systems, which includes all these examples, but modelling the white noise via an Itô integral. See [8] for a discussion of alternative models. In section 3 we discuss stability properties of these systems and in sections 4, 5 and 6 various control and filtering problems, which can be formulated for such systems using the known theory in [9] and [14]. This investigation of semigroup approaches to random vibration problems is by no means complete and several open problems remain, including the boundary noise problem discussed in section 6.

2 SEMIGROUP FORMULATION OF SECOND ORDER STOCHASTIC SYSTEMS

We consider the second order system

$$v_{tt} + \alpha v_t + Av = 0; \; v(0) = v_0, \; v_t(0) = v_1, \; \text{for } \alpha \geq 0 \qquad (2.1)$$

where A is a positive, self adjoint operator on a Hilbert space H with domain D(A).

As in [1] or [9], we obtain a semigroup formulation by introducing $z = \begin{pmatrix} v \\ v_t \end{pmatrix}$, and rewriting (2.1) as

$$\dot{z} = \mathcal{A} z; \; z(0) = z_0, \qquad (2.2)$$

where $z_0 = \begin{pmatrix} v_0 \\ v_1 \end{pmatrix}$ and

$$\mathcal{A} = \begin{pmatrix} 0 & I \\ -A & -\alpha I \end{pmatrix} \text{ is a closed linear operator} \qquad (2.3)$$

on $\mathcal{H} = D(A^{\frac{1}{2}}) \times H$ with $D(\mathcal{A}) = D(A) \times D(A^{\frac{1}{2}})$. If we endow $\mathcal{H}$ with the inner product

$$< z, \bar{z} >_{\mathcal{H}} = < A^{\frac{1}{2}} z_1, A^{\frac{1}{2}} \bar{z}_1 >_H + < z_2, \bar{z}_2 >_H \qquad (2.4)$$

where $z = \begin{pmatrix} z_1 \\ z_2 \end{pmatrix}$, $\bar{z} = \begin{pmatrix} \bar{z}_1 \\ \bar{z}_2 \end{pmatrix}$, then for $z \in D(\mathcal{A})$

$$< z, A z >_H = - \alpha \| z_2 \|^2_H \quad \text{and} \tag{2.5}$$

$$< z, A^* z >_H = - \alpha \| z_2 \|^2_H \tag{2.6}$$

Thus A generates a strongly continuous contractions semigroup S_t on H ([9]). In [16], it is shown that S_t is stable in the sense that

$$\| S_t \|_{L(H)} \leqslant e^{-\omega t} \text{ for some } \omega > 0 \tag{2.7}$$

provided $\alpha > 0$ and $\underline{\omega}(A) = \sup \{\text{Re } \lambda : \lambda \in \sigma(-A)\} < 0$.

In our examples 1 to 4, $H = L_2\ (0,1)$ and there are essentially 4 types of operators

$$A_1 = \frac{\partial^4}{\partial x^4} \qquad D(A_1) = \{h \in H: A_1 h \in H \text{ and } h, h_{xx} = 0 \text{ at } x = 0,1\}$$

$$A_2 = \frac{\partial^4}{\partial x^4} \qquad D(A_2) = \{h \in H: A_2 h \in H \text{ and } h, h_x = 0 \text{ at } x = 0; \; h_{xx}, h_{xxx} = 0 \text{ at } x = 1\}$$

$$A_3 = \frac{\partial^4}{\partial x^4} \qquad D(A_3) = \{h \in H: A_3 h \in H \text{ and } h, h_x = 0 \text{ at } x = 0,1\}$$

$$A_4 = -\frac{\partial^2}{\partial x^2} \qquad D(A_4) = \{h \in H: A_4 h \in H \text{ and } h = 0 \text{ at } x = 0,1\}$$

Using perturbation theory as in [9], we can consider a more general class of systems

$$v_{tt} + \alpha v_t + Av + F(t)v = 0; \; v(0) = v_0, \; v_t(0) = v_1 \tag{2.8}$$

or equivalently on H:

$$\dot{z} = \mathcal{A}(t)z; \; z(0) = z_0 \tag{2.9}$$

where

$$\mathcal{A}(t) = \begin{pmatrix} 0 & I \\ -A & -\alpha I \end{pmatrix} + \begin{pmatrix} 0 & 0 \\ F(t) & 0 \end{pmatrix} \tag{2.10}$$

If $F(t)$ is bounded from $D(A^{\frac{1}{2}})$ to H, then $\mathcal{A}(t)$ generates a mild evolution operator $U(t,s)$ and (2.9) has the mild solution

$$z(t) = U(t,s)z_0 \tag{2.11}$$

Examples of such $\mathcal{A}(t)$ are

$$\mathcal{A}_i(t) = \begin{pmatrix} 0 & I \\ -A_i + f(t)\frac{\partial^2}{\partial x^2} & -\alpha I \end{pmatrix}; \; i = 1,2,3 \text{ and}$$

$$\mathcal{A}_4(t) = \begin{pmatrix} 0 & I \\ -A_4 + f(t)\frac{\partial}{\partial x} & -\alpha I \end{pmatrix} \text{ for } f(t) \text{ bounded in } t.$$

This gives us a natural formulation for example 3 and for examples 1,2,4 where f_0 and $f(t)$ are deterministic. We note that if $f(t) = f$ is time invariant, $U(t,s) = U(t-s,0)$ is in fact a semigroup.

Motivated by the applications discussed in the introduction, we consider the following class of stochastic systems

$$v_{tt} + \alpha v_t + (A + F(t))v + D_1(t)\xi_1 + F_2(t)v\xi_2 + D_2(t)v_t\xi_3 = 0 \tag{2.12}$$

where ξ_i are mutually independent formal 'white noise' type disturbances possibly spatial, $F(t)$ and $F_2(t) \in L_\infty(0,T; \mathcal{L}(D(A^{\frac{1}{2}}),H))$, $D_i(t) \in L_\infty(0,T; \mathcal{L}(K, H))$ and

K = R or H; i = 1,2.

(2.12) allows for very general white noise stochastic disturbances, including the case of f(t) in examples 1,2,4 being a white scalar noise disturbance and in example 3, the case where p(x,t) is a spatial white noise disturbance.

Following the approach in [9], [11] we formulate (2.11) as a stochastic differential equation on $\mathcal{H}$

$$\begin{cases} dz(t) = \mathcal{A}(t)z(t)dt + \begin{bmatrix} 0 \\ D_1(t) \end{bmatrix} dw_1(t) + \begin{bmatrix} 0 & 0 \\ F_2(t) & 0 \end{bmatrix} z(t)dw_2(t) + \\ \qquad + \begin{bmatrix} 0 & 0 \\ 0 & D_3(t) \end{bmatrix} z(t)dw_3(t) \\ z(0) = z_0 \end{cases} \tag{2.13}$$

where $\mathcal{A}(t)$ is given by (2.10), z_0 is a second order random variable on $\mathcal{H}$, $w_1(t)$ is a Wiener process on H or R and $D_1(t) \in L_\infty(0,T;\ \mathcal{L}(H))$ or $L_\infty(0,T;H)$ respectively. $w_2(t)$ and $w_3(t)$ are standard scalar Wiener processes and $w_1(t)$, $w_2(t)$ and $w_3(t)$ are mutually independent. $F_2(t) \in L_\infty(0,T;\ \mathcal{L}(D(A^{\frac{1}{2}}),H))$ and $D_3(t) \in L_\infty(0,T;\ \mathcal{L}(H))$.

The most thorough discussion of solutions of stochastic differential equations is in [10]. Essentially there are 2 possible types of solutions for (2.13) either strong, which means that $z(t) \in D(\mathcal{A})$ w.p.1 and satisfies the integral of (2.13) w.p.1. or a mild solution which satisfies

$$z(t) = U(t,0)\, z_0 + \int_0^t U(t,s) \begin{bmatrix} 0 \\ D_1(s) \end{bmatrix} dw_1(s) + \tag{2.14}$$
$$+ \int_0^t U(t,s) \begin{bmatrix} 0 & 0 \\ F_2(s) & 0 \end{bmatrix} z(s)dw_2(s) + \int_0^t U(t,s) \begin{bmatrix} 0 & 0 \\ 0 & D_3(s) \end{bmatrix} z(s)dw_3(s).$$

Actually [10] treats only the time invariant case $A(t) = A$, but the extension to this case can be carried out along the lines in [11]. Since the operator A here does not generate an analytic semigroup, we do not obtain strong solutions, but mild solutions satisfying (2.14) do exist and are unique under our assumptions. (See [10] or [12]). This is essentially because we have suitably chosen H to make the operators on the noise coefficients linear and bounded. We could also obtain existence results for a wider class of problems with $F_2(t)$ and $D_3(t)$ nonlinear, but satisfying Lipschitzian conditions (see [10] or [12]), however (2.13) covers a sufficiently wide class for most applications.

Now mild solutions do not in general have continuous sample paths, but it is proved in [25] that solutions of (2.14) do. If $A_0 = \begin{pmatrix} 0 & I \\ -A & 0 \end{pmatrix}$, then A_0 generates a group $\bar{S}_t$ and thus for a typical term in (2.14), we can write

$$\int_0^t \bar{S}_{t-s} \begin{pmatrix} 0 & 0 \\ 0 & D_3 \end{pmatrix} z(s)dw_3(s) = \bar{S}_t \int_0^t \bar{S}_{-s} \begin{pmatrix} 0 & 0 \\ 0 & D_3 \end{pmatrix} z(s)dw(s)$$

and from [9] or [11], $\int_0^t \Phi(s)dw(s)$ has continuous sample paths for a general stochastic integrand $\Phi(s)$ adapted to $w(t)$. So if A_0 generates a group (2.14) has continuous sample paths. For general A a perturbation argument establishes continuity for (2.14) (see [25]).

3 STABILITY AND MOMENTS OF SECOND ORDER STOCHASTIC SYSTEMS

A study of stability of infinite dimensional systems was initiated in [13] and simultaneously in [14]. [13] and [14] consider stability properties of the mild solution of the integral equation

$$z(t) = S_t z_0 + \int_0^t S_{t-s} Dz(s)dw_1(s) \tag{3.1}$$

where S_t is a strongly continuous semigroup on H with generator A, $z_0 \in H$, $D \in L(H,L(K,H))$ and $w_1(t)$ is a K-valued Wiener process with covariance W_1. (K is another Hilbert space.)

(3.1) is said to be stable if for each $z_0 \in H$, the mild solution of (3.1) satisfies

$$\int_0^\infty E\{\|z(s)\|^2_H\}ds < \infty \qquad (3.2)$$

which may be shown to be equivalent to

$$E\{\|z(s)\|^2_H\} \leq C e^{-\alpha t}\|z_0\|^2_H \qquad (3.3)$$

for some C, $\alpha > 0$.

In [14] it is proved that (3.1) is stable if and only if there exists a self adjoint, nonnegative operator $P \in L(H)$ which is a solution of

$$2 < Az, Pz >_H + < z, z >_H + \text{trace}\left[(Dz)^* P\, Dz\, W_1\right] = 0 \qquad (3.4)$$

for $z \in D(A)$.

In [13] it is shown that a solution of (3.4) exists under the assumptions that A generates a stable strongly continuous semigroup S_t and

$$\left\|\int_0^\infty S_t^* \Delta\, S_t\, dt\right\|_H < 1 \qquad (3.5)$$

where $< \Delta h, k >_H = \text{trace}\left[(Dh)^* DkW\right]$ for $h,k \in H$

Another type of stability is defined in [14] for

$$z(t) = S_t z_0 + \int_0^t S_{t-s} Dz(s)dw_1(s) + \int_0^t S_{t-s} Fdw_2(s) \qquad (3.6)$$

where S_t, z_0, D, w_1 are as for (3.1) and w_2 is a K_2-valued Wiener process with covariance W_2 and $F \in L(K_2 H)$.

(3.6) is said to be stable if

$$\sup_{t_1>0} \frac{1}{t_1} \int_0^{t_1} E\,\{\| z(s) \|^2_H\} ds < \infty \text{ for each } z_0 \in H \tag{3.7}$$

In fact (3.6) is stable if and only if (3.1) is stable and

$$\lim_{t_1 \to \infty} \frac{1}{t_1} \int_0^{t_1} E\{\| z(s) \|^2_H\} ds = \text{trace} \left[F^* P F W_2 \right] .$$

Furthermore, the Markov process defined by (3.6) has an invariant measure associated with it, that is there exists an invariant measure μ of $P_t(.,.)$ defined by

$$P_t(z,B) = \text{prob}\,\{z(t) \in B \mid z(o) = z_0\}$$

where B is an arbitrary Borel set of H, and

$$\int_H |z|^2 \mu(dz) = \text{trace} \left[F^* P F W_2 \right]$$

(μ is invariant in the sense that $\int P_t(z,B)\,dz = P(B)$, for all t.)

So for stability of (3.6) we can also check (3.5).

Now $\| S_t^* \Delta\, S_t \|_H = \sup_{\| h \|_H = 1} < \Delta S_t h, S_t h >_H$

$$= \sup_{\| h \|_H = 1} \text{trace} \left[(DS_t h)^* DS_t h\ W \right]$$

For the time invariant system (2.14), we have

$$Dz = \begin{pmatrix} 0 & 0 \\ F_2 z_1 & D_3 z_2 \end{pmatrix} , \quad W = \begin{pmatrix} W_1 & 0 \\ 0 & W_2 \end{pmatrix}$$

and trace $(DS_t h)^* DS_t h\, W = \| F_2 (S_t h)_1 \|_H^2 + \| D_2 (S_t h)_2 \|_H^2$

$$\leqslant \| S_t \|^2 \, \| h \|^2_H \; (\| F_2 \|^2 + \| D_2 \|^2)$$

This gives us the rather crude estimate,

$$\left\| \int_0^\infty S_t{}^* \Delta S_t \, dt \right\|_H \leqslant \frac{1}{2\omega} \; (\| F_2 \|^2 + \| D_2 \|^2) \tag{3.8}$$

where we suppose that S_t is stable with $\| S_t \| \leqslant e^{-\omega t}$.

Qualitatively this means that for sufficiently small noise disturbances and sufficiently stable S_t, we get stochastic stability. In [16] estimates of ω are given in terms of α and $\underline{\omega}(A)$, namely

$$\omega \geqslant \frac{2\, \alpha |\underline{\omega}(A)|}{4|\underline{\omega}(A)| + \alpha(\alpha + \sqrt{\alpha^2 + 4|\underline{\omega}(A)|})} \tag{3.9}$$

and $\| F_2 \|$ and $\| D_2 \|$ can be easily estimated as in the following example.

<u>Example 5</u>

$$\frac{\partial^2 v}{\partial t^2} + \alpha \frac{\partial v}{\partial t} + \beta \frac{\partial^2 v}{\partial x^2} \xi + \frac{\partial^4 v}{\partial x^4} = 0$$

$$v(0,t) = 0 = v(1,t) = \frac{\partial^2 v}{\partial x^2}(0,t) = \frac{\partial^2 v}{\partial x^2}(1,t)$$

where ξ is a scalar "white noise" of unit variance.

Then $A = \begin{pmatrix} 0 & I \\ -\dfrac{\partial^4}{\partial x^4} & -\alpha I \end{pmatrix}$ generates a semigroup with $\|S_t\| \leqslant e^{-\omega t}$

and $\omega \geqslant \dfrac{2\alpha\pi^4}{4\pi^4 + \alpha^2 + \alpha\sqrt{\alpha^2 + 4\pi^4}}$

So a sufficient condition for stability is

$$\beta^2 < \frac{4\alpha\pi^4}{4\pi^2 + \alpha^2 + \alpha\sqrt{\alpha^2 + 4\pi^4}}$$

What one would like are results on asymptotic stability as initiated in [13], but the results there are applicable only for A the generator of an analytic semigroup and (3.1) with strong solutions, which is not the situation for second order system. In fact it is possible to prove that if (2.14) is mean square stable in the sense of (3.2), then its sample paths go to zero exponentially fast as $t \to \infty$. As regards the asymptotic stability of the zero solution, as in [13], it can be established at least with respect to finite dimensional initial conditions. These results are established in [25].

Other approaches to the stability of stochastic vibration systems are modal approximations in [6] and [8] and a semigroup approach in [7]. However all these consider coloured noise disturbances and not the Itô equation model discussed here. See also [24] for a analysis of the stability of a stochastic wave equation.

In [23] Pardoux follows a more direct approach in studying existence and uniqueness of stochastic systems and for (3.6) with $F = 0$ he establishes an energy inequality which can be used to establish pathwise stability results, along the lines in [13]. However $F = 0$ is a severe limitation in vibration problems for it means that in example 1, 2 and 3 that $f_0 = 0 = m$.

4 FORMULATION OF FILTERING PROBLEMS

In [9] can be found a general theory for the solution of the filtering problem for systems with bounded operators.

This includes signals described by (2.13) and observations given by

$$y(t) = \int_0^t Cz(s)ds + w_5(t) \tag{4.1}$$

where $w_5(t)$ is a k-dimensional Wiener process and $C \in L(H,R^k)$. It is of interest to see whether or not problems with point observations are permissible. That $C : H \to R^k$ given by

$$C \begin{pmatrix} h_1 \\ h_2 \end{pmatrix} = h_1(x_0) \text{ for } 0 < x_0 < 1. \tag{4.2}$$

This corresponds to "noisy" measurements of $z(x_0,t)$. In fact, $h_1 \in D(A^{\frac{1}{2}})$, since $H = D(A^{\frac{1}{2}}) \times H$ and for all our operators, A_i; $i = 1,2,3,4$, we can verify that

$$\left| C \begin{pmatrix} h_1 \\ h_2 \end{pmatrix} \right| \leq \text{constant } \|h\|_H \tag{4.3}$$

and so point observations of v are covered by the general theory in [9]. Point observations of the velocity v_t are not, however.

5 STOCHASTIC CONTROL PROBLEMS FOR PERFECT OBSERVATIONS

Recently the stochastic regulator problem for quadratic cost and a linear state with control and state dependent noise was extended to the infinite dimensional case by Ichikawa in [14]. For our second order systems a typical signal would be

$$dz(t) = A(t)z(t)dt + \begin{pmatrix} 0 \\ D_1 \end{pmatrix} dw_1(t) + \begin{pmatrix} 0 & 0 \\ F_2 & 0 \end{pmatrix} zdw_2(t) + \begin{pmatrix} 0 & 0 \\ 0 & D_3 \end{pmatrix} zdw_3(t)$$

$$+ \begin{pmatrix} 0 \\ B_1 \ u(t) \end{pmatrix} dt + \begin{pmatrix} 0 \\ B_2 \ u(t) \end{pmatrix} dw_4(t) \tag{5.1}$$

where $A(t)$, D_1, F_2, D_3, w_1, w_2 and w_3 are as for (2.13), the control $u(t)$ has values in a Hilbert space U, $B_1 \in L(U,H)$, $B_2 \in L(U, L(K,H))$ and $w_2(t)$ is a K-valued Wiener process independent of all the other Wiener processes.

The cost functional to be minimized is

$$J(u) = E\{< Gz(t_1), z(t_1) >_H + \int_0^{t_1} [< Mz(t), z(t) >_H$$

$$+ < Nu(t), u(t) >_U]dt\} \tag{5.2}$$

Of interest to us here is that G and M can include penalties for point sensing of the displacement $v = z_1(t)$ as these operators are still in $L(H)$ (cf. (4.3)). Also feedback controls of the type

$$u(t) = (C\ O)\ z(t) \tag{5.3}$$

with C given by (4.2) are admissible if $B_1 \in H$ and $B_2 \in L(K,H)$. See [14] for the relevant theory.

6 SEPARATION PRINCIPLE

The theory for the separation principle can be found in [9] and it holds here for second order systems of type (5.1) with $F_2 = D_3 = O = B_2$ observation (4.1) and quadratic cost (5.2). The comments in § 4 and § 5 concerning point observations and controls are valid in this case too.

7 FORMULATION OF BOUNDARY NOISE AND BOUNDARY CONTROL

The problems formulated thus far have fitted nicely into the "bounded" semigroup theory of [9]. However, when one wishes to formulate problems where the control action or the noise enters on the boundary, the mathematics is more difficult. For the deterministic boundary control problem several approaches have been tried, namely [20], [18], [17] and [10]. [10], [18] are semigroup approaches and [17] is a cosine operator approach but none of these actually solve the control problem, which had already been done in [20] by a more direct approach. Because of the "second order" nature of the Itô integral formulation, the formulation of boundary noise problems are more difficult, although it is possible to formulate boundary noise problems for some parabolic systems [21], [22]. For second order systems I have failed to obtain a suitable formulation along the lines of [10] or [18] using the Itô integral representation for the white noise. It appears to me that a more promising approach to this problem would be to use the white noise model of [19].

8 REFERENCES

1. Balas, M. J. "Modal Control of Certain Flexible Dynamic Systems", *SIAM J. Control and Optimization,* **16**, 450-462, (1978).

2. Elishakoff, I. "Interaction of Flutter and Random Vibration in Plates", B. L. Carkson Ed., Stochastic Problems in Dynamics, Pitman (1977).

3. Fryba, L. "Response to Bridges to Moving Random Loads", Ibid.

4. Ariaratnam, S. T. "Dynamic Stability of a Column under Random Loading", Dynamic Stability of Structures, Proc. Int. Conf. Pergamon Press, N.Y., 267-284, (1967).

5. Lepore, J. A. and Shah, H. C. "Dynamic stability of axially loaded columns supported by columns subjected to stochastic excitation", *AIAA Journal,* **6**, 1515-1521, (1968).

6. Plaut, R. H. and Infante, E. F. "On the stability of some continuous systems subject to random excitation", *A.S.M.E. Trans. Journal of Appl. Mechanics,* 623-627, (1970).

7. Wang, P. K. C. "On the almost sure stability of linear stochastic distributed parameter systems", *A.S.M.E. Trans. Journal of Appl. Mechanics,* 182-186, (1966).

8. Kozin, F. "Stability of Linear Stochastic Systems", Proc. of Int. Symposium on Stability of Stochastic Dynamical Systems, Lecture Notes in Mathematics 294, Springer Verlag (1972).

9. Curtain, R. F. and Pritchard, A. J. "Infinite Dimensional Linear Systems Theory", Lecture Notes in Control and Information Sciences", **8**, Springer Verlag, (1978).

10. Chojnowska-Michalik, A. "Stochastic Differential Equations in Hilbert Spaces and their Applications", Ph.D. Thesis, Institute of Mathematics, Polish Academy of Sciences, (1976).

11. Curtain, R. F. "Stochastic Evolution Equations with General White Noise Disturbance", *J. Math. Anal. and Appl.*, **60**, 570-595, (1977).

12. Ichikawa, A. "Linear Stochastic Evolution Equations in Hilbert Space", Control Theory Centre Report No. 51, University of Warwick, (1976), (To appear in J. Diff. Eqns.).

13. Haussman, U. G. "Asymptotic Stability of the Linear Itô Equation in Infinite Dimensions", *J. Math. Anal. and Appl.*, **64**, (1978).

14. Ichikawa, A. "Dynamic Programming Approach to Stochastic Evolution Equations", Control Theory Centre Report No. 60, University of Warwick, (1977).

15. Curtain, R. F. and Falb, P. L. "Itô's lemma in Infinite Dimensions", *J. Math. Anal. and Appl.*, **31**, 434-448, (1970).

16. Pritchard, A. J. and Zabczyk, J. "Stability of Infinite Dimensional Systems", Control Theory Centre Report No. 70, (1977).

17. Triggiani, R. "A Cosine Operator Approach to Modelling Boundary Input Hyperbolic Systems", Proc. 8th. IFIP Conference on Optimization, Würzburg, W. Germany, Sept. (1977).

18. Zabczyk, J. "A Semigroup Approach to Boundary Controls", Proc. 2nd IFAC Conference on Distributed Parameter Systems, Warwick, UK, (1977).

19. Balakrishnan, A. V. "Applied Functional Analysis", Springer Verlag, (1976).

20. Lions, J. L. "Optimal Control of Systems Governed by Partial Differential Equations", Springer Verlag, (1971).

21. Curtain, R. F. "Linear Stochastic Control for Distributed Systems with Boundary Control", Control Theory Centre Report 46, University of Warwick, (1976).

22. Curtain, R. F. "A Semigroup Approach of the LQG Problem for Infinite Dimensional Systems", *IEE Circuits and Systems*,

25, 713-720, (1978).

23. Pardoux, E. Doctoral Thesis , L'université de Paris Sud, Centre d'Orsay, (1975).

24. Yavin, Y. "On the Instability of an oscillatory distributed parameter system", *Int. J. System Science*, **4**, (1973).

25. Curtain, R. F. "Asymptotic Stability for Second Order Stochastic Partial Differential Equations", (To appear).

3.5 OPTIMISATION WITHIN NESTED RENEWAL PROCESSES

J. Ansell,* A. Bendell** and S. Humble***

(Sheffield City Polytechnic, UK)

1 INTRODUCTION

Renewal processes are a well established and well documented class of stochastic processes which have proved to be extremely useful in the solution of many practical problems, especially within the field of reliability engineering. The main results are given in [3]. In [2] we considered the class of stochastic processes generated by nesting renewal processes within other renewal processes in order to model the damage sustained by, and the consequent replacement of, motor vehicle components. Such nested renewal processes are also of interest in the study of group-arrival/batch-service queues, storage systems, epidemics and computer software. They are related to, but not identical with, alternating, superimposed and multivariate wear and shock processes, as well as the original general formulation of ordinary renewal processes by Smith [7] which allowed for blows of negative size. The authors' interest in these processes originated in the study of some practical engineering problems, and in this paper we discuss the more basic properties of the processes, categorize the optimisation problems that arise in their application areas, and consider the identification of appropriate objective functions. Whilst we are not concerned here with the establishment of relational properties nor the proving of existence theorems, it is apparent that interesting problems of this type arise in relation to nested renewal processes and it is also hoped that this paper will encourage others to investigate these processes further.

In Section 2 of the paper the main properties of nested renewal processes are discussed, whilst in Section 3 optimisation within such processes is considered and optimisation

---

* Now at University of Keele

** Now at Dundee College of Technology

*** Now at Royal Military College of Science, Shrivenham

criteria are identified within the various application areas. For deteriorating units the optimal time and damage to replacement are obtained, whilst for group-arrival/batch-service systems optimum batch sizes and inter-service periods are identified. In connection with storage systems we discuss the identification of optimum dispatch policies, whilst in the context of epidemics we consider the selection of optimum preventive policies.

An illustration of a nested renewal process is given in Fig. 1. Shocks occur to a component randomly in time in an ordinary renewal process, each shock causing a random amount of damage. Damages are identically and independently distributed, and damages resulting from shocks are accumulated. In addition to this cumulative process there is a second ordinary renewal process in time, the effect of which is to restart the cumulative renewal process at zero accumulated shocks and consequently zero cumulative damage. This represents component replacement, which in this paper will be assumed to be due to failure.

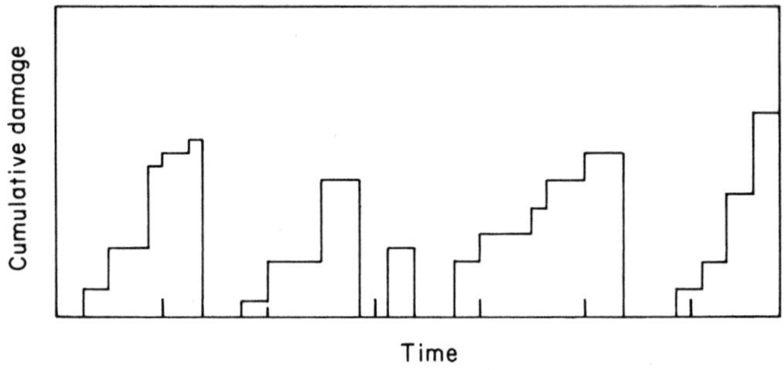

Fig. 1

This second-order renewal process can be extended by the addition of a third renewal process in time representing system, e.g. vehicle replacement, the effect of which is to restart the second order process from the state it was in at time zero. Consequently the third order renewal process contains two competing ordinary renewal processes which each restart the cumulative renewal process. However, whilst the occurrence of a third-order renewal implies that a second-order renewal also occurs, the converse does not hold. In a similar way the extension of the second-order nested process to a third-order one can be further extended to the nth order, by sequentially introducing new time-dependent renewal processes around the earlier ones. These might represent, for example, the replacement mechanisms of progressively larger sub-assemblies.

2 NESTED RENEWAL PROCESSES

Let $g(t)$ be the density function of time to a shock or first-order renewal, and let $h(T)$ be the density function of time to a second-order renewal or replacement due to failure. Then the elapsed time t is returned to zero as soon as a shock occurs, and similarly t and T are returned to zero following a failure. If $f(x)$ is the density function for damage x resulting from a shock and $p(x|t)$ represents the density function for accumulated damage x at time t, then the system can be described by the differential equation

$$\frac{dp(x|t)}{dt} = -\ p(x|t)\int_0^t ds \sum_{m=0}^{\infty} Z^m(t-s)\left[h(s)+H(s)\sum_{k=1}^{\infty} g^k(s)\right]$$
$$+\int_0^x dy\ p(y|t)\int_0^t ds\ Z^m(t-s)H(s)\sum_{k=1}^{\infty} g^k(s)f(x-y),\quad x>0,\ t>0 \tag{1}$$

where

$$\frac{dp(0|t)}{dt} = -\ p(0|t)\int_0^t ds \sum_{m=0}^{\infty} Z^m(t-s)H(s)g(s) + \sum_{m=0}^{\infty} h^m(t),\quad t>0$$

$$\left.\frac{dp(x|t)}{dt}\right|_{t=0} = g(0)f(x),\quad x>0$$

$$\left.\frac{dp(0|t)}{dt}\right|_{t=0} = -g(0)$$

$$Z^m(t) = \int_0^t Z^{m-1}(t-s)h(s)ds,\quad m=1,2,3,\ldots$$

$$Z^0(t) = \delta(t),\quad p(x|0) = \delta(x)$$

$$H(t) = \int_t^{\infty} h(s)ds,$$

g^k, h^k represent the kth convolutions of g and h respectively, $h^0(t) = \delta(t)$, $h^1(t) = h(t)$, $g^1(t) = g(t)$, and δ is the (Dirac) "density" of the 1-point distribution on 0.

It follows that the probability that k shocks have been accumulated at time t is

$$p(k|t) = \int_0^t ds \left[\sum_{i=0}^{\infty} h^i(t-s)H(s) \left\{ \int_0^s du \; g^k(s-u)G(u) \right\} \right] \tag{2}$$

where $G(u) = \int_u^{\infty} g(t)\,dt$,

and $g^0(t) = \delta(t)$.

The density of accumulated damage at time t is then obtained as

$$p(x|t) = \sum_{k=0}^{\infty} p(k|t) f^k(x), \tag{3}$$

where $f^k(x)$ is the kth convolution of $f(x)$ and $f^0(x) = \delta(x)$, $f^1(x) = f(x)$.

Since

$$\sum_{k=0}^{\infty} p(k|t) = 1 \text{ for all } t,$$

the Laplace transform of $p(k|t)$ can be written in the form

$$p^*(k,r) = L(k,r) \Big/ \left[r \sum_{k=0}^{\infty} L(k,r) \right], \tag{4}$$

where

$$L(k,r) = \mathcal{L}\left\{ H(t) \int_0^t du \; g^k(t-u)G(u) \right\}.$$

Similarly the single and double Laplace transforms of $p(x|t)$ can be written as

$$p^*(x,r) = L(x,r) \Big/ \left[r \int_0^{\infty} L(x,r)\,dx \right]$$

$$p^{**}(s,r) = L^*(s,r) \big/ \left[r\, L^*(0,r) \right], \tag{5}$$

where

$$L(x,r) = \sum_{k=0}^{\infty} L(k,r) f^k(x)$$

$$L^*(s,r) = \mathcal{L}\left\{ L(x,r) \right\} .$$

Laplace transforms are particularly useful in obtaining asymptotic results since

$$\lim_{t\to\infty} p(k|t) = \lim_{r\to 0} r\, p^*(k,r)$$

and (6)

$$\lim_{t\to\infty} p^*(s,t) = \lim_{r\to 0} r\, p^{**}(s,r)$$

Examples are given in [2].

The mean number of shocks accumulated at time t is

$$E(k|t) = \sum_{k=0}^{\infty} k\, p(k|t) = \frac{\partial}{\partial Z} \pi(Z,t) \Big|_{Z=1}, \tag{7}$$

where $\pi(Z,t)$ is the generating function

$$\pi(Z,t) = \sum_{k=0}^{\infty} Z^k\, p(k|t) .$$

For ordinary renewal processes it is well known that $E(k|t)$ is monotonically non-decreasing in t. However for nested renewal processes $E(k|t)$ provides a comparison of the rates of occurrence of shocks, and replacements due to failure, so that this is no longer necessarily the case. For example, assuming g(t) to be an exponential density with parameter λ and $h(\tau)$ to be Gamma $(3,\beta)$ we have

$$E(k|t) = \lambda[2-3\exp(-\beta t)]/\beta+2\lambda\exp(-3\beta t/2)\sin(3\beta^2 t/4-\psi)$$

where $\psi = \tan^{-1}(-\beta/2)+\tan^{-1}(-3\beta/2)+\tan^{-1}(-3\beta/14)$.

Thus for this case the value of $E(k|t)$ oscillates as t increases. The mean accumulated damage at t is

$$E(x|t) = E(k|t)\int_0^\infty xf(x)\,dx = E(k|t)\mu_D \tag{8}$$

where μ_D is the expected damage per shock. Similarly the higher order moments about the mean can be deduced in the usual manner from the generating function or its Laplace transform. Other properties of these processes are given in [2].

To generalize our above results for second order processes to n^{th} order ones where $n \geqslant 3$ we now let $h_i(\tau)$ represent the density function corresponding to the ith replacement policy $i = 1,2,\ldots,n-1$. In this case, the probability distribution function for the accumulated number of shocks at t, and the density of the accumulated damage at t are now

$$p^{(n)}(k|t) = \int_0^t ds \sum_{\ell=0}^{\infty} h_{n-1}^{\ell}(t-s)H_{n-1}(s)p^{(n-1)}(k|s)$$

$$p^{(n)}(x|t) = \int_0^t ds \sum_{\ell=0}^{\infty} h_{n-1}^{\ell}(t-s)H_{n-1}(s)p^{(n-1)}(x|s), \tag{9}$$

where

$$H_n(t) = \int_t^\infty h_n(s)\,ds$$

and

$$p^{(2)}(k|t) = p(k|t),\ p^{(2)}(x|t) = p(x|t).$$

As in (4) and (5) the Laplace transforms of $p^{(n)}(k|t)$ and $p^{(n)}(x|t)$ can again be written in compact forms.

In obtaining the above results it was assumed that the failure density h(.) is a function of time. If instead it is a function of accumulated damage x, equation (3) is replaced by

$$p(x|t) = \sum_{k=0}^{\infty} f^k(x)H(x)\int_0^t du \left[\int_0^u ds\ g^k(u-s)G(s)\right]\sum_{i=0}^{\infty} W^i(t-u) \quad (10)$$

where

$$W^i(t) = \int_0^t ds\ W^{i-1}(t-s)\sum_{k=1}^{\infty} g^k(s)\left[\int_0^{\infty} dx \int_0^{\infty} dy f^{k-1}(x-y)f(y)\left\{H(x-y)-H(x)\right\}\right],$$

$i=1,2,\ldots$

$W^0(t) = \delta(t)$

In [2] we discuss the case of competing time-dependent and damage-dependent mechanisms.

3 OPTIMISATION

We are concerned here with units subject to the cumulative damage, failure and replacement cycle that we have described as a second-order nested renewal process. Whilst replacement automatically takes place upon failure, cost considerations may mean that it is worthwhile to replace a unit at some time and damage prior to failure. The problem is to identify the optimal time and cumulative damage thresholds for control limit type preventative replacement policies to be applied to such units. These control limit type policies are such that no action is taken if the time, or alternatively the cumulative damage, is less than the specified threshold, and a preventative replacement is made if the appropriate variable is equal to or exceeds the threshold. Since nested renewal processes can also be applied to problems in queueing, storage and epidemiology, we also consider the identification of optimum batch size and inter-service period policies for group-arrival/batch-service queues, optimum dispatch policies for certain storage systems, and optimum preventative policies for epidemics. As we indicate below, certain of these problems have been considered previously by certain authors in related, although not identical, formulations to ours. However, these emphasize somewhat different aspects of the mathematical structure, since we are primarily concerned with the classification of the optimisation problems that arise in the application of nested renewal processes.

3.1 Cumulative damage and deteriorating units

Allowing for the failure density h(.) to be either a function of time t or cumulative damage x, and for the preventative replacement threshold also to be in terms of time or cumulative damage, we have in fact four distinct optimisation problems. The selection of optimal preventative policies is based on expected cost criteria, and here we assume that the cost of a preventative replacement c_1, as well as the cost of a replacement on failure c_2, is a constant independent of time and damage at failure. For preventative replacement to be worthwhile $c_1 < c_2$. In each of the four cases there are a number of plausible formulations of the expected-cost-per-unit-time function to be minimized in order to obtain the optimal replacement threshold. For example, for the case where h(.) is a function of x and the problem is to identify the optimum damage to preventative replacement, X, one possible criterion is to choose X to minimize the objective function

$$\psi(X) = \lim_{t\to\infty} t^{-1} c(t,X) \tag{11}$$

where c(t,X) represents the expected cost up to time t

$$c(t,X) = \int_0^t ds \sum_{i=0}^{\infty} V^i(t-s,X)\left[c_1\omega_1(s,X) + c_2\omega_2(s,X)\right],$$

and

$$\omega_1(s,X) = \sum_{m=1}^{\infty} g^m(s)\int_0^X f^{m-1}(x)H(x)F(X-x)\,dx$$

$$\omega_2(s,X) = \sum_{m=1}^{\infty} g^m(s) \int_0^X dx \int_0^x dy f^{m-1}(x-y) f(y) \left[H(x-y)-H(x)\right]$$

$$v^i(t,X) = \int_0^t ds\, v^{i-1}(t-s,X) \left[\omega_1(s,X) + \omega_2(s,X)\right], \; i=1,2,\ldots$$

$$v^0(t,X) = \delta(t).$$

This type of expected cost per unit time criterion has been used previously in related problems by a number of authors, e.g. [4,5,6,8], and it follows from standard results in renewal theory that (11) can be rewritten as the ratio of the expected cost of a replacement to the expected life length of a unit,

$$\psi(X) = \frac{c_1\int_0^\infty \omega_1(t,X)\,dt + c_2\int_0^\infty \omega_2(t,X)\,dt}{\int_0^\infty t\left[\omega_1(t,X) + \omega_2(t,X)\right]dt}. \tag{12}$$

An alternative optimisation criterion is to choose X to minimize the expected cost per unit time over one cycle,

$$\phi(X) = c_1 \int_0^\infty t^{-1}\omega_1(t,X)\,dt + c_2 \int_0^\infty t^{-1}\omega_2(t,X)\,dt. \tag{13}$$

Neither of these optimisation criteria is ideal. From a physical viewpoint perhaps the most sensible procedure would be to choose X to minimize $c(T,X)/T$ or $c(T,X)$ for a given finite time period T, but even for the simpler of our problems such a procedure is prohibitively difficult analytically. The optimal policy over an infinite time horizon, which can be derived from considering (11) or (12), is by no means necessarily optimal in the real situation in which one desires to obtain a practical operating policy based on a finite time constraint. However, supposing that it was decided a priori that the system is to be run for a fixed number of cycles rather than a fixed length of time, the optimal policy will be obtained from (13). Such a policy would arise naturally if there were a fixed number of spare units available, and is in any event more attractive than that derivable from (12) since it takes account of the correlation between the values of $\omega_i(t,X)(i=1,2)$ and t. Consequently, in this paper we shall use objective functions of the type in (13). We start with the simplest cases.

3.1.1 Time threshold $[T]$ *and failure density in time* $[h(t)]$

Since failure and preventative replacement are independent of cumulative damage x, x is of no interest, the system reduces to an ordinary renewal process, and our optimisation criterion is just to choose the time threshold T to minimize

$$\phi_1(T) = c_1 H(T)/T + c_2 \int_0^T h(t)/t \, dt. \tag{14}$$

This relatively trivial case has been considered extensively in the literature, although usually in terms of an objective function of the type in (11) and (12) (see e.g. [3,6]). For example, if the time to failure density is Gamma $(2,\beta)$ the optimal time to preventative replacement using (14) is

$$T = \left[c_1 + (4c_2-3c_1)^{\frac{1}{2}} c_1^{\frac{1}{2}}\right] / \left[2(c_2-c_1)\beta\right],$$

whilst using the more usual criterion it is the solution of the equation

$$(c_2-c_1)e^{-\beta T} = c_2+\beta T(2c_1-c_2)$$

or equivalently

$$T = \frac{c_2}{\beta(c_2-2c_1)} - \frac{c_2-c_1}{\beta(c_2-2c_1)} e^{-c_2(c_2-\lambda c_1)^{-1}} \sum_{i=0}^{\infty} \frac{(i+1)^i}{(i+1)!}$$

$$\left[\frac{(c_2-c_1)}{c_2-2c_1} e^{-c_2(c_2-2c_1)^{-1}}\right]^i .$$

Thus, for this simple example at least, the new optimisation criterion does not lead to more complicated solutions than the criterion more usually applied and, with it, replacement tends to take place earlier.

3.1.2 *Time threshold [T] and failure density in cumulative damage [h(x)]*

Such a case will arise if damage is non-measurable, even though failures are dependent upon accumulated damage. The special case of this model in which the accumulated damage by time t is, in the absence of failure, gamma has been previously considered [1] using an optimisation criterion of the type (12). Our criterion is to choose T to minimize

$$\phi_2(T) = \frac{c_1}{T}\int_0^T ds \sum_{m=1}^{\infty} g^m(T-s)G(s)\int_0^{\infty} f^m(x)H(x)\,dx$$

$$+ c_2\int_0^T t^{-1}\sum_{m=1}^{\infty} g^m(t)\,dt\int_0^{\infty}dx\int_0^x dy\,f^{m-1}(x-y)f(y)\left[H(x-y)-H(x)\right]. \tag{15}$$

For example if h(x) is exponential with parameter β and g(t) is Gamma (2,λ) then the optimal time T is a solution to the equation

$$0 = \frac{c_2\left[1-f^*(\lambda)\right]}{\left[f^*(\lambda)\right]^{\frac{1}{2}}}\sinh\left\{\beta T\left[f^*(\lambda)\right]^{\frac{1}{2}}\right\} + \frac{c_1(\beta T-1)}{T}\left[\frac{\sinh\left\{\beta T\left[f^*(\lambda)\right]^{\frac{1}{2}}\right\}}{\left[f^*(\lambda)\right]^{\frac{1}{2}}}\right.$$

$$\left. - \cosh\left\{\beta T\left[f^*(\lambda)\right]^{\frac{1}{2}}\right\}\right].$$

3.1.3 *Damage threshold [X] and failure density in cumulative damage [h(x)]*

Optimisation problems related to this one have been considered by a number of authors, e.g. [4,5,8], although again usually based on the optimisation criteria of (11) and (12). Our objective function $\phi_3(X)$ is given in (13) with $\omega_i(t,X)$ i=1,2 defined in (11).

3.1.4 Damage threshold [X] and failure density in time [h(t)]

Such an optimisation problem might arise if records of replacements were not kept, but an easily measurable auxiliary variable (cumulative damage) was available as the basis of a replacement policy. In this case our optimal policy is to choose X to minimize

$$\phi_4(X) = c_1 \int_0^\infty dt\ t^{-1} H(t) \sum_{m=1}^{\infty} g^m(t) \int_0^X f^{m-1}(x) F(X-x)\,dx$$
$$+ c_2 \int_0^\infty dt\ t^{-1} h(t) \sum_{m=0}^{\infty} \int_0^t ds\, g^m(t-s) G(s) \int_0^X f^m(x)\,dx. \qquad (16)$$

As an example suppose that f(x) is exponential with parameter μ, g(t) is Gamma $(2,\lambda)$ and h(t) is Gamma $(2,\beta)$. Then the optimum X is a solution of

$$\frac{\beta}{\lambda}\left[\frac{c_2\beta}{c_1\lambda}\left(2 + \frac{\beta}{\lambda}\right) + 2 + \frac{2\beta}{\lambda} + \frac{\beta^2}{\lambda^2}\right] \exp\left[\frac{\lambda^2 \mu X}{(\lambda+\beta)^2}\right] =$$
$$- \frac{(\lambda+\beta)^4}{\lambda^4 (\mu X)^{3/2}} \int_0^{\lambda\mu^{\frac{1}{2}} X^{\frac{1}{2}} (\lambda+\beta)^{-1}} (Z^2 - \mu X) e^{Z^2}\, dZ.$$

3.1.5 Variable costs

If one or both of the costs of replacement c_1 and c_2 depend upon the time and/or cumulative damage at failure, then of course the cost term or terms just move inside the corresponding integral in (13), (14), (15) or (16). For example for the case of 3.1.1 where h(t) is Gamma $(2,\beta)$ and where the cost of replacement on failure is now

$$c_2(t) = ct^n,\ n=0,1,2,\ldots$$

the optimal time T to preventative replacement is a solution of the equation

$$c_1 n!(1+\beta T+\beta^2 T^2) = c\beta^2 T^{n+2}.$$

3.2 Group-arrival/batch-service queues

Consider a service facility such that input requiring service arrives in groups or batches, with time-between-batch density g(t). In general the sizes of batches are continuous with density f(x), and the arriving input is allowed to accumulate for a certain period T before the accumulated input is serviced. Alternatively, service may take place when the quantity of accumulated input awaiting service first exceeds some threshold X. Services are instantaneous but at a fixed cost c_1 (independent of the size of the batch being serviced), whilst storage costs c_2 per unit stored per unit time. This queueing system, which may for example correspond to the accumulation of transactions awaiting action by head office at a bank branch, constitutes a second-order nested renewal process in which the survivor function H(.) takes one or other of the forms

$$H(t) = \begin{cases} 1, & t<T \\ 0, & t\geqslant T \end{cases},$$

$$H(x) = \begin{cases} 1, & x<X \\ 0, & x\geqslant X \end{cases}. \qquad (17)$$

Our problem is to identify that value of T or X that will minimize the costs of processing the inputs, i.e. storage costs plus service costs. We may choose T or X to minimize the expected processing cost per unit time or the expected processing cost per item. Thus it would appear that as in Subsection 3.1 we have four distinct optimisation problems. In each of these we shall for convenience use an approximation for the expected storage cost of the same type as is frequently employed in stock control problems. If we interpret the second-order nested renewal process in Fig. 1 as illustrating the build up and removal by servicing of waiting input then it is apparent that it is a reflection of the type of graph that frequently arises in stock control problems to represent the falling off of stocks due to demand and their consequent replenishment. Consequently, if in a period of length t input accumulates from 0 to an amount x we shall treat the expected storage cost as $c_2\, t\, x/2$, so that the expected storage cost per unit time is $c_2 x/2$, and the expected storage cost per item is $c_2 t/2$. Of course, if as here x and/or t vary over different cycles the values of x and/or t in the last two expressions are replaced by their expected values. The effect of the approximation should be small if the sizes of the arriving batches are small relative to the sizes of the batches being serviced, i.e. if many group arrivals constitute a service batch.

Based on minimizing the expected cost per unit time, our procedure is to choose the time between services T to minimize

$$\phi_5(T) = \frac{c_1}{T} + \frac{c_2}{2}\int_0^\infty x\int_0^T ds \sum_{m=0}^{\infty} g^m(T-s)G(s)f^m(x)\,dx, \tag{18}$$

or to choose the minimum batch size serviced, X, to minimize

$$\phi_6(X) = c_1\int_0^\infty t^{-1}\int_0^X dx \sum_{m=1}^{\infty} f^{m-1}(x)F(X-x)\,dx g^m(t)\,dt$$
$$+ \frac{c_2}{2}\int_0^X dx \int_{X-x}^{\infty} dy \sum_{m=1}^{\infty} (x+y) f^{m-1}(x) f(y). \tag{19}$$

If, on the other hand, our criteria is to minimize the expected cost per item, then we wish to choose X to minimize

$$\phi_7(X) = c_1\int_0^X dx\int_{X-x}^{\infty} dy\,(x+y)^{-1} \sum_{m=1}^{\infty} f^{m-1}(x) f(y)$$
$$+ \frac{c_2}{2}\int_0^\infty t \int_0^X \sum_{m=1}^{\infty} f^{m-1}(x)F(X-x)\,dx g^m(t)\,dt. \tag{20}$$

The criterion of choosing the time between services to minimize the expected cost per item processed is eliminated, since the corresponding cost function is always infinite due to the inclusion in it of G(T) divided by zero. As an example of these queueing problems suppose that g(t) is Gamma $(2,\lambda)$ and f(x) is Gamma $(2,\mu)$, then the value of T which minimizes (18) is a solution of the equation

$$2\mu c_1/c_2 = (1+\lambda T)\lambda T^2 e^{-\lambda T},$$

whilst the value of X which minimizes (19) is a solution of

$$\frac{c_2}{2}\left[\sinh(\mu X)-\cosh(\mu X)+2-\mu X\right] = c_1\lambda\mu\left[\int_0^{\mu X} Z^{-1}\sinh(Z)\,dZ+\mu X\sinh(\mu X)-\right.$$

$$\left.-\cosh(\mu X)\right].$$

3.3 *Storage*

Input to a store arrives in batches, and is allowed to accumulate either for a period of time T before dispatch, or until the capacity of the store X is first exceeded whereupon (instantaneous) dispatch takes place. If only one of these dispatch processes is present, the optimisation problem can be formulated identically to that of the queues of Subsection 3.2. If both are present, a number of new optimisation problems arise. For example, we may wish to determine the optimum time between dispatches for a store of a given size X, subject to clearance by dispatch at fixed intervals T, which balances the cost of a dispatch c_1 against the storage cost per unit per unit time c_2.

3.4 *Epidemics*

Consider an epidemic in which the time between new outbreaks has density g(t), the extent of new infection at a new outbreak (in general assumed continuous) has density f(x), and infection resulting from outbreaks accumulates additively. A cost c_2 is sustained for each unit which remains infected per unit time. Suppose also that the epidemic resulting from new outbreaks terminates naturally according to the density h(.), which is either a function of the accumulated extent of the infection x or of the length of time since the last epidemic terminated t. In such circumstances the authorities may wish to take preventative policies in order to end the epidemic and we assume they are instantaneous, one hundred percent effective and cost a constant amount c_1 whatever the extent of the infection. In particular these preventative measures may be taken when the accumulated extent of infection reaches some level X or at regular intervals T. It is reasonable to assume that the authorities will apply these preventative policies so as to minimize the expected cost of their policies per unit time or per unit infected. For example, for the case where h(.) is a function of t and the objective is to minimize the cost of preventative measures per unit time they may choose T to minimize

$$\phi_8(T) = \frac{c_1}{T} + \frac{c_2}{2}\int_0^T dt \sum_{n=0}^{\infty} h^n(T-t)\left[H(t)\int_0^t ds \sum_{m=0}^{\infty} g^m(t-s)G(s)\right. \tag{21}$$

$$\left. + \int_0^t dsh(s)\int_0^s dZ \sum_{m=0}^{\infty} g^m(s-Z)G(Z)\right] . \int_0^{\infty} xf^m(x)\,dx.$$

4 REFERENCES

1. Abdel-Hameed, M.S. "Optimal replacement policies for devices subject to a Gamma wear process", In The Theory and Application of Reliability, C.P. Tsokas and I.N. Shimi, (eds.), Academic Press, New York, 397-412, (1977).

2. Ansell, J., Bendell, A. and Humble, S. "Nested renewal processes", *Adv. Appl. Prob.*, (1980).

3. Cox, D.R. "Renewal Theory", London, Methuen, (1962).

4. Feldman, R.N. "Optimal replacement with semi-Markov shock models", *J. Appl. Prob.*, **13**, 108-117, (1976).

5. Feldman, R.N. "The maintenance of systems governed by semi-Markov shock models", In The Theory and Applications of Reliability, C.P. Tsokas and I.N. Shimi, (eds.), Academic Press, New York, 215-226, (1977).

6. Hjorth, U. "A comparison of some parametric models for the estimation of maintenance times from small samples", 11th European Meeting of Statisticians, Oslo, (1978).

7. Smith, W.L. "Regenerative stochastic processes", *Proc. Roy. Soc. A.*, **232**, 6-31, (1955).

8. Taylor, H.M. "Optimal replacement under additive damage and other failure models", *Naval Res. Logist. Quart.*, **22**, 1-18, (1975).

3.6 STABILITY AND CONTROL OF STOCHASTIC EVOLUTION EQUATIONS

A. Ichikawa

(Faculty of Engineering, Shizuoka University, Hamamatsu 432, Japan)

1 INTRODUCTION

In [3] we have studied stability and regulator problems in infinite dimensions using semigroup theory and have succeeded in extending Wonham's finite dimensional results to evolution equations using the dynamic programming approach. In general, stochastic evolution equations do not have strong solutions and so Ito's lemma cannot be applied. But with the aid of approximating smooth systems we can overcome this difficulty.

This paper is an extension of [3] to the case where the state-dependent noise term involves unbounded operators. This is done using intermediate spaces defined by fractional powers of the negative of the generator A. Most of the results in [3] are valid without significant change. So only those proofs which require different arguments are included in this paper and we refer the reader to [3] for details.

2 STOCHASTIC EVOLUTION EQUATIONS

Let X, H_i, i=1,2, be Hilbert spaces. Let H_i be separable. Consider the stochastic differential equation in X,

$$\begin{aligned} dx(t) &= Ax(t)dt + D(x(t))dw_1(t) + Fdw_2(t), \\ x(0) &= x_0, \end{aligned} \tag{2.1}$$

where A is the infinitesimal generator of a strongly continuous semigroup T(t) on X, $w_i(t)$ are mutually independent Wiener processes in H_i with covariance operator W_i, i=1,2 [1], $F \in L(H_2,X)$ $D \in L(X_\alpha, L(H_1,X))$, and X_α will be defined later.

The integral equation corresponding to (2.1) is

$$x(t) = T(t)x_0 + \int_0^t T(t-r)D(x(r))dw_1(r) + \int_0^t T(t-r)Fdw_2(r). \tag{2.2}$$

In general (2.1) requires more restrictive conditions than (2.2). So we try to establish a solution of (2.2). In [3] we took $D \in L(X,L(H_1,X)) = L(X\times H_1,X)$, but here we introduce $X_\alpha \subset X$. Thus D is unbounded in X. Our assumption is

C1: $(-A)^\alpha$, $0 \le \alpha \le 1$ is well-defined and

$$|(-A)^\alpha T(t)| \le \frac{c}{t^\alpha},$$

where c is a generic constant.

By X_α we denote $\mathcal{D}((-A)^\alpha)$ with graph norm $\| \ \|_\alpha$. The assumption that $(-A)^\alpha$ exists is not very restrictive since we can replace $-A$ by $-A + k$ for some $k > 0$ [7]. We denote by (Ω,Σ,σ) the underlying probability space.

Proposition 2.1 Assume C1. Then for each $0 \le \alpha < \frac{1}{2}$ there exists a unique solution of (2.2) in the class $L_2((0,T)\times\Omega, X_\alpha) \cap C([0,T], L_2(\Omega,X))$ which is adapted to $\sigma_t\{w_i(\cdot)\}$, $i=1,2$, the σ-algebra generated by $w_i(s)$, $0 \le s \le t$.

Since we deal with $L_2((0,T)\times\Omega,X_\alpha)$ we need a stochastic integral of the following type. Let $f(t) \in L_2(0,T)$ and $\Phi(t) \in L(H_1,Y)$ be adapted to $\sigma_t\{W_1(\cdot)\}$ such that $\int_0^T E|\Phi(t)|^2 dt < \infty$, where Y is a Hilbert space. Then we can define a stochastic integral

$$\int_0^t f(t-r)\ \Phi(r)dw_1(r) \tag{2.3}$$

in $L_2((0,T)\times\Omega, Y)$. In fact let $f_n(t) \in C(0,T)$ be a sequence which is convergent to f(t) in $L_2(0,T)$. Then

$$y_n(t) = \int_0^t f_n(t-r)\ \Phi(r)\,dw_1(r)$$

is well-defined and is in $C([0,T], L_2(\Omega,Y))$. Now

$$E|y_n(t)-y_m(t)|^2 \le \text{trace } W_1\int_0^t |f_n(t-r)-f_m(t-r)|^2\ E|\Phi(r)|^2\ dr.$$

So this yields

$$\int_0^T E|y_n(t)-y_m(t)|^2 dt \le \text{trace } W_1\int_0^T\int_0^t |f_n(t-r)-f_m(t-r)|^2 E|\Phi(r)|^2 drdt$$

$$\le \text{trace } W_1\int_0^T\int_r^T |f_n(t-r)-f_m(t-r)|^2 dt\ E|\Phi(r)|^2 dr$$

$$\le \text{trace } W_1\int_0^T |f_n(t)-f_m(t)|^2 dt \int_0^T E|\Phi(r)|^2 dr$$

$$\longrightarrow 0$$

Hence $y_n(t)$ is convergent in $L_2((0,T)\times\Omega, Y)$ and we define the stochastic integral (2.3) by this limit. Using this type of stochastic integral, Proposition 2.1 can be established by contraction arguments in $L_2((0,T)\times\Omega, X_\alpha)$ see Theorem 2.1 [4]. An extension of this proposition to the case of Lipschitz operators D is also possible [3].

Ito's lemma, given below, is still valid in our new set up.

Theorem 2.1. Suppose that the solution x(t) of (2.2) also satisfies (2.1). Let $g(t,x):[0,t_1]\times X \to R'$ be a continuous map satisfying:

(i) g(t,x) is differentiable in t for each $x \in \mathcal{D}(A)$ and the derivative $g_t(t,x)$ is continuous in t with estimate

$$|g_t(t,x)| \le c(1+|x|)(1+|x|+|Ax|)\ ,$$

(ii) $g(t,\cdot)$ is twice Fréchet differentiable on X for each t,

(iii) $g_x(t,x)$, $g_{xx}(t,x)$ are continuous in (t,x).

Then $z(t) = g(t,x(t))$ has the stochastic differential

$$dz(t) = \{g_t(t,x(t)) + \langle g_x(t,x(t)), Ax(t)\rangle$$

$$+ \frac{1}{2} \text{ trace } D(x(t))W_1D^*(x(t))g_{xx}(t,x(t))$$

$$+ \frac{1}{2} \text{ trace } FW_2F^*g_{xx}(t,x(t))\}dt$$

$$+ \langle g_x(t,x)), D(x(t))dw_1(t) + Fdw_2(t)\rangle.$$

Next we introduce a linear equation

$$\frac{d}{dt} \langle P(t)x,x\rangle + 2\langle Ax,P(t)x\rangle + \langle [M+\Delta(P(t))]x,x\rangle = 0, \quad x \in \mathcal{D}(A)$$

$$P(t_1) = G \tag{2.4}$$

and its integrated version

$$P(t)x = \int_t^{t_1} T^*(r-t)\,[M + \Delta(P(r))]\; T(r-t)x\,dr$$

$$+ T^*(t_1-t)GT(t_1-t)x, \quad x \in X \tag{2.5}$$

where $M \geq 0$, $G \in L(X)$ and $\langle \Delta(R)x,y \rangle = \text{trace } D^*(y)RD(x)W_1$, $x,y \in X_\alpha$, $R \in L(X)$. By a standard method [3] we can establish a unique solution satisfying (2.4), (2.5) in the class of linear self-adjoint nonnegative strongly continuous operators on X. $P(t)$ is useful as we see below.

<u>Theorem 2.2</u> Let $x(t)$ be the solution of (2.2). Then

$$\int_s^{t_1} E\langle Mx(t), x(t)\rangle dt + E\langle Gx(t_1), x(t_1)\rangle$$

$$= E\langle P(s)x(s),x(s)\rangle + \int_s^{t_1} \text{trace } F^*P(t)FW_2dt. \tag{2.6}$$

As in [3] we note that there is a unique solution $x(t,\lambda)$ to

$$dx(t) = Ax(t)dt + \lambda R(\lambda,A)\left[D(x(t))dw_1(t) + Fdw_2(t)\right]$$
$$x(0) = x_0 \varepsilon \mathcal{D}(A) \tag{2.7}$$

where $0 < \lambda \varepsilon \rho(A)$ and $R(\lambda,A)$ is the resolvent of A. This is due to the fact $A\lambda R(\lambda,A) = \lambda-\lambda^2 R(\lambda,A) \varepsilon L(X)$. Note also that $x(t,\lambda)$ satisfies the integral equation associated with (2.7). Applying Ito's lemma to $< P(t)x(t,\lambda), x(t,\lambda) >$, and the limit process using the following lemma, will establish Theorem 2.2.

Lemma 2.1 $x(t,\lambda) \to x(t)$ in $C([0,t_1], L_2(\Omega,x)) \cap L_2((0,t_1)\times\Omega, X_\alpha)$ as $\lambda \to \infty$, where $x(t)$ is the solution of (2.2).

Proof. Since

$$x(t,\lambda) - x(t) = \int_0^t T(t-r)\lambda R(\lambda,A)D(x(r,\lambda)-x(r))dw_1(r)$$

$$+ \int_0^t T(t-r)\left[\lambda R(\lambda,A)-I\right]Fdw_2(r).$$

we have

$$E\|x(t,\lambda)-x(t)\|_\alpha^2 \le 3E\left\|\int_0^t T(t-r)\lambda R(\lambda,A)D(x,(r,\lambda)-x(r))dw_1(r)\right\|_\alpha^2$$

$$+ 3E\left\|\int_0^t T(t-r)\left[\lambda R(\lambda,A)-I\right]D(x(r))dw_1(r)\right\|_\alpha^2$$

$$+ 3E\left\|\int_0^t T(t-r)\left[\lambda R(\lambda,A)-I\right]Fdw_2(r)\right\|_\alpha^2$$

$$\le c\int_0^t \frac{1}{(t-r)^{2\alpha}} E\|x(r,\lambda)-x(r)\|_\alpha^2\, dr + \varepsilon(t,\lambda)$$

where $\varepsilon(t,\lambda)$ is the sum of last two terms above. Now let $k>0$ and calculate

$$\int_0^{t_1} e^{-kt} E\|x(t,\lambda)-x(t)\|_\alpha^2\, dt \le c\int_0^{t_1}\int_0^{t} \frac{e^{-k(t-r)}}{(t-r)^{2\alpha}}\, e^{-kr} E\|x(r,\lambda)-x(r)\|_\alpha^2\, drdt$$

$$+ \int_0^{t_1} e^{-kt}\, \varepsilon(t,\lambda)dt$$

$$\le c\left\|\frac{e^{-kt}}{t^{2\alpha}}\right\|_{L_1(0,t_1)} \int_0^{t_1} e^{-kt} E\|x(t,\lambda)-x(t)\|_\alpha^2\, dt$$

$$+ \int_0^{t_1} e^{-kt}\, \varepsilon(t,\lambda)dt.$$

For sufficiently large k,

$$c\left\|\frac{e^{-kt}}{t^{2\alpha}}\right\|_{L_1(0,t_1)} = \ell < 1$$

and $\int_0^{t_1} e^{-kt} E\|x(t,\lambda)-x(t)\|_\alpha^2\, dt \le \frac{1}{1-\ell}\int_0^{t_1} e^{-kt}\, \varepsilon(t,\lambda)dt.$

But $\lambda R(\lambda,A) \to I$ strongly as $\lambda \to \infty$. Therefore

$$\int_0^{t_1} e^{-kt}\, E\|x(t,\lambda)-x(t)\|_\alpha^2\, dt \to 0 \text{ as } \lambda \to \infty,$$

which implies

$$\int_0^{t_1} E\|x(t,\lambda)-x(t)\|_\alpha^2\, dt \to 0 \text{ as } \lambda \to \infty.$$

We also have the estimate

$$E|x(t,\lambda)-x(t)|^2 \leq c \int_0^t E\|x(r,\lambda)-x(r)\|_\alpha^2 \, dr$$

$$+ 3E\left| \int_0^t T(t-r)\left[\lambda R(\lambda,A)-I\right]D(x(r))\,dw_1(r)\right|^2$$

$$+ 3E\left| \int_0^t T(t-r)\left[\lambda R(\lambda,A)-I\right]F\,dw_2(r)\right|^2$$

$\to 0$ as $\lambda \to \infty$. □

3 STOCHASTIC STABILITY

First wè set $F = 0$ in (2.2) and consider

$$x(t) = T(t)x_0 + \int_0^t T(t-r)D(x(r))dw_1(r). \tag{3.1}$$

Definition 3.1 The system (3.1) (or (A,D)) is stable if the solution $x(t)$ satisfies $\int_0^\infty E|x(t)|^2 \, dt < \infty$ for each $x_0 \in X$.

Theorem 3.1 The following statements are equivalent:

(i) (A,D) is stable,

(ii) there exists $0 \leq P \in L(X)$ such that

$$2 < Ax, Px > + < \Delta(P)x, x > = -< x, x > \text{ for any } x \in \mathcal{D}(A). \tag{3.2}$$

If $\bar{d} = |D|^2$ trace W_1

is sufficiently small, then (i), (ii) are equivalent to

(iii) $E|x(t)|^2 \leq ae^{-\delta t}|x_0|^2$ for some $a \geq 1$, $\delta > 0$.

Proof We need only to show (ii)$\Longrightarrow$ (iii). The rest of the proof is identical to that of Theorem 3.1 in [3] and we use

Theorem 2.2. First note that (ii) implies the exponential stability of T(t) [3]. Thus

$$|(-A)^{\alpha}T(t)| \le b \frac{e^{-\beta t}}{t^{\alpha}} \quad \text{for some } b,\beta > 0.$$

From (3.1) we have the estimate

$$E\|x(t)\|_{\alpha}^{2} \le b^{2} \frac{e^{-2\beta t}}{t^{2\alpha}} |x_0|^2 + \bar{d} \int_0^t \frac{e^{-2\beta(t-r)}}{(t-r)^{2\alpha}} E\|x(r)\|_{\alpha}^{2} dr.$$

If $\bar{d}$ is sufficiently small we still have an estimate

$$E\|x(t)\|_{\alpha}^{2} \le c \frac{e^{-2\bar{\beta}t}}{t^{2\alpha}} |x_0|^2 \quad \text{for some } 0 < \bar{\beta} < \beta.$$

We also have $|T(t)| \le c_1 e^{-\beta t}$, $c_1 \ge 1$ and

$$E|x(t)|^2 \le c_1^2 e^{-2\beta t}|x_0|^2 + \bar{d} \int_0^t e^{-2\beta(t-r)} E\|x(r)\|_{\alpha}^{2} dr$$

$$\le c_1^2 e^{-2\beta t}|x_0|^2 + \bar{d} \int_0^t e^{-2\beta(t-r)} \frac{e^{-2\bar{\beta}r}}{r^{2\alpha}} |x_0|^2 dr$$

$$\le (c_1^2 e^{-2\beta t} + c\, t^{1-2\alpha} e^{-2\bar{\beta}t}) |x_0|^2.$$

Hence

$$E|x(t)|^2 \le a\, e^{-\delta t}|x_0|^2 \quad \text{for some } a \ge 1,\ \delta > 0. \ \square$$

<u>Corollary 3.1</u> If $\bar{d}$ is sufficiently small then (3.2) has at most one solution. Let x(t) be the solution of (2.2) and suppose (A,D) is stable. Then

$$\lim_{t\to\infty} \frac{1}{t} \int_0^t E|x(r)|^2 dr = \text{trace } F^*PFW_2,$$

where P is the solution of (3.2).

Corollary 3.2 If there is a $0 \leq P \in L(X)$ such that

$$2 \langle Ax,Px\rangle + \langle\Delta(P)x,x\rangle \leq -b\langle x,x\rangle,$$

then (A,D) is stable.

The next theorem is concerned with invariant measures of the Markov process x(t) obtained from (2.2).

Theorem 3.2 If (A,D) is stable, then there exists an invariant measure μ of the Markov process x(t) and

$$\int_X |\xi|^2 \mu(d\xi) = \text{trace } F^*PFW_2$$

where P is the solution of (3.2).

Proof See [3]. □

4. REGULATOR PROBLEMS

In [3] we considered quadratic problems for the control system

$$x(t) = T(t)x_0 + \int_0^t T(t-r)D(x(r))dw_1(r) + \int_0^t T(t-r)Fdw_2(r)$$
$$+ \int_0^t T(t-r)Bu(r)dr + \int_0^t T(t-r)C(u(r))dw_3(r). \tag{4.1}$$

Here $w_3(t)$ is a Wiener process in a separable Hilbert space H_3 with covariance operator W_3, u(t) is a control with values in a Hilbert space U, $B \in L(U,X)$, $C \in L(U,L(H_3,X))$ and $w_3(t)$ is independent of $w_i(t)$, i=1,2. We can check that all the results in [3] concerning (4.1) are still valid. For example, consider the minimization problem for the quadratic cost functional

$$C(u) = E\langle Gx(t_1),x(t_1)\rangle + \int_0^{t_1} E\{\langle Mx(t),x(t)\rangle + \langle Nu(t),u(t)\rangle\}dt, \tag{4.2}$$

where $G \geq 0$, $M \in L(X)$, $0 < N \in L(U)$ and $N^{-1} \in L(U)$ and admissible controls are $u(t)$'s which are adapted to $\sigma_t\{w_i(\cdot), i=1,2,3\}$ and satisfy $\int_0^{t_1} E|u(t)|^2\, dt < \infty$.

Theorem 4.1 The optimal control is given by the feedback law

$$\bar{u} = -[N + \Gamma(Q(t))]^{-1}B^*Q(t)x \tag{4.3}$$

and the minimum cost is

$$C(u) = \langle Q(0)x_0,x_0\rangle + \int_0^{t_1} \text{trace } F^*Q(t)FW_2dt, \tag{4.4}$$

where $Q(t) \geq 0$ is the unique solution of the Riccati equation

$$\frac{d}{dt}\langle Q(t)x,x\rangle + 2\langle Ax,Q(t)x\rangle + \langle Mx,x\rangle + \langle\Delta(Q(t))x,x\rangle$$

$$- \langle Q(t)B[N+\Gamma(Q(t))]^{-1}B^*Q(t)x,x\rangle = 0, \quad x \in \mathcal{D}(A) \tag{4.5}$$

$$Q(t_1) = G$$

and $\langle \Gamma(Q(t))u,v \rangle = \text{trace } C^*(v)Q(t)C(u)W_3$, $u,v, \in U$.

Now we turn to optimal stationary control problems. We take the class of feedback controls $K(x): X \to U$ with

$$|K(x) - K(y)| \leq c|x-y|,$$

and $u=K(x)$ is admissible if the Markov process given by (4.1) with $u=K(x)$ has an invariant measure μ_k such that

$$\int_X |x|^2\mu_k(dx) < \infty.$$

Our control problem is to minimise

$$C(u) = \int_M \{< Mx,x > + < NK(x),K(x) >\} \mu_k(dx)$$

over all admissible controls.

Theorem 4.2 Suppose that $(A-BK_1,D,C)$ is stable for some $K_1 \varepsilon L(X,U)$ and that for some $J \varepsilon L(X)$, $A-JM^{\frac{1}{2}}$ generates a stable semigroup $\tilde{T}(t)$ with $|\tilde{T}(t)| \leq \tilde{a}e^{-\tilde{\beta}t}$ and

$$\frac{\tilde{a}^2 d^2 \text{ trace } W_1}{\tilde{\beta}} < 1, \ d = |D|.$$

Then the optimal control exists and is given by the feedback law

$$\bar{u} = -[N + \Gamma(Q)]^{-1} B^*Qx,$$

and the minimum cost is $C(\bar{u}) = \text{trace } F^*QFW_2$, where $0 \leq Q$ is the unique solution of the Riccati equation

$$2< Ax,Qx > + <\{M+\Delta(Q)-QB[N+\Gamma(Q)]^{-1}B^*Q\}x,x > =0, \ x \varepsilon \mathcal{D}(A).$$

Proofs of these results utilize dynamic programming and the key point is to use approximating systems of type (2.7) for which we can apply Ito's lemma. The results are then obtained by the limit process.

5 FINAL REMARKS

Pardoux [6] covers a class of generators A which arise from sesquilinear forms and are coercive. His results correspond to $\alpha = 1/2$. But coercivity implies that (A,D) is stable [2] under C1. In fact, we can use Corollary 3.2. Hence for his problems stability almost follows from existence.

Reference [4] contains examples of parabolic equations and delay equations which can be covered by the evolution system in this paper. A stable semigroup of a hyperbolic equation is found in [5]. This and the class of delay equations are contained in the case $\alpha = 0$. The control results can be applied to various examples considered in [2], [4], [6], [8].

6 REFERENCES

1. Curtain, R.F. and Falb, P.L. "Ito's lemma in infinite dimensions", *J. Math. Anal. Appl.*, **31**, 434-448, (1970).

2. Haussmann, U.G. "Asymptotic stability of the linear Ito equation in infinite dimensions", *J. Math. Anal. Appl.*, **65**, 219-335, (1978).

3. Ichikawa, A. "Dynamic programming approach to stochastic evolution equations", *SIAM J. Control Opt.*, **17**, 152-174, (1979).

4. Ichikawa, A. Linear stochastic evolution equations in Hilbert space", *J. Diff. Eqns.*, **28**, 266-277, (1978).

5. Ichikawa, A. and Pritchard, A.J. "Existence, uniqueness and stability of nonlinear evolution equations", *J. Math. Anal. Appl.*, **68**, 454-476, (1979).

6. Pardoux, E. "Thesis", Université Paris, Sud (1975).

7. Tanabe, H. "Evolution Equations", Iwanami, Tokyo (1975).

8. Zabczyk, J. "On stability of infinite dimensional stochastic systems", Banach Center Publications, *Probability Theory*, **5**, 273-281, (1979).

CHAPTER 4

ALGORITHMS

4.1 A DISCRETE APPROXIMATION TO OPTIMAL STOCHASTIC CONTROLS

U. G. Haussmann

(Department of Mathematics, University of British Columbia, Vancouver, Canada, V6T 1W5)

1 INTRODUCTION

Consider the optimal control problem

$$\min_u E\{\int_0^T c(t,x(t))dt + c_0[x(T)]\} \tag{1.1}$$

subject to

$$\begin{aligned} dx &= f(t,x(t),u(t,x(t))dt + \sigma(t,x(t))dw^u, \ 0 \le t \le T, \\ x(0) &= x_0, \quad u(t,x(t)) \in \Gamma, \end{aligned} \tag{1.2}$$

where u is to be an admissible control as defined below, E denotes expectation, $x(\cdot)$ is an $\mathbb{R}_d$ valued stochastic process, w^u is a standard $\mathbb{R}_d$ valued Wiener process, $c(\cdot,\cdot)$ and $c_0(\cdot)$ are $\mathbb{R}_1$ valued, f is $\mathbb{R}_d$ valued and σ is $d \times d$ matrix valued. x_0 is a fixed point in $\mathbb{R}_d$. Γ is the set of control points. We assume throughout

(A1) c, c_0, f, σ are bounded and continuous,

(A2) $\sigma(t,x)^{-1}$ exists and is bounded,

(A3) Γ is a compact convex subset of a Euclidean space,

(A4) $f(t,x,\Gamma)$ is convex for each (t,x), although (A4) is for convenience only. It can be dropped as pointed out at the end. Cost rates depending explicitly on the control can also be treated, c.f. the remarks at the end of the article.

As admissible controls we take Borel measurable functions $u:[0,T] \times \mathbb{R}_d \to \Gamma$. Hence we have a completely observable Markovian control problem. Moreover (1.2) has a unique weak solution for any admissible control [1], and optimal controls exist [2].

Early attempts to solve the problem numerically consisted of either approximating it by a linear regulator problem, or by solving the Bellman equation numerically, possibly using policy improvement, c.f. [3], [4] § 6.6. The first method lacks rigorous justification, and the second one is cumbersome numerically and again is only valid under further smoothness assumptions. Recently Kushner [5] has proposed an interesting new approach. The idea is to approximate x by a discrete time and state Markov chain ξ, and then to choose u to solve a corresponding discrete control problem involving ξ (much easier numerically). He does require, however, added Lipschitz continuity of f and c, but not (A2). Our approach is somewhat different. We again approximate x by ξ but then find a condition on (ξ,u) which, in the limit, reduces to the condition of the maximum principle. The condition on (ξ,u) can then be used to generate numerically the approximate optimal costs.

In section two we give some preliminaries on the maximum principle from [2]. In section three we describe the discretization procedure, and in section four we establish the convergence in the limit to the continuous processes. Finally in section five we show that the costs generated by the discrete Markov chains converge to the optimal cost for (1.1), (1.2).

2 THE STOCHASTIC MAXIMUM PRINCIPLE

According to [1], a solution to (1.2) induces, or is, a measure P^u on $C[0,T]$, the space of all $\mathbb{R}_d$ valued continuous functions defined on $[0,T]$. If π_t is the projection map $C[0,T] \to \mathbb{R}_d$ defined by $\pi_t x = x(t) \equiv x_t$, then the expected cost-to-go under (1.2) given $x_s = y$, $V^u(s,y)$, is a conditional expectation (at y) on $C[0,T]$ under P^u given π_s, i.e.

$$V^u(s,y) = E^u\{\int_s^T c(t,x_t)dt + c_0(x_T) \mid \pi_s\}(y).$$

According to [2], there is a Borel function H^u defined on $[0,T] \times \mathbb{R}_d$ and a martingale M, orthogonal to w^u, such that

$$V^u(t,x_t) - V^u(s,x_s) = -\int_s^t c(\tau,x_\tau)d\tau + \int_s^t H^u(\tau,x_\tau)\cdot dw^u(\tau) \quad \text{a.e.}$$
$$+ M_t^u - M_s^u, \qquad (2.1)$$

with $\int_0^T E|H^u|^2 dt < \infty$. Moreover u is optimal if and only if

$$H^u(t,x)\cdot\sigma(t,x)^{-1}f(t,x,u(t,x)) = \inf_{\mu\in\Gamma} H^u(t,x)\cdot\sigma(t,x)^{-1}f(t,x,\mu) \qquad (2.2)$$

where a.e. refers to a measure m defined by

$m(A) = \int_0^T \left[\int_{A_T} (\bar{P}_0\pi_t^{-1})(dx)\right] dt$ with $A_t = \{x: (t,x) \in A\}$ and with $\bar{P}$ being the measure induced by (1.2) with $f \equiv 0$. Note that P^u is equivalent to $\bar{P}$. The maximum principle is also discussed in [6] in greater generality. Our aim is to find a discrete version of (2.2) which reduces in the limit to (2.2).

3 THE MARKOV CHAIN APPROXIMATION

We set $a(t,x) = \frac{1}{2}\sigma(t,x)\sigma(t,x)^*$ (* denotes transpose), and we <u>assume</u>

$$a_{ii}(t,x) - \sum_{\substack{j=1 \\ j\neq i}}^{d} |a_{ij}(tx)| \geq 0 \text{ all } i,t,x. \qquad (3.1)$$

This condition can be relaxed somewhat, c.f. [5] p. 118, p. 125. Choose any h, Δ positive such that T/Δ is an integer N, and

$$h^2/\Delta \geq \sup_{t,x,\mu} \{h \sum_{i=1}^{d} |f_i(t,x,\mu)| + 2 \sum_{i=1}^{d} a_{ii}(t,x) - \sum_{\substack{i,j=1 \\ i\neq j}}^{d} |a_{ij}(t,x)|\}. \tag{3.2}$$

Since f and σ are bounded, (3.2) can always be satisfied for any small h if Δ is sufficiently small. (x_0,h,Δ) now define a grid on $[0,T] \times \mathbb{R}_d$ if we start at x_0 and increment by Δ along the t axis, and by h along all the d other axes. Call it $G(x_0,h,\Delta)$. For $u(\cdot,\cdot)$ defined on $G(x_0,h,\Delta)$, we define the following transition probabilities for the Markov chain $\{\xi_n^{h\Delta u}\}_{n=0}^{N}$, $\xi_0^{h\Delta u} = x_0$.

$$\Pr\{\xi_{n+1} = y | \xi_n = x\}$$

$$= \Delta a_{ij}^{\pm}(n\Delta,x)/h^2 \text{ if } y = x + e_i h \pm e_j h,\ i \neq j,$$

$$= \Delta a_{ij}^{\mp}(n\Delta,x)/h^2 \text{ if } y = x - e_i h \pm e_j h,\ i \neq j,$$

$$= \Delta\{h\, f_i^{\pm}(n\Delta,x,\ u(n\Delta,x)) + a_{ii}(n\Delta,x) - \sum_{\substack{j=1 \\ j\neq i}}^{d} |a_{ij}(n\Delta,x)|\}/h^2 \quad \text{if } y = x \pm e_i h, \tag{3.3}$$

$$= 1 - \Delta\{h \sum_{i=1}^{d} |f_i(n\Delta,x,u(n\Delta,x))| + 2 \sum_{i=1}^{d} a_{ii}(n\Delta,x) - \sum_{\substack{i,j=1 \\ i\neq j}} |a_{ij}(n\Delta,x)|\}/h^2 \quad \text{if } y = x,$$

$= 0$ otherwise,

where as usual a function g decomposes into two non-negative parts $g = g^+ - g^-$, and e_i is the ith standard basis element in $\mathbb{R}_d$. (3.1) and (3.2) ensure that (3.3) defines a probability.

The chain $\{\xi_n^{h\Delta u}\}_{n=0}^N$ lives on the grid $G(x_0,h,\Delta)$, and can be represented as $(f^u(t,x) \equiv f(t,x,u(t,x)))$

$$\xi_{n+1}^{h\Delta u} = \xi_n^{h\Delta u} + f^u(n\Delta,\xi_n^{h\Delta u})\Delta + \beta_n^{h\Delta u}, \tag{3.4}$$

where $\{\beta_n^{h\Delta u}\}_{n=0}^{N-1}$ is a sequence of orthogonal random vectors satisfying

$$\begin{aligned} &E\{\beta_n^{h\Delta u}|\xi_n^{h\Delta u}\} = 0 \\ &E\{\beta_n^{h\Delta u}(\beta_n^{h\Delta u})^*|\xi_n^{h\Delta u}\} = \Delta\ \Sigma_n^{h\Delta u}(\xi_n^{h\Delta u}) \end{aligned} \tag{3.5}$$

and $\Sigma_n^{h\Delta u}(x) = 2a(n\Delta,x) + h\ \text{diag}\{|f_i^u(n\Delta,x)|\} - \Delta\ f^u(n\Delta,x)f^u(n\Delta,x)^*$.

We can interpolate $\xi^{h\Delta u}$ to a process defined on $[0,T]$ by setting $\xi^{h\Delta u}(t) \equiv \xi_{I(t)}^{h\Delta u}$ if $I(t) = [t/\Delta]$. Let us homogenize the process by writing $\zeta^{h\Delta u}(t) \equiv (I(t)\Delta,\xi^{h\Delta u}(t))$. Let us define two further processes.

$$F^{h\Delta u}(t) = \sum_{k=0}^{I(t)-1} \tilde{f}^u(k\Delta,\xi_k^{h\Delta u})\Delta$$

$$B^{h\Delta u}(t) = \sum_{k=0}^{I(t)-1} \tilde{\beta}_k^{h\Delta u},$$

where $\tilde{f}^{u*} = (1,f^{u*})$ and $\tilde{\beta}^* = (0,\beta^*)$. Then

$$\zeta^{h\Delta u}(t) = \tilde{x}_0 + F^{h\Delta u}(t) + B^{h\Delta u}(t)$$

if $\tilde{x}_0^* = (0,x_0^*)$.

4 CONVERGENCE

From [5], §7.2, §9.2, it follows that if $(h,\Delta) \to (0,0)$ through a suitable subsequence, then $(\zeta^{h\Delta}(t), F^{h\Delta}(t), B^{h\Delta}(t))$, all defined on some probability space (Ω, A, Q) by Skorokhod embedding, converge uniformly in t, w.p.1 to $(t, x(t), \int_0^t \bar{f}(s,\omega)ds, \int_0^t \sigma(s,x_s)dw)$ with

$$x(t) = x_0 + \int_0^t \bar{f}(s,\omega)ds + \int_0^t \sigma(s,x_s)dw(s) \tag{4.1}$$

where $\bar{f}$ is some measurable function and where w is a Brownian motion. Moreover $f^u(\zeta^{h\Delta u}(t,\omega))$ converges weakly in $L_1(dt \times dQ)$ to $\bar{f}(t,\omega)$. Since $f^u, \bar{f}$ are bounded then the convergence is also weak in $L_2(dt \times dQ)$.

By [7], theorem 4.3, $\bar{f}$ and w can be assumed to be x adapted, and since w is Brownian motion relative to the σ-algebras generated by x, then future incremenets of w are independent of the past of x. Hence $\int_t^{t+\delta} \sigma(s,x_s)dw$ is $\{x_s : t \leq s \leq t+\delta\}$ measurable. From (4.1) it follows that $\int_t^{t+\delta} \bar{f}(s,\omega)ds$ is similarly measurable. Dividing by δ and taking δ to zero implies that $\bar{f}$ is x adapted (we may have to change $\bar{f}$ on a (t,ω) null set); hence $\bar{f}(t,\omega) = g(t,x(t,\omega))$ for some Borel measurable g. Now x satisfies

$$x(t) = x_0 + \int_0^t g(s,x_s)ds + \int_0^t \sigma(s,x_s)dw. \tag{4.2}$$

Since for each t,x, $f(t,x,\Gamma)$ is convex, then g can be taken as $g(s,x) = f(s,x,\bar{u}(s,x))$ with $\bar{u}$ admissible, by the McShane-Warfield implicit function lemma, c.f. [5], §9.2.

Let us be more precise about u defined on $G(x_0,h,\Delta)$. u does vary with the grid, and so we should write $u^{h\Delta}$. By

$f^u(\zeta^{h\Delta u})$ we mean $f(\zeta^{h\Delta u}, u^{h\Delta}(\zeta^{h\Delta u}))$.

We set $V_0^{h\Delta u}(x_0) = E\{\int_0^T c(\zeta^{h\Delta u}(s))ds + c_0(\xi^{h\Delta u}(T))\}$, and $V^{h\Delta u}(t) = E\{\int_t^T c(\zeta^{h\Delta u}(s))ds + c_0(\xi^{h\Delta u}(T)) | \zeta^{h\Delta u}(t)\}$. By weak convergence $V_0^{h\Delta u}(x_0) \to V^{\bar{u}}(0,x_0)$, and hence if we define $K^{h\Delta u}(T)$ by

$$c_0(\xi^{h\Delta u}(T)) = V_0^{h\Delta u}(x_0) - \int_0^T c(\zeta^{h\Delta u}(s))ds + K^{h\Delta u}(T)$$

then from (2.1) it follows that

$$K^{h\Delta u}(T) \to \int_0^T H^{\bar{u}}(\tau,x_\tau) \cdot dw^{\bar{u}} + M_T^{\bar{u}} \equiv K(T) \quad \text{w.p.l.}$$

We collect the main points of this section in

Lemma 4.1. <u>There exists an admissible control</u> $\bar{u}$, <u>a subsequence</u> $(h,\Delta) \to 0$ <u>and a probability space</u> $(\Omega,\mathcal{A},Q)$ <u>carrying</u> $\zeta^{h\Delta u}(\cdot)$, $B^{h\Delta u}(\cdot)$, $K^{h\Delta u}(T)$, $x(\cdot)$, $B(\cdot)$, $K(T)$ <u>such that</u>

(i) x <u>satisfies</u> (1.2) <u>with</u> $u = \bar{u}$ <u>and</u> $\int_0^t \sigma dw^u = B(t)$,

(ii) <u>for</u> ω <u>not in some null set</u>, $(\zeta^{h\Delta u}(t), B^{h\Delta u}(t), K^{h\Delta u}(T)) \to (t, x(t), B(t), K(T))$ <u>uniformly in</u> t,

(iii) $f(\zeta^{h\Delta u}(t), u^{h\Delta}(\zeta^{h\Delta u}(t))) \to f(t,x(t),\bar{u}(t,x(t)))$ <u>weakly in</u> $L_2(dt \times dQ)$.

(iv) $V_0^{h\Delta u}(x_0) \to V^{\bar{u}}(0,x_0)$.

5 THE MAIN THEOREM

Let us assume that the convergent subsequence is labelled by m, i.e. the superscript $h\Delta u$ is replaced by m.

Theorem 5.1 <u>If u^m satisfies</u>

$$E\{V^m(t+\Delta)\beta_k^m|\zeta^m(t) = z\}\cdot\left[\sigma(z)\sigma(z)^*\right]^{-1}f(z,u^m(z)) = \inf_{\mu\in\Gamma} E\{V^m(t+\Delta)\beta_k^m|\zeta^m(t) = z\}\cdot\left[\sigma(z)\sigma(z)^*\right]^{-1}f(z,\mu) \tag{5.1}$$

<u>for all</u> $z = (k\Delta,x) \in G_m$, $t = k\Delta$, <u>then</u>

$$H^{\bar{u}}(z)\cdot\sigma(z)^{-1}f(z,\bar{u}(z)) = \inf_{\mu\in\Gamma} H^{\bar{u}}(z)\cdot\sigma(z)^{-1}f(z,\mu) \quad \text{a.e. } (m) \tag{5.2}$$

Proof: Let v be any continuous admissible control. Write

$$\phi^m(t) = \left[\sigma(\zeta^m(t))\sigma(\zeta^m(t))^*\right]^{-1}\left[f^m(\zeta^m(t)) - f^v(\zeta^m(t))\right]$$

$$\phi(t) = \left[\sigma(t,x(t))\sigma(t,x(t))^*\right]^{-1}\left[f^{\bar{u}}(t,x(t)) - f^v(t,x(t))\right],$$

$$D^m(t) = \sum_{k=0}^{I(t)-1}\phi^m(k\Delta)\cdot\beta_k^m = \int_0^t\phi^m(s)\cdot dB^m(s) + \rho^m(t),$$

$$D(t) = \int_0^t\phi(s)\cdot dB(s) = \int_0^t\phi(s)\cdot\sigma(s,x_s)dw,$$

where $E|\rho^m(t)|^2 \leq O(\Delta_m)$ uniformly in m.

Observe that (5.1) implies (if $\ell_k^m \equiv E\{K^m(T)|\xi_0^m,\ldots,\xi_k^m,\xi_{k+1}^m\} - E\{K^m(T)|\xi_0^m,\ldots,\xi_k^m\}$

$$\begin{aligned} EK^m(T)D^m(T) &= E\{\sum_{k=0}^{N-1}\ell_k^m D^m(T)\} \\ &= E\{\sum_{k=0}^{N-1}\ell_k^m\phi^m(k\Delta)\cdot\beta_k^m\} \\ &= E\{\sum_{k=0}^{N-1}\phi^m(k\Delta)\cdot E\{V^m((k+1)\Delta)\beta_k^m|\xi_k^m\}\} \leq 0 \end{aligned} \tag{5.3}$$

where we have used the fact the conditional means of β_k^m and ℓ_k^m are zero, and that $\ell_k^m = V^m((k+1)\Delta) - V^m(k\Delta) + \Delta c(\zeta_k^m)$.

Now

$$|EK^m(T)D^m(T) - EK(T)D(T)|$$

$$\leq (E|K^m(T) - K(T)|^2)^{1/2}(E|D^m(T)|^2)^{1/2}$$

$$+ \left|EK(T)\int_0^T \phi^m(t)\cdot d(B^m - B)\right|$$

$$+ \left|EK(T)\int_0^T [\phi^m(t) - \phi(t)]\cdot dB\right| + (K\Delta_m E|K(T)|^2)^{1/2}.$$

Since $E|D^m(T)|^2$ is bounded in m, and since $K^m(T,\omega)$ is bounded in m, ω by (A1), then lemma 4.1 (ii) implies that the first term goes to zero. So does the last. Since $\phi^m(t,\omega)$ is also bounded in m,t,ω, then again lemma 4.1 (ii) yields that the second term converges to zero. The third term equals (since M isorthogonal to w)

$$\left|E\int_0^T [\phi^m(t) - \phi(t)]\cdot d < K, B >_t\right|$$

$$= \left|E\int_0^T [\phi^m(t) - \phi(t)]\cdot\sigma(t,x_t)H^{\bar{u}}(t,x_t)dt\right|$$

which again converges to zero by lemma 4.1 (iii). Note $K(t) = \int_0^t H^{\bar{u}}\cdot dw + M_t^{\bar{u}}$.

Now (5.3) implies

$$0 \geq EK(T)D(T)$$
$$= E\int_0^T H^{\bar{u}}\cdot\sigma^{-1}(f^{\bar{u}} - f^v)dt. \qquad (5.4)$$

If v is any admissible control, we can approximate it in $L_1(dt \times dx)$ by a continuous admissible control (since Γ is convex) using local averages. Since L_1 convergence implies convergence in measure, since transition densities exist, c.f. [1], and since σ^{-1} and f are bounded and $H^{\bar{u}}$ is $L_2(dt \times dQ)$, then $E \int H^{\bar{u}} \cdot \sigma^{-1} f^v dt$ can be approximated arbitrarily well by the same expression but using a continuous control $\tilde{v}$, and so (5.4) holds for all admissible v. This implies (5.2).

Corollary: $V_0^{h\Delta u}(x_0) \to \hat{V}$, the optimal cost, for any sequence $(h,\Delta) \to 0$.

Proof: By (5.2) $\bar{u}$ is an optimal cost. Any other subsequence will yield another optimal cost $\bar{u}'$, but $V^{\bar{u}} = V^{\bar{u}'} = \hat{V}$ since the optimal cost is unique.

Remarks. The condition (5.1) defines $u^{(m)}$ implicitly, not explicitly, since to compute the conditional expectation one requires $u^{(m)}(k\Delta,x)$. However a simple iterative procedure yielded good results in two numerical examples. In fact for a three dimensional predicted miss problem, where the optimal $\hat{u}$ can be computed analytically, we found that $u^{(m)}(t,x) = \hat{u}(t,x)$, for most $(t,x) \in G_m$, so that even the optimal controls were generated, see appendix. In fact it follows that if $\{u^{(m)}\}$ converges then it converges to an optimal control. More computational work will be done to see whether the controls do in fact usually converge. The size of the problem is still a limiting factor, and we can only handle $d \leq 3$.

If the cost rate is of the form c(t,x,u), then one can transform it into the c_0 term by augmenting the state by one dimension, and considering the partially observable control problem where u is independent of x_{d+1} the augmented component. It can be shown that (~ denotes the augmented variable), $\tilde{H}^u(t,\tilde{x}) = (H^u(t,x),1)$ and that (2.2) becomes

$$E\{H^u(t,x)\cdot\sigma(t,x)^{-1}f(t,x,u(t,\tilde{x})) + c(t,x,u(t,\tilde{x}))\,|\,x(t) = x\}$$

$$= \inf_{\mu\in\Gamma} E\{H^u(t,x)\cdot\sigma(t,x)^{-1}f(t,x,\mu) + c(t,x,\mu)\,|\,x(t) = x\}. \qquad (2.2)'$$

Since the right side is independent of x_{d+1} this is equivalent to treating the (d+1) dimensional completely observable problem with

$$\tilde{H}^u(t,x)\cdot\sigma(t,x)^{-1}\tilde{f}(t,x,u(t,\tilde{x})) = \inf_{\mu\in\Gamma} \tilde{H}^u(t,x)\cdot\tilde{\sigma}(t,x)^{-1}f(t,x,\mu) \qquad (2.2)$$

to which the above theory applies. Moreover the optimal u is independent of x_{d+1}. Note that $\tilde{f} = \binom{f}{c}$, $\tilde{\sigma} = \begin{pmatrix}\sigma & 0\\ 0 & 1\end{pmatrix}$.

Finally we observe that the convexity requirement (A4) can be dropped. We convexify the problem as in [2]. Now if $\Gamma(t,x) = f(t,x,\Gamma)$ and $\hat{\Gamma}(t,x) = \mathrm{co}\Gamma(t,x) = \overline{\mathrm{co}\Gamma(t,x)}$ then the right side of (5.1) is

$$\inf_{\phi\in\Gamma(t,x)} E\{V\beta|z\}\cdot(\sigma\sigma^*)^{-1}\phi = \inf_{\phi\in\hat{\Gamma}(t,x)} E\{V\beta|z\}\cdot(\sigma\sigma^*)^{-1}\phi$$

so that u^m satisfies the optimality condition for the convexified problem. Now f^v represents any $\hat{\Gamma}(t,x)$ valued drift (v is now just an identifying parameter), but it can still be approximated by a continuous drift since Γ is convex. Hence the drift g(t,x) solves the convexified problem and thus also the original problem, with the same H, $\hat{V}$. Thus $V_0^{h\Delta u}(x_0)$ converges to the optimal cost.

Appendix. Consider the "predicted miss" problem

$$\min E\{|k\cdot x(T)|^2\}$$

subject to

$$dx = (Ax+Bu)dt + C\,dw,$$

$$|u_i(t)| \leq 1,$$

where $u^* = (u_1,u_2) \in \mathbb{R}_2$, $x \in \mathbb{R}_3$, $w \in \mathbb{R}_3$, $k \in \mathbb{R}_3$. It is known, [6], that the optimal control law is $\hat{u}(t,x) = -\text{sgn}[B(t)^*s(t)s(t)^*x]$ where sgn is the signum function and s satisfies

$$\frac{ds}{dt} = - A^*s, \quad s(T) = k.$$

An algorithm based on the main theorem was applied to this problem when

$$k^* = (3,-2,1), \quad A = \begin{pmatrix} .2 & 0 & 0 \\ 0 & .2 & 0 \\ .1 & 0 & .1 \end{pmatrix}, \quad B = \begin{pmatrix} 1 & 0 \\ -1 & 0 \\ 0 & 1 \end{pmatrix},$$

$$C = \begin{pmatrix} .5 & .25 & .25 \\ -.25 & .5 & .25 \\ -.25 & -.25 & .5 \end{pmatrix}, \quad x(0)^* = (1,1,1), \quad T = .1 .$$

Although (A1) is not satisfied, the algorithm still converged. We summarize the results obtained on an IBM system 370. Three iterations were performed:

Iteration	h	Δ	Cost	C.P.U. time (seconds)
1	.5	.05	2.947	.262
2	.25	.025	2.702	1.13
3	.2	.013	2.675	15.2

In the table, "Cost" refers to $V^{(m)}(0)$ for the mth iteration. The last column refers to the time required by the central processing unit to solve the problem. No further iterations were performed due to memory limitations (the program uses a $4 \times 26 \times 26 \times 26$ array as well as smaller ones). More complex problems would require quite a bit more CPU time -- of the order of minutes.

The table indicates that indeed the $V^{(m)}(0)$ appear to converge. Moreover it was noted that for iteration one, $u^{(1)} = \hat{u}$ at all the gridpoints. This was no longer true at a very few new grid points for iterations two and three. These results are encouraging, but it seems unlikely that the convergence of $u^{(m)}$ holds for more general control problems.

6 REFERENCES

1. Strook, D. W. and Varadhan, S. R. S. "Diffusion Processes with Continuous Coefficients, I. II," *Comm. Pure Appl. Math.*, **22**, 345-400, 479-530, (1969).

2. Bismut, J. M. "Théorie probabiliste du contrôle des diffusions", Memoirs A.M.S., **4**, Providence, R.I., (1976).

3. Bismut, J. M. "An Approximation Method in Optimal Stochastic Control", *SIAM J. Control Optimization*, **16**, 122-130, (1978).

4. Fleming, W. H. and Rishel, R. W. "Deterministic and Stochastic Optimal Control", Springer Verlag, New York, (1975).

5. Kushner, H. J. "Probanility Methods for Approximations in Stochastic Control and Elliptic Equations", Academic Press, New York, (1977).

6. Haussmann, U. G. "On the Stochastic Maximum Principle", *SIAM J. Control Optimisation*, **16**, 236-251, (1978).

7. Wong, E. "Representation of Martingales, Quadratic Variation and Applications", *SIAM J. Control*, **9**, 621-633, (1971).

4.2 ON STOCHASTIC VARIABLE-METRIC METHODS

L. C. W. Dixon

(The Numerical Optimisation Centre, The Hatfield Polytechnic, UK)

and

L. James

(Marconi - Elliott Avionics Systems Ltd., Boreham Wood, UK)

1 INTRODUCTION

This paper is motivated by the desire to be able to solve two distinct but closely related problems, viz the stochastic optimisation problem and the system parameter estimation problem. Both are shown to be soluble within the structure of the variable metric approach. Experimental evidence of the efficiency of the latter method is presented.

2 THE STOCHASTIC OPTIMISATION PROBLEM

Let us consider the problem of finding the value x* such that $F(x) > F(x^*)$ all $x \in S \in R^N$, when the only information available at x is an approximation

$$F_n(x) = F(x) + n. \tag{1}$$

For convenience we will assume that n is drawn from a normal distribution $N(0,\sigma)$ and will distinguish two sub-problems. In Problem IA the error n at a point x is repeatable, e.g. is due to rounding error in the computation of F(x), whilst in Problem IB a different value of n may be obtained at each measurement at x.

Many optimisation algorithms are iterative and generate a sequence of estimates $x^{(k)}$ of x* by selecting a direction $p^{(k)}$ and step size a_k at each iteration,

$$x^{(k+1)} = x^{(k)} - ap^{(k)}. \tag{2}$$

The convergence of such algorithms has been thoroughly examined for the case in which $\sigma = 0$ and the gradient vector

$$g_i = \partial F/\partial x_i \tag{3}$$

is assumed available at $x^{(k)}$, see for instance Wolfe [1].

The extension of Wolfe's theory to the case where the gradient at $x^{(k)}$ is estimated by central differences

$$\overline{g}_i = (F(x + he_i) - F(x - he_i))/2h \tag{4}$$

is considered in Dixon [2]. There the algorithm considered is basically one in which p is chosen to satisfy Condition I and a is then chosen to satisfy Conditions II and III;

$$\text{I} \quad \tilde{f}' \geq \epsilon_1 \|p\| \cdot \|\tilde{g}\| \tag{5}$$

$$\text{II} \quad F(x - ap) - F - a\tilde{f}' \geq \epsilon_3 \, a\tilde{f}' \tag{6}$$

$$\text{III} \quad F - F(x - ap) \geq \epsilon_4 \, a\tilde{f}' \tag{7}$$

where $\tilde{f}'$ is the linear estimate of the change in F due to a step $a = 1$ in the direction p (usually $\tilde{f}' = \tilde{g}^T p$).

In that paper it is shown that if such an algorithm is applied to a "well behaved function" then the following result holds.

Theorem 1

There exists ϵ_0^* (σ, h) such that if a terminating rule $\|\tilde{g}\| < \epsilon_0$ is included, and if $\epsilon_0 > \epsilon_0^*$ then the algorithm will terminate in a finite number of iterations.

Difficulties arise when it is necessary to obtain a solution more accurately than that guaranteed by ϵ_0^*. Practical experience has shown that this is frequently necessary when optimising on-line or using analogue computers. When the slope $\tilde{f}'$ is not correct there need be no value of a satisfying (6) and (7) and then depending on the strategy adopted deterministic algorithms may either terminate permaturely or continue indefinitely.

In this paper we extend these results for a particular class of algorithm in which p is calculated by solving

$$Gp = g \tag{8}$$

where G is an approximation to the Hessian of F at $x^{(k)}$. The problem of obtaining a suitable step size is treated first, whilst the discussion of the method for obtaining G is deferred to section 3.2. In these circumstances the step a = 1 should frequently be acceptable and we will introduce a simple line search based on the Armijo search for deterministic problems. In this search a is chosen from the integral powers of b, b^j (0 < b < 1, j pos or neg), the value chosen is the least (most negative) value of j such that $a = b^j$ satisfies (7) but $a = b^{j-1}$ fails (7). It can be shown that, with this particular line search, convergence follows even if Condition II is not satisfied, as both the Armijo Rule and Condition II ensure

$$a\|p\| > a_0 \, g^T p/\|p\| \tag{9}$$

for some value of a_0. In addition it follows from Kushner [3] that if (9) is satisfied then (7) can be weakened to

$$F - F(x - ap) \geq \epsilon_4 \, a\tilde{f}' - \beta_k \tag{10}$$

$$\text{provided } E \sum_k^{\infty} |\beta_k| < \infty. \tag{11}$$

We will now modify the Armijo line search by introducing a deadband $\pm\beta_k$, a concept first introduced by Farah [4].

2.1 Stochastic line search

For each function evaluation along the line we perform two tests

$$\text{TEST 1} \quad F - F(x - ap) \geq \epsilon_4 \, a\tilde{f}' + \beta_k \tag{12}$$

$$\text{TEST 2} \quad F - F(x - ap) < \epsilon_4 \, a\tilde{f}' - \beta_k \tag{13}$$

If TEST 1 is satisfied in any of the steps below the point $x - ap$ is accepted as $x^{(k+1)}$ and the next iteration started.

<u>Line Search Logic</u>

STEP 1 Evaluate F at $j = 0$, $a = 1$. If TEST 2 is satisfied go to step 2 else to step 3

STEP 2 Evaluate F at $j = 1$ and $j = 2$ and go to step 4

STEP 3 Evaluate F at $j = +1$ and $j = -1$ and go to step 4

STEP 4 As test 1 has not been satisfied, revise the estimates of F, $\tilde{g}$, p and G at $x^{(k)}$ using the values of the function just obtained (as described in section 2.2 below).

<u>Note 1</u> This involves at most 3 different values of j per line search and satisfies (9) provided

$$\beta_k > a_0 \left[\frac{g^T p}{\|p\|}\right]^2 \tag{14}$$

<u>Note 2</u> As β_k is being used to measure the uncertainty in F, $F(x - ap)$ and $\tilde{f}'$, whenever it is decided to improve the accuracy of these estimates by K repeated measurements then the value of β should be reset to β/K. This is of course not applicable to Problem IA.

<u>Note 3</u> This line search is presented as an alternative to that described in Section 5 of Dixon [2]. The Kalman Filter approach outlined in Dixon [5] and given below implies that the least squares line search outlined in [2] is no longer necessary.

2.2 Kalman Filter approach

In addition to being able to calculate a step size "a" we need to select a search direction p. At each point $x^{(k)}$ we have an estimate of the function value F, gradient g, and Hessian G of the objective function and calculate p from $Gp = +g$. To apply Kalman Filter theory to estimating new values of G and g, we form a state vector X

$$X = \begin{pmatrix} F \\ \tilde{g} \\ \underline{G} \end{pmatrix} \quad \text{where } \underline{G}_{n(i - i) + j} = G_{i,j} \tag{15}$$

which has a dynamic system given by

$$X^+ = \begin{pmatrix} F + \tilde{g}^T s + \frac{1}{2} s^T G s \\ \tilde{g} + Gs \\ \underline{G} \end{pmatrix} X + v \tag{16}$$

for a move $\Delta x = s$. This can be put in the form for Kalman Filter theory

$$X^+ = AX + v \text{ and } Y = CX + w, \tag{17}$$

where Y is the trial function evaluation at a point Δx from $x^{(k)}$ and CX is $F + \tilde{g}^T \Delta x + \frac{1}{2} \Delta x^T G \Delta x$. Kalman Filter theory then tells us how to update X after new line search measurements Y and each successful step s. This is outlined very briefly in [5].

To contrast this approach with the standard deterministic variable metric formula we shall consider the simpler system that arises if accurate measurements are made on the gradient vector g. This was first investigated by Thomas [6]. Then the state vector $X = \underline{G}$ and the measurement vector Y can be treated as $y = \Delta g$. The general Kalman Filter selects the new value of X^+ by solving

$$\min_{X^+} (X^+ - AX)^T \tilde{P}^{-1} (X^+ - AX) + (Y - CX^+) R^{-1} (Y - CX^+) \tag{18}$$

for appropriate matrices $\tilde{P}$ and R.

For this simplified problem this becomes

$$\min_{\underline{G}} \left[\Delta\underline{G}^T \tilde{P}^{-1} \Delta\underline{G}\right] + \left[(y - Gs)^T R^{-1} (y - Gs)\right] \tag{19}$$

a quadratic form whose solution depends on the relative magnitude of two error matrices.

The deterministic variable metric formula can be derived from similar problems i.e. Broyden's (1965) 1st Rank One Formula is obtained

$$\text{By } \underset{\underline{G}}{\text{Min}} \left[\Delta\underline{G}^T \, \Delta\underline{G} \right] \quad \text{s.t.} \quad Gs = y \tag{20}$$

which is the limit of the Kalman Filter algorithm as $R \to 0$ and $\tilde{P} = I$. It is more normal in variable metric theory to use the fact that $\Delta\underline{G}^T\Delta\underline{G} = \|\Delta G\|_F^2$, the Frobenius Norm of the matrix ΔG and to write this problem as $\underset{\Delta G}{\text{Min}} \|\Delta G\|_F^2$ s.t. $(G_O + \Delta G)s = y$. Similarly Powell's symmetric Rank 2 formula is obtained from

$$\begin{aligned} &\underset{\Delta G}{\text{Min}} \|\Delta G\|_F \quad \text{s.t.} \ \Delta Gs = y - G_O s \\ &\qquad\qquad \text{and} \quad \Delta G^T = \Delta G, \end{aligned} \tag{21}$$

the Davidon - Fletcher - Powell formula for $H = G^{-1}$ by inverting the solution of

$$\begin{aligned} &\underset{\Delta G}{\text{Min}} \|W^{-1}\Delta G W^{-1}\| \quad \text{s.t.} \quad \Delta Gs = y - G_O s \\ &\qquad\qquad\qquad\qquad \Delta G^T = \Delta G \end{aligned} \tag{22a}$$

where $W = (H^+)^{\frac{1}{2}}$,

and the more efficient Broyden-Fletcher-Shanno formula for H by solving

$$\begin{aligned} &\underset{\Delta H}{\text{Min}} \|W \, \Delta H W\| \quad \text{s.t.} \quad \Delta Hy = s - H_O y \\ &\qquad\qquad\qquad\qquad \Delta H^T = \Delta H. \end{aligned} \tag{22b}$$

It will be seen that the Kalman Filter approach modifies the standard variable metric formula by generating the optimal covariance matrix P, and also by relaxing the Quasi-Newton relationship $Gs = y$ which should not be satisfied exactly even on a deterministic (non-quadratic) objective function. When the function evaluations are subject to noise the need for such a relaxation is more apparent, but experiments reported by Thomas [6] and Spedicato [8] indicate that Kalman Filter generated algorithms of type (19) are very robust when applied to solving deterministic optimisation problems and the solution of non-linear simultaneous equations respectively. Analytic formula cannot be given readily for the Kalman Filter variable metric approach derived from (15) but it is simple to calculate the estimates iteration by iteration.

2.3 Constrained problems

It is now a fairly simple matter to extend this approach to constrained problems. There are a number of measures that have a minimum at the solution of a constrained problem.

$$\text{Min } F \qquad \text{s.t.} \quad e_i(x) = 0. \tag{23}$$

These include the augmented Lagrangian

$$A = F - \lambda^T e + \frac{1}{r} e^Y e \quad (\text{variables } x, \lambda) \tag{24}$$

the exact penalty function

$$E = F - e^T (N^T N)^{-1} N^T g + \frac{1}{r} e^T e \quad (\text{variables } x) \tag{25}$$

or the gradient of the Lagrangian

$$D = \| g - N\lambda \|^2 + \| e \|^2 \quad (\text{variables } x, \lambda) \tag{26}$$

The above theory can then be applied to any of these measures to solve the constrained problem (23).

Inequality constraints can be included both by special treatment of upper and lower bounds and/or by the inclusion of positive slack variables.

3 SYSTEM PARAMETER ESTIMATION

Let us now turn to the related problem of identifying the parameters of an input/output system

$$\begin{aligned} y_k + a_1 y_{k-1} + a_2 y_{k-2} &= b_1 u_{k-1} + b_2 u_{k-2} \\ \text{say } y_k + Ay_k &= Bu_k \end{aligned} \tag{24}$$

when the only available data are a string of pairs of input data and measured output m_k.

It is well known that the simple least squares approach of

$$\text{Min} \sum^{K} e_k^{\,2} \;;\quad e_k = m_k + Am_k - Bu_k \tag{25}$$

leads to biassed solutions unless all the errors are system errors of type v, Equ. (17). This is true whether the off line or on line approach is adopted.

Similarly it is well known that the complete maximum likelihood approach of defining a general additional filter

$$\xi_k + C\xi_k = e_k + De_k \tag{26}$$

and applying a general purpose optimisation routine to $\text{Min} \sum_k \xi_k^2$ is very expensive as at each function evaluation we need both to examine the complete data stream and solve a large set of simultaneous equations for ξ_k.

Stott and James [9] have shown that by adopting the GLS objective function introduced by Clarke [10], in which

$$\xi_k = e_k + De_k \tag{27}$$

both these operations can be avoided, and a very efficient algorithm results. In their approach the data are precondensed into two matrices which contain all the information needed to evaluate $\sum_{k=1}^{K} \xi_k^2$ for any values of $x^T = (a, b, d)$.

<u>Note</u> $\xi_k = e_k + De_k = (I + D)(I + A)m_k + (I + D)Bu_k$

$$\text{Now define } I + A^+ = (I + D)(I + A)\,;\; B^+ = (I + D)B \tag{28}$$

and $x^+ = (A^+, B^+)^T$

so that these filters are longer than the original filter.

Then the objective function is a quadratic function of A^+, B^+ and if the matrix Z is the matrix of past data that would be required to do a least squares fit to those parameters then

$$F_K = \frac{1}{K}\left[\Sigma\, m_k^2 + x^{+T} Z_K^T m + \frac{1}{2}\, x^{+T}(Z_K^T Z_K)x^+\right]. \tag{29}$$

Off line the data are preanalysed into the matrices $Z^T Z$ and $Z^T m$ and a variable metric method used to minimise F by first calculating x^+ and then substituting in this formula. The longer the data stream the more effective this method becomes compared with not precondensing the data.

On line we may treat the problem as that of finding the parameters $x^T = (a, b, d)$ which minimise the objective

$$F_\infty = \lim_{K\to\infty} \frac{1}{K}\left[\sum_{k=1}^{K} m_k^2 + x^{+T} Z_K^T m + \frac{1}{2} x^{+T}(Z_K^T Z_K)\, x^+\right] \qquad (30)$$

Then evaluations of F_K made as K increases are only approximate values of F_∞ so the problem could be treated as an example of problem 1. However in this case F_K can be calculated accurately and cheaply for different values of x and therefore a more efficient recursive estimation package can be proposed based on the variable metric principle and the condensed data approach.

The Recursive Variable Metric Algorithm developed by James as part of his work for an M.Phil consists of the following three stages:-

(1) Generate K_1 data points and form condensation matrices

$(Z_{K_1}^T \; Z_{K_1})$, $Z_{K_1}^T m$.

(Note K_1 ought to be sufficiently large for F_{K_1} to have a non singular Hessian.) Then set D = O and predict initial values of A, B and select an initial variable metric matrix H_O and set $k = K_1$.

(2) Perform k_i variable metric iterations on F_K calculating F_K and the gradient using (29). Obviously if the variable metric algorithm converges in less than k_i iterations we move to step 3.

(3) Accept k_r more data points, updating the condensation matrices and the Hessian matrix by the Sherman Morrison

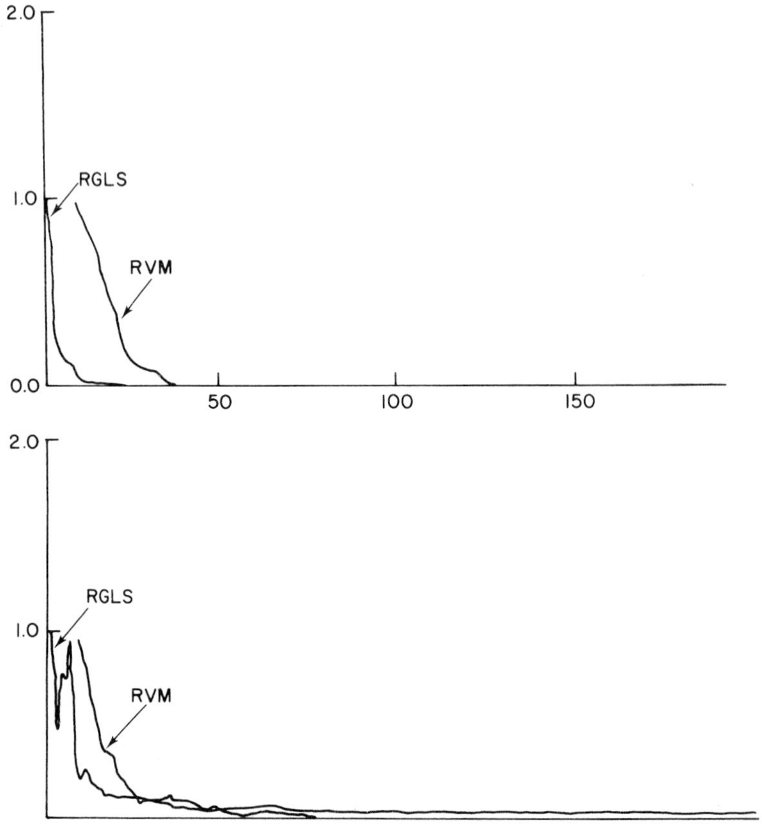

Fig. 1 Results for system 1; $a_1 = -1.5$, $a_2 = 0.7$, $b_1 = 1.0$, $b_2 = 0.5$

Upper graphs: Noise/signal ratio = 0.2
Lower graphs: Noise/signal ratio = 1.0

Methods compared R.G.L.S = recursive generalised least squares
R.V.M. = recursive variable metric

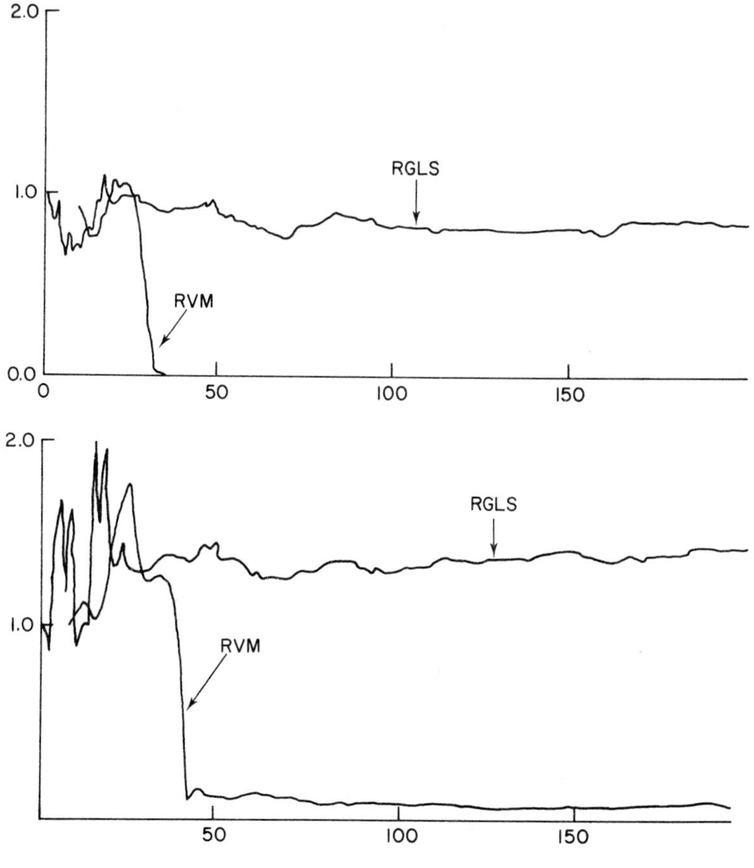

Fig. 2 Results for system 2; $a_1 = +1.6$, $a_2 = 0.64$, $b_1 = 1.0$, $b_2 = 0.5$

Upper graphs: Noise/signal ratio = 0.2
Lower graphs: Noise/signal ratio = 1.0

Methods compared R.G.L.S = recursive generalised least squares
R.V.M = recursive variable metric

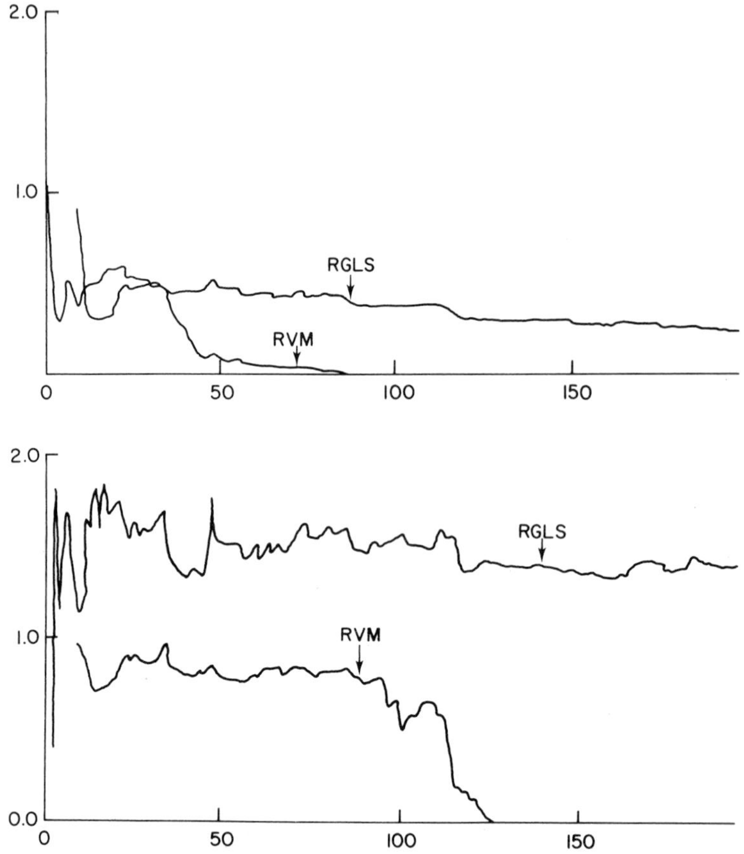

Fig. 3 Results for system 3; $a_1 = 0.1$, $a_2 = 0.02$, $b_1 = 1.0$, $b_2 = 0.5$

Upper graphs: Noise/signal ratio = 0.2
Lower graphs: Noise/signal ratio = 1.0

Methods compared R.G.L.S = recursive generalised least squares
R.V.M = recursive variable metric

formula and return to stage 2 with $K = K + k_r$.

In Figs. 1 - 3 the performance of RVM method is contrasted with that of the Hastings-James and Sage [11] recursive generalised least squares algorithm. Both of these approaches use the same performance index (29) and the difference in performance is due to different iteration strategies. In these results the RVM algorithm was run with $K_1 = 9$, $k_i = k_r = 1$, and the deterministic BFS variable metric algorithm (22b) was used at stage 2 with safeguards satisfying (5), (6) and (7), the vertical axis is the norm $\|x-x_c\|$ where x_c is the correct set of parameters that were used to generate the data. Details of the systems tested are indicated on the figures, it will be seen that RGLS only really worked on one of the three systems whilst RVM satisfactorily located the correct parameter values efficiently both for a noise/signal ratio of 0.2 and 1.0. We claim this indicates that it is a very powerful method for finding on line estimates of system parameters rapdily and accurately.

4 CONCLUSIONS

A new Kalman Filter/Variable metric approach to the general stochastic optimisation problem has been described which allows for the updating of the estimated values of F, g and G for all function evaluations. This approach has the advantages that the estimates optimally take into account uncertainty in the evaluations.

A new recursive identification algorithm is also described based on the variable metric principle and computational evidence is given indicating the efficiency of the algorithm.

5 ACKNOWLEDGEMENTS

The authors wish to acknowledge the contribution made to these developments by Dr. R. Coleman and Dr. S. Stott, during many useful discussions. The second author, L. James, also wishes to acknowledge the financial support of the S.R.C. during the period this work was undertaken.

6 REFERENCES

1. Wolfe, P. "Convergence conditions for ascent methods", *SIAM Review*, **11**, 226-235, (1969) and *SIAM Review*, **13**, 185-8, (1971).

2. Dixon, L. C. W. "Paper 4" in Oettli and Ritter, editors, "Optimisation and Operations Research - Oberwolfach 1975", Springer-Verlag, Berlin, 35-55, (1976).

3. Kushner, H. "Stochastic Approximation Algorithms for local optimisation of functions ...", *IEEE Trans AC,* **AC 17**, 5, 646-54, (1972).

4. Farah, J. L. "Unconstrained Local optimisation of stochastic functions", Imperial College, London, Report No. 76/37.

5. Dixon, L. C. W. "On on-line Variable Metric Methods", N.O.C. T.R. 79, presented at "IX International Symposium of Mathematical Programming", Budapest, (1976).

6. Thomas, S. W. "Sequence estimation techniques for Quasi Newton Algorithms", Cornell University T.R. 75/227.

7. Broyden, C. "A class of methods for solving nonlinear equations", *Maths. of Computation,* **19**, 577-593, (1965).

8. Spedicato, E. and Greenstadt, J. "On some classes of variationallly derived Quasi Newton methods for systems of nonlinear algebraic equations", unpublished, (1978).

9. Stott, S. and James, L. Electronic Letters, **11**, 14, 311, (1975).

10. Clarke, D. W. "Generalised - Least Squares estimation of the parameters of a dynamical model", IFAC symp. "Identification in Automatic Control System", Prague Paper 3.17, (1967).

11. Hastings-James, R. and Sage, M. W. "Recursive Generalised least squares procedure for on line identification of process parameters", *Proc. IEE,* **116**, 2057-62, (1969).

4.3 ON THE STABILITY AND CONVERGENCE OF SELF-TUNING CONTROLLERS

P. J. Gawthrop

(Department of Engineering Science, Oxford University, UK)

1 INTRODUCTION

Convergence properties of self-tuning algorithms [2,3,7,24] have been considered by Ljung [14,15] and Kushner [9] using rather general theorems, to some extent specifically developed for such problems. Stability methods for model-reference algorithms have also been used for some time [10,17]. Recently, the close relationship between the two types of algorithm has been noted [5,16].

This paper combines input-output stability methods [20-23,27] with the martingale convergence theorem [4] to give a type of convergence of the self-tuning controller w.p.1. The method is based on some model-reference ideas of Landau [11,12], and thus is another example of the close links between the model-reference and self-tuning control and estimation procedures. The results are, however, incomplete in the sense that the theorems depend on non-trivial properties of the system input and output sequences.

2 LEAST-SQUARES PREDICTION AND CONTROL

A canonical representation [1] of single-input, single-output, finite dimensional, randomly disturbed, discrete-time systems is given by:

$$A(q^{-1})y(t) = q^{-k}B(q^{-1})u(t) + C(q^{-1})\xi(t) \qquad (2.1)$$

where: $y(t)$ is the system output (assumed available for measurement),
$u(t)$ is the control input to the system,
$\xi(t)$ is a stationary innovations process,
t is the (integer) time index,
k is the (integer) input time delay,
$A(q^{-1}) = a_0 + a_1q^{-1} + \ldots + a_nq^{-n}$

q^{-1} is the backwards shift operator,
B, C and all polynomials used hereafter are of the same form.

Without loss of generality, it is assumed that:

$$a_o = 1,\ b_o \neq 0,\ c_o = 1 \tag{2.2}$$

Further, it follows from the representation theorem [1], that the polynomial $C(q^{-1})$ is such that:

$$C(q^{-1}) = 0 \quad \Rightarrow |q| \leqslant 1 \tag{2.3}$$

Define the sequence $\phi(t)$, obtained from the sequences $y(t)$ and $u(t)$ by:

$$\phi(t) = \frac{P_1(q^{-1})}{P_2(q^{-1})} + \frac{q^{-k}Q_1(q^{-1})}{Q_2(q^{-1})} \tag{2.4}$$

It is a standard result [19,25] that the least-squares prediction of $\phi(t+k)$ in terms of $y(i)$ and $u(i)$ for $-\infty < i < t$ is given by

$$C(q^{-1})\phi^*(t+k|t) = F(q^{-1})y(t) + G(q^{-1})u(t) \tag{2.5}$$

F and G are obtained from A, B and C using standard manipulations.

The prediction error $e(t+k)$ is a moving average process of order k:

$$\begin{aligned} e(t+k) &\triangleq \phi(t+k) - \phi^*(t+k|t) \\ &= \sum_{i=0}^{k-1} \gamma_1 \zeta(t+k-i) \end{aligned} \tag{2.6}$$

Define the vectors:

$$\theta = \{f_o, f_1, \ldots, f_n;\ g_o, g_1, \ldots, g_n\}^T \tag{2.7}$$

$$x(t) = \{y(t), y(t-1), \ldots, y(t-n); u(t), u(t-1), \ldots, u(t-n)\}^T \tag{2.8}$$

equation 2.6 may be written as:

$$C(q^{-1})\phi^*(t+k|t) = x^T(t)\theta \qquad (2.9)$$

The minimum variance control law [1,3,7] associated with a particular form of ϕ is, for the purposes of this paper, defined as:

$$x^T(t)\theta = 0 \qquad (2.10)$$

From the various definitions, it follows that it is always possible to manipulate u(t) to satisfy this equation.

The stability of this control law in closed-loop with the system 2.1 is not guaranteed; this is a side issue as far as this paper is concerned.

Extensions to include setpoints and known disturbances are given elsewhere [3,7]; they do not affect the theory here.

3 SELF-TUNING

If the coefficients of A, B and C (equation (2.1) are unknown, θ (equation 2.7) is also unknown and thus the control law 2.10 is not realisable.

Self-tuning methods [2,3,7,24] have been developed to cope with systems where θ is unknown but constant. The basic idea is to estimate θ using linear least-squares [6] (as if C did not appear in equation 2.9), and use the resultant estimate in equation 2.10 as if it were θ.

That is, an estimate of θ is generated by:

$$\hat{e}(t) = \phi(t) - x^T(t-k)\hat{\theta}(t-1) \qquad (3.1)$$

$$\hat{\theta}(t) = \hat{\theta}(t-1) + S^{-1}(t-k)x(t-k)\hat{e}(t) \qquad (3.2)$$

$$S(t) = S(t-1) + x(t)x^T(t) \qquad (3.3)$$

with initial conditions:

$$\theta(0) = \theta_0 \qquad (3.4)$$

$$S(0) = S_0, \text{ a positive definite matrix.} \qquad (3.5)$$

The control law is given by:

$$x^T(t)\hat{\theta}(t) = 0 \tag{3.6}$$

Remarks.

1. In practice, equation 3.3 is not used directly, but rather S^{-1} is directly updated.

2. Equation 3.6 is realisable if $\hat{g}_0 = 0$; the analysis assumes that this is true.

3. Exponential forgetting [6] is sometimes used in the estimation algorithm; however as its use precludes convergence it is not considered here. Its use does not affect the stability result.

The purpose of the self-tuner is to drive ϕ^* to zero; the problem addressed here is to determine under what conditions this aim is achieved.

4 METHOD OF ANALYSIS

The approach used here owes its outline to Landau [11,12], but the details are rather different, and the conclusions somewhat stronger. The three main steps are as follows:

1. For each realisation of the sequences involved, write the adaptive algorithm as a feedback system.

2. Show that the feedback system is the interconnexion of positive (or passive) [20-23,27] subsystems, and hence deduce input-output stability.

3. Combine step 2 with the martingale convergence theorem to deduce convergence w.p.1.

The results presented here are unfortunately defective in that step 2 has not been accomplished for k>1, and step 3 gives the result that:

$$\underset{N\to\infty}{\mathrm{Lt}}\ \frac{1}{N}\sum_{t=0}^{N}\left[\phi^*(t+k|t)\right]^2 \Rightarrow 0 \quad \text{w.p.1} \tag{4.1}$$

Step 1 proceeds as follows. Define:

$$\tilde{\theta}(t) = \hat{\theta}(t) - \theta \tag{4.2}$$

$$\tilde{e}(t) = \hat{e}(t) - e(t)$$
$$= \phi^*(t|t-k) - x^T(t-k)\hat{\theta}(t-1) \quad (4.3)$$

Equations 3.1 and 3.2 become:

$$\tilde{\theta}(t) = \tilde{\theta}(t-1) + S^{-1}(t-k)x(t-k)\left[\tilde{e}(t)+e(t)\right] \quad (4.4)$$

$$v(t) = x^T(t-k)\tilde{\theta}(t-1) \quad (4.5)$$

$$\tilde{e}(t) = \frac{1}{C} x^T(t-k)\theta - x^T(t-k)\hat{\theta}(t-1)$$

$$= -\frac{1}{C}\left[v(t) + (C-1)x(t-k)(\hat{\theta}(t-1)-\hat{\theta}(t-k))\right] \quad (4.6)$$

$$= -\frac{1}{C}\left[v(t) + w(t)\right]$$

$$\text{where } w(t) = (C-1)x^T(t-k)\sum_{i=1}^{k-1}\left[\hat{\theta}(t-1) - \hat{\theta}(t-i-1)\right] \quad (4.7)$$

$$= (C-1)\sum_{i=1}^{k-1} x^T(t-k)S^{-1}(t-i-1)x(t-i-1)e(t-i)$$

The self-tuning algorithm may thus be written as the interconnexion of three systems:

Σ_1: The time-varying system of equations 4.4-4.5.

Σ_2: The transfer function 1/C defined in equation 4.6.

Σ_3: The time-varying system given by equation 4.7.

Note that if k=1, w(t)=0 and Σ_3 may be omitted.

5 STABILITY

Having written the self-tuning algorithm as an interconnected system, it is natural to analyse its stability using positivity concepts [18,20-23,27]. As the quantity of interest is a scalar output (ϕ^*) rather than a state ($\hat{\theta}$) input-output methods are chosen.

The basic problem is to manipulate equations 4.2-4.7 (without the noise term e) into such a form that the resultant subsystems are positive, the interconnexions are neutral, and the resultant system is well-posed [20-23,26,27].

It has not yet been possible to include Σ_3 in the analysis and hence from now on k is assumed to be unity.

The positivity of the subsystems are examined in the following lemmas.

Lemma 5.1 (Positivity of Σ_1)

Consider the system Σ_1' with output $v'(t)$ and input $e'(t)$ obtained by adding negative scalar feedback f, and positive scalar direct link d, to Σ_1. That is Σ_1' is given by equations 4.4 and 4.5 but with v and e replaced by:

$$v'(t) = v(t) + de(t) \tag{5.1}$$

$$e'(t) = e(t) + fv(t) \tag{5.2}$$

Then Σ_1' is strictly positive in the sense that the dissipation rate [22] is given by:

$$\delta(t) = v'(t)e'(t) - r^2 e > 0; \; r^2 > 0 \tag{5.3}$$

if the following conditions hold.

1. $\sigma(t) \triangleq x^T(t)S^{-1}(t)x(t) < \sigma_0 < 1$ for all $t>0$ (5.4)

2. $f = \frac{1}{2}(\sigma_0 + 1)$ (5.5)

3. $d = \frac{1}{2} \dfrac{\sigma_0}{1-f^2}$ (5.6)

Remarks.

1. It can be shown [8] from the definition of σ in equation 5.4 that $0<\sigma<1$; hence the theorem requires some independent condition on the sequence σ. The importance of σ is one of the main results, but the analysis of this sequence is beyond the scope of this paper.

2. Inclusion of a forgetting factor [6] does not affect this result; indeed it increases the dissipation rate.

Proof (outline).

Σ'_1 is given by:

$$\theta'(t) = \left[1 - fS^{-1}(t-k)x(t-k)x^T(t-k)\right]\theta'(t-1) + S^{-1}(t-k)x(t-k)e'(t) \qquad (5.7)$$

$$v'(t) = x^T(t-k)\theta'(t-1) + de'(t) \qquad (5.8)$$

Define the "storage function" [22]:

$$\theta'^T(t)S(t-k)\theta'(t) \geqslant 0 \qquad (5.9)$$

and the "supply rate" [22]: $v'(t)e'(t)$ (5.10)

It follows that the "dissipation rate" [22] is given by:

$$\delta(t) = v'(t)e'(t) - \theta'^T(t)S(t-k)\theta'(t) + \theta'^T(t-1)S(t-k-1)\theta'(t-1) \qquad (5.11)$$

After some manipulations (similar to those of section 11, reference [18]), and using equation 3.3,

$$\delta(t) = (p(t)x^T(t-k)\theta'(t-1) + g(t)e'(t))^2 + r^2(t)e'^2(t) \qquad (5.12)$$

where: $p^2(t) = 2f - 1 - f^2\sigma(t)$ (5.13)

$$p(t)g(t) = f\sigma(t) \qquad (5.14)$$

$$g^2(t) = 2d - \sigma(t) - r^2 \qquad (5.15)$$

Substituting conditions 1-3 into equations (5.13-5.15) it follows that p, g and r are real and $r > 0$. Hence equation 5.12 implies the result.

Lemma 5.2

Consider the system Σ_2', with output $e'(t)$ and input $v'(t)$ obtained by adding a negative direct-link term f, and a positive feedback term d around Σ_2. That is as given by equation 4.6 but with input and output given by equations 5.1 and 5.2. Then Σ_2' is positive if the transfer function:

$$\frac{\frac{1}{C} - f}{1-d(\frac{1}{C} - f)} \tag{5.16}$$

is positive real.

Remark.

The limiting case as $\sigma \longrightarrow 0$, $f \longrightarrow 1/2$, $d \longrightarrow 0$, gives the condition:

$$\frac{1}{C} - \frac{1}{2} \tag{5.17}$$

is positive real. This is a condition obtained by Ljung [15].

Proof.

The transfer function (5.16) gives the input-output relation of Σ_2'. The fact that a positive real transfer function is a positive system is a standard result [18,21].

To deduce stability from positivity, it is necessary to show that the feedback system in question is well-posed [21,26]. Well-posedness guarantees a unique solution to the system (for given inputs and initial conditions), and its insensitivity to certain modelling errors. The required result is contained in the following lemma.

Lemma 5.3 (Well-posedness)

From theorem 4.1 of reference [21], well-posedness of linear feedback systems follows if the product of the direct link terms of the operators is less than unity.

Σ_1 contains no direct link term, and thus the unmodified system is well-posed.

The product of the direct link terms of Σ'_1 and Σ'_2 is given by:

$$\Delta = d\,\frac{C(0)-f}{1-d(C(0)-f)} \qquad (5.18)$$

From equations 2.2, 5.5 and 5.6:

$$\Delta = \frac{\sigma_0}{2-\sigma_0} \qquad (5.19)$$

From expression 5.4, $\sigma_0 < 1$; this implies the required result.

Lemmas 5.1 and 5.2 hold for all k>0; however as no results have been shown for Σ_3, the following theorem is restricted to the case where k=1.

Theorem 5.1

If the system time delay k=1, then the feedback system, equations 4.4-4.6, is input-output stable [21]:

$$\sum_1^N (\text{output})^2 \leqslant \text{constant}.\sum_1^N (\text{input})^2 \qquad (5.20)$$

if conditions 5.4-5.7 and 5.16 are satisfied.

Proof.

From corollary 4.3.4, reference [21], the results of lemmas 5.1-5.3 imply this result.

6 CONVERGENCE

Up to this point, processes such as x(t) have been rather loosely provided with the single argument t. In this section it is necessary to be rather more precise. Each realisation of the random sequence $\xi(t)$ will be indexed by ω, and thus for a given set of initial conditions, processes such as e(t), x(t), S(t), etc. will be written: $e(t;\omega)$, $x(t;\omega)$, $S(t;\omega)$ etc. This extra notation will be suppressed when the meaning is clear.

The fact that input-output stability of the self-tuner implies a form of convergence w.p.1 is exposed in the following theorem.

Theorem 6.1

Given the following conditions:

1. The conditions of theorem 5.1 are satisfied.

2. The sequence x is uniformly bounded in the mean square

then the self tuning controller (equations 3.1-3.6) operating on system 2.1, converges in the sense that:

$$\operatorname*{Lt}_{N\to\infty} \frac{1}{N} \sum_{t=0}^{N} (\phi^*(t+k|t;\omega))^2 = 0 \quad \text{w.p.1} \tag{6.1}$$

Proof.

Consider the sequence of vectors:

$$z(t;\omega) = S(i-1)x(i-1)\xi(i) \tag{6.2}$$

This sequence has been studied previously [8], as has a related scalar version [13]; and has been shown to be a vector martingale of bounded variance. Hence, from the martingale convergence theorem [4], it follows that:

$$\operatorname*{Lt}_{t\to\infty} z(t;\omega) = z^{\infty}(\omega) \quad \text{w.p.1} \tag{6.3}$$

$$\text{where: } ||z^{\infty}(\omega)|| < \infty \tag{6.4}$$

As Σ_1 (equation 4.4,4.5) is linear and contains the integrator term $1/(1-q^{-1})$; for each realisation indexed by ω, it may be rewritten as:

$$\theta''(t) = \theta''(t-1) + S^{-1}(t-k)x(t-k)\tilde{e}(t) \tag{6.5}$$

$$v(t) = x^T(t-k)\theta''(t-1) + v''(t) \tag{6.6}$$

$$v''(t) = x^T(t-k)\left[z(t) - z^{\infty}\right] \tag{6.7}$$

$$\theta''(0) = \hat{\theta}(0) - \theta + z^{\infty} \tag{6.8}$$

The input $e(t;\omega)$ has thus been replaced by the input $v''(t;\omega)$ and the system Σ_1 by Σ_1'' without affecting the stability results of previous sections or the process $\tilde{e}(t;\omega)$. Further, as k=1, equations 4.3 and 3.6 give that:

$$\tilde{e}(t;\omega) = \phi^*(t|t-k;\omega) \tag{6.9}$$

Hence, given that the system comprising Σ_1'' and Σ_2 is input-output stable:

$$\sum_{t=0}^{N} (\phi^*(t|t-k;\omega))^2 < M \sum_{t=0}^{N} (v''(t;\omega))^2 + M' \tag{6.10}$$

where M is a positive constant.

Dividing by N:

$$\frac{1}{N} \sum_{t=0}^{N} (\phi^*(t|t-k;\omega))^2 < \frac{M}{N} \sum_{t=0}^{N} (v''(t;\omega)) + \frac{M'}{N} \tag{6.11}$$

Equations 6.7 and 6.3, combined with condition 2, imply that:

$$\underset{N\to\infty}{\text{Lt}} \frac{1}{N} \sum_{t=0}^{N} v''^2(t;\omega) = 0 \quad \text{w.p.1} \tag{6.12}$$

It may be verified that 6.12 implies that the right-hand side of inequality 6.11 tends to zero as $N \longrightarrow \infty$ and thus the result is proved.

Remark.

This theorem may be extended to cases where k>1 by considering the k martingales generated by the moving-average process e(t).

7 CONCLUSION

Conditions for a type of convergence w.p.1 of the self-tuning controller have been derived for the particular case of unity input delay (k=1). These conditions involve two quantities: the polynomial C (equation 2.1) and $\sigma(t)$ (equation 5.4).

The quantity σ provides a link between the properties of the actual control loop, reflected in values of x(t), and the convergence properties of the adaptive mechanism. Roughly speaking, σ close to zero indicates "steady-state" conditions in system 2.1; it also gives the weakest condition (expression 5.17) on the polynomial C. The complete analysis of the self-tuning algorithm would involve examination of the properties of σ.

This method is currently being extended to include: input delay k>1, systems of unknown order, and non-linear forms of equation 2.1.

8 REFERENCES

1. Åström, K. J. "Introduction to stochastic control theory", Academic Press, (1970).

2. Åström, K. J. and Wittenmark, B. "On self-tuning regulators", *Automatica,* **9**, 185-199, (1973).

3. Clarke, D. W. and Gawthrop, P. J. "Self-tuning controller", *Proc. IEE,* **122**, 9, 929-934, (1975).

4. Doob, J. L. "Stochastic processes", Wiley, (1953).

5. Egardt, B. "A unified approach to model-reference adaptive systems and self-tuning regulators", Rept. TFRT 7413, Div. automatic control, Lund Institute of Technology, (1978).

6. Eykhoff, P. "System identification", Wiley, (1974).

7. Gawthrop, P. J. "Some interpretations of the self-tuning controller", *Proc. IEE,* **124**, 10, 889-894, (1977).

8. Goodwin, G. C. and Payne, R. L. "Dynamic system identification: Experiment design and data analysis", Academic Press, (1977).

9. Kushner, H. J. "Convergence of recursive adaptive and identification procedures via weak convergence theory", *IEEE Trans. autom. control,* **AC–22**, 6, 921-930, (1977).

10. Landau, I. D. "A survey of model-reference adaptive techniques - theory and applications", *Automatica,* **10**, 353-379, (1974).

11. Landau, I. D. "Unbiased recursive identification using model-reference adaptive techniques", *IEEE Trans. autom. control,* **AC–21**, 194-202, (1976).

12. Landau, I. D. "An addendum to "unbiased recursive identification using model reference adaptive techniques"", *IEEE Trans. autom. control,* **AC–23**, 97-99, (1978).

13. Ljung, L. "Consistency of the least-squares identification method", *IEEE Trans. autom. control,* **AC–21**, 5, 779-781, (1976)

14. Ljung, L. "Analysis of recursive stochastic algorithms", *IEEE Trans. autom. control,* **AC–22**, 4, 551-575, (1977).

15. Ljung, L. "On positive real transfer functions and the convergence of some recursive schemes", *IEEE Trans. autom. control,* **AC–22**, 4, 539-550, (1977).

16. Ljung, L. and Landau, I. D. "Model-reference adaptive systems and self-tuning regulators - some connections", IFAC Conference, Helsinki, (1978).

17. Parks, P. C. "Lyapunov redesign of model-reference adaptive control systems", *IEEE Trans.,* **AC–11**, 362-367, (1966).

18. Popov, V. M. "Hyperstability of control systems", Springer-Verlag, (1973).

19. Whittle, P. "Prediction and regulation by linear least-square methods", E.U.P., (1963).

20. Willems, J. C. "Stability, instability, invertibility and causality", *SIAM J. Control,* **7**, 4, 645-671, (1969).

21. Willems, J. C. "The analysis of feedback systems", Research monograph No. 62, MIT press, (1971).

22. Willems, J. C. "Dissipative dynamic systems", I "General theory", II "Linear systems with quadratic supply rates", *Arch. Rational mech. anal.,* **45**, 5, 321-395, (1972).

23. Willems, J. C. "Mechanisms for stability and instability in feedback systems", *Proc. IEEE,* **64**, 1, 24-35, (1976).

24. Wittenmark, B. "A self-tuning regulator", Rept. 7311, Div. automatic control, Lund Institute of Technology, (1973).

25. Yaglom, A. M. "An introduction to the theory of stationary random functions", Translator R. A. Silverman, Dover (1973).

26. Zames, G. "Realizability conditions for nonlinear feed-back systems", *IEEE Trans. circuit theory,* **11**, 186-194, (1964).

27. Zames, G. "On the input-output stability of time-varying nonlinear feedback systems", Part I "Conditions derived using concepts of loop-gain, conicity and positivity", *IEEE Trans. autom. control,* **11**, 228-239, (1966).

CHAPTER 5

INFORMATION

5.1 INFORMATIONAL ASPECTS OF STOCHASTIC CONTROL

H. S. Witsenhausen

(Bell Laboratories, New Jersey, USA)

1 INTRODUCTION

The crucial importance of the information structure in stochastic control problems is particularly evident in the non-classical case. After reviewing a few general concepts, the special case of the two-agent signalling problem is considered below. This is one of the simplest non-classical situations and is also related to the (Shannon) Information Theory. There has been hope of obtaining good bounds for this problem from the so-called "generalized rate-distortion theory". Also, in the case where all variables range over finite sets, a direct solution appears to be within computable reach. The inherent difficulty of non-classical problems is illustrated by the arguments that will be given to show that both of the above expectations are overly optimistic.

2 NON-CLASSICAL INFORMATION PATTERNS

In a situation where a finite number of decisions are to be made, each of these decisions is viewed as being made by an "agent" (usually in fact a computer or device). For each agent, there is a prescription of how the decision is to depend on the data available to the agent. This is called the "control law" or "decision function" for that agent. The information pattern is the specification of the information available to the agents. For any given agent, the data he receives can only depend on (i) random variables "chosen by nature", which can be considered functions of a generic element ω of a probability space and (ii) decisions taken by other agents.

Note that when the set of all control laws, (this may be called the <u>design</u>) has been selected, then (but only then) the decisions become random variables, that is, functions of ω. For this reason the data of type (ii) above are more difficult to exploit as compared to the case in which only type

(i) data exist (the "static team" case). This situation is mitigated in that there may be reductions to static form of an initially dynamic problem; however, these reductions may shift difficulties to another aspect of the design optimization.

The problem is "sequential" [7] when the agents can be ordered in such a way that the data available to any one agent depend only on ω and on the decisions of preceding agents. It has long been known in game theory that non-sequential problems arise in the real world. The same is true in the one-player game known as stochastic control. (For instance, two branches of the government can influence the world and each other without being fully aware of what their colleagues are doing, and without a fixed predetermined time-interleaving of their various decisions.)

Restricting our attention to sequential problems, several subcategories can be distinguished (see [7] for more details), such as the static problems (which are sequential in a trivial way). Most intensively studied have been the problems with "classical" information pattern. These are the ones where an ordering of the agents exists which satisfies not only the sequential condition previously stated but in addition the condition that each agent knows (i.e. is in a position to reconstruct) what the preceding agent knew. This assumption amounts to saying that all control is exerted from a central location which has perfect memory.

As there is considerable practical interest in situations where centralization of decision making is not complete, and also, to a lesser extent, in situations where memory is limited, the study of non-classical, dynamic sequential problems is currently one of the most pertinent research frontiers in the decision-making art. However, even the simplest problems of this type are of astounding resilience as we will exemplify below.

3 THE SIGNALLING PROBLEM

The concept of signalling, in general situations, is somewhat vague and not necessarily helpful. However there is an extremely simple-seeming type of problem which catches the signalling concept in its purest form. This form is inherited from the Shannon information theory, with the important difference that no repetitions (hence no block coding and no delay) are considered.

There are two agents. Nature picks a random ω which has two independent components $\omega = (\omega_1, \omega_2)$. The first agent observes ω_1 and then makes decision u. The second agent observes y, a function of u and ω_2 after which he makes decision v. The cost whose expectation is to be minimized is a function J of ω_1 and v (plus perhaps a function of u). Thus agent two needs information about ω_1 and can only get it through the "signal" u chosen by agent one. The function of u and ω_2 characterizes a "channel".

Clearly the problem is sequential, dynamic and non-classical. It is possibly the oldest such, but is now the subject of renewed attention, most notably by Kastner, Ho and Wong [4,5]. The scalar linear-quadratic-Gaussian case has a simple solution (see [7]): the easily determined best linear design, meets the Shannon rate-distortion bound [1], and is therefore optimal. This was one of the facts on which was based the assertion in [6] that the counterexample to linearity of the solution, given there, (which is not a signalling problem as defined above) was the simplest possible such. It was already clear then that rate-distortion theory had limited value in stochastic control because (i) the signalling situation is a very special one and (ii) the Shannon bound, while tight on a block coding basis, can only be tight on a one-shot basis if a test channel extremizing the rate-distortion calculation (usually there is only one such) is equivalent to a cascade of channels one of which is a copy of the given channel.(*) However since then, a generalized rate-distortion theory has been developed [2,8,9] which gives more flexibility in bounding one-shot problems as suggested in [7]. This possibility is examined more closely in the next sections.

4 THE GENERALIZED DATA-PROCESSING THEOREM

The key to the derivation of rate distortion bounds is the data-processing theorem. As the treatment in the only readily accessible reference [8] is somewhat confusing, a simple proof of this crucial result is given here.

(*) Thus one can construct any number of source, channel, criterion triples for which the rate-distortion bound is sharp on a single use basis.

Let q be a real function defined and convex on $(0,\infty)$ with $q(x) = o(x)$ as $x\to\infty$. For a pair of discrete random variables X,Y with joint probability distribution p_{xy}, one defines the generalized mutual information

$$I^q(X;Y) = \sum_{x,y} p_{xy}\, q\left(\frac{p_x p_y}{p_{xy}}\right)$$

where p_x and p_y denote the marginal probabilities

$$p_x = \sum_y p_{xy} \text{ and } p_y = \sum_x p_{xy}.$$

This definition only applies, at first, to pairs for which all p_{xy} are positive. However, the condition $q(x) = o(x)$ implies that the extension by continuity to the case of $p_{xy} \geq 0$ is possible and is accomplished by dropping terms with $p_{xy} = 0$ from the summation in the definition of I^q.

The generalized data-processing theorem states the following: If the discrete random variables X,Y,Z form a Markov chain (that is, if X and Z are conditionally independent, given Y) then $I^q(X;Z) \leq \min(I^q(X;Y);\ I^q(Y;Z))$.

[In the Shannon theory, $q(x) = -\log x$ and the theorem can be found in all standard texts.]

As the Markov property is symmetric with respect to sequence reversal (Z,Y,X is Markov if, and only if, X,Y,Z is) it is enough to prove that $I^q(X;Z) \leq I^q(X;Y)$, and, by continuity of I^q, one may assume that all joint probabilities p_{xyz} are strictly positive. Then (with the usual notation in which $p_{xy} = \Pr\{X=x,\ Y=y\}$, etc..)

$$I^q(X;Y) = \sum_{xy} p_{xy}\, q\left(\frac{p_x p_y}{p_{xy}}\right)$$

$$= \sum_{xyz} p_{xyz}\, q\left(\frac{p_x p_y}{p_{xy}}\right)$$

$$= \sum_{xz} p_{xz} \sum_{y} \frac{p_{xyz}}{p_{xz}} q\left(\frac{p_x p_y}{p_{xy}}\right)$$

$$\geq \sum_{xz} p_{xz} q\left(\frac{p_x}{p_{xz}} \sum_{y} \frac{p_{xyz} p_y}{p_{xy}}\right)$$

using successively: the definition; $p_{xy} = \sum_z p_{xyz}$;

$\sum_y \frac{p_{xyz}}{p_{xz}} = 1$ and Jensen's inequality for the convex function q.

Using the Markov property, one has

$$\sum_y \frac{p_{xyz} p_y}{p_{xy}} = \sum_y p_y\, p_{z|xy} = \sum_y p_y\, p_{z|y} = \sum_y p_{yz} = p_z .$$

Hence,

$$I^q(X;Y) \geq \sum_{xz} p_{xz} q\left(\frac{p_x p_z}{p_{xz}}\right) = I^q(X;Z),$$

completing the proof.

5 GENERALIZED RATE-DISTORTION BOUNDS

Lower bounds on the minimum expected cost achievable in a signalling problem are obtained in two steps. For simplicity, we will assume that the cost is a function of only ω_1 and v. Dependence on u can be taken care of, but this might only obscure the arguments.

The first step is to determine the channel capacity. This is the maximum, over all probability distributions on the set of possible values of the variable u, of the mutual information $I^q(u;y)$ between this variable and the output variable y observed by agent 2. Let C^q be the number so obtained.

The second step is to consider all test channels from ω_1 to v (that is all stochastic matrices of transition probabilities $p_{v|\omega_1}$) such that $I^q(\omega_1;v) \leq C^q$ and to minimize the resulting expected cost.

That is one minimizes

$$\sum_{\omega_1,v} p_{\omega_1} p_{v|\omega_1} J(\omega_1,v)$$

over all $p_{v|\omega_1} \geq 0$ satisfying $\sum_v p_{v|\omega_1} = 1$ and

$$\sum_{\omega_1,v} p_{\omega_1} p_{v|\omega_1} q\left(\frac{p_{\omega_1}}{p_{v|\omega_1}}\right) \leq C^q$$

If the value of the minimum is M^q then this is a lower bound on the minimum cost because for any design, the cascade of channels consisting of (i) the decision function of agent one, (ii) the actual channel and (iii) the decision function of agent two, make up a composite channel from ω_1 to v. All the channels that occur this way for all pairs of decision functions are contained in the set over which the minimization of step two is taken. For indeed, the sequence (ω_1,u,y,v) is always a Markov chain so that, for any design, by the data processing theorem, one has

$$I^q(\omega_1,v) \leq I^q(u,y) \leq C^q.$$

This inclusion implies that $M^q \leq E\{J(\omega_1,v)\}$ for any design.

Since for any convex q with $q(x) = o(x)$ one obtains a different bound, there is hope that reasonably close bounds can be obtained, perhaps even the exact answer. In fact, if one goes to a further "vector" generalization of rate distortion theory as in [8], it is shown there that the bound can then always be made sharp.(†) However, this does not apply to the scalar q considered in [9] and above.

(†) In the sense that the supremum of the bounds over all choices of the auxiliary convex function of n variables is the exact minimum. However, the proof is not constructive.

6 LACK OF CONVEXITY

The principal limitation of the above approach seems not to have yet been recognized. It is of course known that by going from -log to general convex q one gives up the easy extension to block coding over repeated source drawings and channel uses. This is of no consequence to the control application. However, the -log function has another essential property that is also lost for general q, namely the following.

The joint distribution of two random variables X,Y can be specified by giving the vector p of marginal probabilities of X and the stochastic matrix T whose entries are the conditional probabilities of Y given X. The generalized mutual information $I^q(X;Y)$ is a certain function F(p,T) of p and T. When q = -log it is easily shown, and well known, that (i) F is concave in p for fixed T and (ii) F is convex in T for fixed p. The first property makes the capacity step the maximization of a concave function on a convex set (the probability simplex). The second property makes the rate-distortion step the minimization of a linear function over the convex set $\{T: F(p,T) \leq C\}$, or alternatively, if one minimizes mutual information for a given level D of distortion as parameter, it becomes the minimization of the convex function F(p,•) of T over the convex polytope of the T satisfying the nonnegativity, unit sum and distortion level conditions.

Unfortunately, with general convex q both (i) and (ii) may be lost. As a result, the two steps that must be carried out become extremization problems of about the same magnitude as the direct solution of the signalling problem that they will bound.

To illustrate the far-reaching nature of this difficulty, we need only consider the determination of C^q for a binary symmetric channel with cross-over probability β. This amounts to maximizing over p in $[0,1]$ the expression

$$F(p) = (1-p)\left[(1-\beta)q\left(1 - p + \frac{\beta}{1-\beta}p\right) + \beta q\left(1 - p + \frac{1-\beta}{\beta}p\right)\right]$$

$$+ p\left[\beta q\left(p + \frac{1-\beta}{\beta}(1-p)\right) + (1-\beta)q\left(p + \frac{\beta}{1-\beta}(1-p)\right)\right].$$

By symmetry $F(p) = F(1-p)$ so that $F'(\frac{1}{2}) = 0$. However, for such choices of q as $q(x) = x^{-2}$ or $q(x) = e^{-\lambda x}$, $(\lambda>4)$ and small β, one has $F''(\frac{1}{2}) > 0$ so that $p=\frac{1}{2}$ is a local minimum. The "obvious" conclusion that the capacity of the BSC is attained for equally probable use of the two inputs is thus false in general.

7 THE SIMPLEST SIGNALLING PROBLEM

Perhaps the simplest signalling problem is the one characterized by the following specifications:

a) the source selects the value of random variable $X(=\omega_1)$ among n equally probable alternatives denoted $\{1,\ldots,n\}$
b) agent 1 selects among k possible signals $\{1,\ldots,k\}$
c) agent 2 receives this signal exactly (thus the channel is the noiseless k-ary channel and ω_2 is not needed).
d) agent 2 selects among m possible actions V in $\{1,\ldots,m\}$
e) the cost for X=i, V=j is given by the n by m matrix $[c_{ij}]$.

The problem is thus to partition the X alphabet into k subsets $A_1,\ldots,A_k$ so as to minimize

$$\frac{1}{n}\sum_{s=1}^{k}\min_{1\le j\le m}\sum_{i\in A_s} c_{ij}, \tag{*}$$

where use has been made of the fact that the X values are equiprobable. (Note that any X distribution can be reduced to the equiprobable case by replacement of c_{ij} by $p_i c_{ij}$).

The problem is trivial when $k \ge \min(n,m)$ for if $k \ge n$ one can signal the exact X-value, while if $k \ge m$ agent 1 can determine and send the optimal V value for each X. In either case, one achieves expected cost

$$\frac{1}{n}\sum_{i=1}^{n}\min_{1\le j\le m} c_{ij}.$$

When $k < \min(n,m)$ one straightforward solution algorithm is to try in turn each of $\binom{m}{k}$ subsets of k decisions. One of these sets must be (or contain) a true optimal decision set. If B is one of these k-sets of V-values, the best result achievable using only V's in B is

$$\frac{1}{n} \sum_{i=1}^{n} \min_{j \in B} c_{ij} \qquad (**)$$

which is achieved by partitioning the values of X according to the $j \in B$ which achieves the above minimum for X=i. In other words, the minimum of (*) over all partitions is equal to the minimum of (**) over all B of size k.

Unfortunately if e.g. one has n=m=2k, then the effort required in this method increases geometrically with k.

Hopes to find a substantially more efficient, yet still infallible, algorithm are dashed by the fact that even with c_{ij} values restricted to {0,1}, the problem for general k,n,m is NP complete (*) [3]. [For any fixed value of k, the $\binom{m}{k}$ algorithm above works in polynomial time.]

For free k one may be tempted to try for B a set containing the k indices j which give $\sum_{i=1}^{n} c_{ij}$ its k smallest values. However it is easy to make up examples in which this strategy misses the optimum by a large factor.

8 A SPECIFIC EXAMPLE

Consider the special case of the signalling problem in which $n = m > k > 1$ and $c_{ij} = 1-\delta_{ij}$. Thus agent two tries to guess X and pays one unit if in error. This is a zero-delay source coding problem or "quantization" problem, which has an exact solution easy to determine. Indeed, all of the $\binom{n}{k}$ k-sets B give the same value in (**): the minimum expected costs $J^* = \frac{n-k}{n}$. This can be achieved by <u>any</u> partition into

---

(*) The problem of exact covering by 3-sets (long known to be NP complete) can be stated as a signalling problem of the type considered here.

nonvoid sets A_s, s=1,...,k with agent two picking some arbitrary fixed element in A_s when receiving signal s.

One can now determine the Shannon rate-distortion bound for this problem (that is, use q = -log). The capacity of the noiseless k-ary channel is log k. (The symmetry and the concavity with respect to the input probabilities imply that the uniform distribution is optimal). As for the second step, let t_{ji} be the transition probabilities of a test channel from X = i to V = j. Besides being nonnegative, the t_{ji} are subject to $\sum_j t_{ji} = 1$ for each i and to the capacity constraint

$$\sum_{i,j} \frac{1}{n} t_{ji} \log \frac{n\, t_{ji}}{\sum_{i'} t_{ji'}} \leq \log k.$$

The bound B is the minimum over all such t_{ji} of

$$\sum_{ij} \frac{1}{n} t_{ji} c_{ij} = 1 - \frac{1}{n} \sum_i t_{ii}$$

Try a solution with all $t_{ji} > 0$. Adjoining the unit sum and capacity constraints with Lagrange multipliers λ_i and μ gives the first order conditions

$$\lambda_i + \mu(\log t_{ji} - \log \sum_{i'} t_{ji'}) + \delta_{ij} = 0$$

for all i,j pairs. Consider the choice, for $0 \leq \theta \leq 1$ of

$$t_{ji} = (1-\theta)\delta_{ij} + \theta/n.$$

For each k, θ can be chosen to satisfy the capacity constraint with equality, a condition which reduces to

$$\log n - \frac{n-1}{n} \theta \log(n-1) - h\left(\frac{n-1}{n} \theta\right) = \log k$$

with $h(x) = -x \log x-(1-x)\log(1-x)$ as usual. Then, for $\theta < 1$, the multipliers $\lambda_i = \lambda$ and μ can be chosen such that

$\lambda + \mu \log(\theta/n) = 0$ and $\lambda + \mu \log(1-(n-1)\theta/n) + 1 = 0$, so that all first order conditions are satisfied. By the convexity of the second auxiliary problem, the first order conditions are sufficient for optimality, giving the bound

$$B = 1 - \frac{1}{n}\sum_i t_{ii} = \frac{n-1}{n}\theta$$

In summary, the bound is given for each k, by the transcendental equation

$$\log n - B \log(n-1) - h(B) = \log k$$

(this even holds trivially for $k=1$, $B=(n-1)/n$) and the resulting inequality $B \leq J^*$ holds with equality for the extreme cases $k=1$ and $k=n$.

Letting n go to infinity with fixed k/n yields an asymptotic expression for the bound:

$$B \rightarrow \frac{\log n - \log k}{\log n}$$

which shows that for fixed $k/n = \alpha$, B goes to 0 while $J^* = 1-\alpha$ remains constant.

In conclusion, while the Shannon bound is easily determined, it is ineffective, in this case, for large n.

9 REFERENCES

1. Berger, T. "Rate Distortion Theory", Prentice Hall, (1971).

2. Csiszar, I. "A class of measures of informativity of observation channels", *Periodica Mathematica Hungarica*, **2**, 191, (1972).

3. Garey, M. R. and Johnson, D. S. Private Communication

4. Ho, Y. C., Kastner, M. P. and Wong, E. "Teams, Signalling, and Information Theory", *IEEE Trans. on Autom. Contr.*, **AC-23**, 305-312, (1978).

5. Kastner, M. P. "Information and signalling decentralized decision problems", Harvard Univ., Div. Appl. Sci. Techn. Rep. 669, Sept., (1977).

6. Witsenhausen, H. S. "A counterexample in stochastic optimum control", *SIAM J. Contr.*, **6**, 131-147, (1968).

7. Witsenhausen, H. S. "The intrinsic model for discrete stochastic control: some open problems", Lecture Notes in Econ. and Math. Systems, **107**, 322-335, Springer, (1975).

8. Zakai, W. and Ziv, J. "A generalization of the rate-distortion theory and applications", C.I.S.M. Courses and Lectures, **219**, 87-123, Springer, (1975).

9. Ziv, J. and Zakai, M. "On functionals satisfying a data-processing theorem", *IEEE Trans.*, **IT–19**, 275-283, (1973).

5.2 SOME REMARKS ON THE CONCEPT OF INFORMATION STATE

J. C. Willems

(Mathematics Institute, University of Groningen, Netherlands)

1 INTRODUCTION

One of the problems which enters in a natural way in dynamic decision problems with uncertainty is the automatic reduction and updating of the consecutive measurements for use in the future decisions. The variables which one needs to store in this process form what is called the *information state*. In this paper a formal definition of this concept will be given. In general the ideas and results are in a somewhat preliminary form and are therefore only briefly sketched. A more complete paper will be forthcoming. The framework in which we treat our problems is very much like in the work of Witsenhausen [1]. Another relevant paper by the same author (where much more general problems are treated) is [2]. A formal definition of the concept of information state may be found in [3]. In a sense the present paper elaborates and generalizes the approach taken there.

Let us first recall the definition of a *dynamical system in input/output form*. This is defined by $\{T,U,\mathcal{U},Y,\mathcal{Y},F\}$ with $T \subset \mathbb{R}$ the time axis, U the input alphabet (the set of input values), $\mathcal{U} \subset U^T$ the input space, Y the output alphabet (the set of output values), $\mathcal{Y} \subset Y^T$ the output space, and $F:\mathcal{U} \to \mathcal{Y}$ the system function which is assumed to be *non-anticipating*, that is to say, that if $u_1, u_2 \in \mathcal{U}$ happen to have the property $u_1(t) = u_2(t)$ for $t \leq t'$ then it will likewise hold that $(Fu_1)(t) = (Fu_2)(t)$ for $t \leq t'$. An *uncertain* dynamical system in input/output form is now defined as $\Sigma := \{\Omega,T,U,\mathcal{U},Y,\mathcal{Y},F\}$ with Ω the *uncertainty set*, $T,U,\mathcal{U},Y$ and $\mathcal{Y}$ as before, and $F:\Omega \times \mathcal{U} \mapsto \mathcal{Y}$ such that $F(\omega,\cdot)$ is non-anticipating for all $\omega \in \Omega$.

[the symbol := denotes set definition, $\mapsto$ denotes a mapping]

Example: Let $\{\Omega, A, P\}$ be a probability space, $x_0 \in \mathbb{R}^n$ a random vector, and w a q-dimensional Wiener process on $[t_0, t_1]$. Under the usual suitable smoothness assumptions on a, b, c, d and U the stochastic differential equation:

$$dx = a(x,u(t),t)dt + b(x,u(t),t)dw, \quad x(t_0) = x_0;$$

$$dy = c(x,u(t),t)dt + d(x,u(t),t)dw, \quad y(t_0) = 0;$$

will define an uncertain dynamical system with $T = [t_0, t_1]$, etc.

In stochastic control the uncertainty set is taken to be a probability space as in the above example. In this framework one consequently needs to add certain measurability assumptions. For example in Definition 1 one should assume that Y is a measurable space and that $F(\omega,u)$ is, as a Y-valued function as T, a well-defined stochastic process for all u. The reader will have no difficulty adding, where needed, such conditions which we will consequently conveniently pass over.

In a dynamic control situation where on-line decisions are being made one has given an uncertain dynamical system $\Sigma = \{\Omega, T, U, \mathcal{U}, Y, \mathcal{Y}, F\}$ which models the way the observations y depend on the uncertainty ω and the (past) control u. The decision law in its turn is defined by a map which processes the observations y and chooses the control variables on the basis of these. This decision law could for instance have been obtained as the solution of an optimization problem, but that need not concern us here. The fact that the observations can depend on the control used is an essential feature of feedback control and results in the complication that it becomes rather delicate to define the class of decision laws which are allowed. We propose the following definition of admissibility: Let Σ be an uncertain dynamical system. Then a map G: $\mathcal{Y} \to \mathcal{U}$ is said to define an *admissible control law* for Σ if:

(i) G is non-anticipating;
(ii) the equations $y = F(\omega,u)$; $u = Gy$ have a unique solution for all $\omega \in \Omega$. Let $\omega \mapsto u =: F_u(\omega)$, $y =: F_y(\omega)$ be the solution thus obtained. Thus $F_u, F_y : \Omega \to \mathcal{U}, \mathcal{Y}$;

(iii) if $(F_u(\omega_1)(t)) = (F_u(\omega_2)(t)) =: v(t)$ for $t \leq t'$

and $(F(\omega_1,v))(t) = (F(\omega_2,v))(t)$ for $t \leq t'$

then it must also hold that $(F_y(\omega_1))(t) = (F_y(\omega_2))(t)$ for $t \leq t'$.

Conditions (i) and (ii) are standard. However (iii) is quite essential: it expresses the fact that the information about ω in the obtained observation $F_y(\omega)$ must actually have been delivered by the system, i.e., by $F(\omega,u)$. It is easy to construct an example of a control law which satisfies (i) and (ii) but not (iii).

2 ADMISSIBLE STOCHASTIC CONTROL LAWS

Consider the stochastic system:

$$dx = Ax\,dt + B\,dw + G\,u(t)dt, \quad x(t_0) = x_0;$$

$$dy = Cx\,dt + D\,dw, \quad y(t_0) = 0;$$

which is studied, for example, in continuous time LQG stochastic control problems. Let us assume that the usual assumptions on w and x_0 are satisfied and that the time interval of interest is $[t_0,t_1]$. Assume that there are m controlled variables and p measured variables. The dynamical law from u to y takes the form $y = Lu + f$ with L a deterministic bounded linear operator from $L_2((t_0,t_1);\mathbb{R}^m)$ into $L_2((t_0,t_1);\mathbb{R}^p)$ and f a stochastic process on $[t_0,t_1]$ which is easily computed from x_0 and w. The uncertainty thus enters the measurements additively through the function f. A problem which has received some attention in the literature is the question of determining an as broad as possible class of admissible feedback control laws for this case. In terms of the definition of admissible control we would call a non-anticipating map $G: L_2((t_0,t_1);\mathbb{R}^p) \mapsto L_2((t_0,t_1);\mathbb{R}^m)$ an admissible control law if the map $I - LG$ has a non-anticipating inverse on $L_2((t_0,t_1);\mathbb{R}^p)$. It is possible [4] to prove the optimality of the well-known optimal linear control law for this rather broad class of admissible control laws. This class includes the locally

Lipschitz laws considered for example by Wonham [5] and others. Several authors [6,7] have proposed to take as admissible controls for this case those which can be obtained by applying a non-anticipating map on f. While it indeed yields a broader class of controls than those obtained above this approach has the æsthetic disadvantage that it is hard to justify the assumption intuitively from the "experimental" set-up of the problem and the more fundamental disadvantage that it brings in controls which cannot be obtained from feedback laws. It is indeed easily verified that for the system $y(t) = \int_{t_0}^{t} u(\sigma)d\sigma + f(t)$ it will be impossible to obtain the relationship $u(t) = -f(t) + f(t-1)$ from a non-anticipating feedback control law. In fact, it is easy to convince oneself that the mere assumption that u could be any non-anticipating function of f is inadequate and that it is rather fortunate that this has not led to difficulties in these treatments of the LQG-theory.

3 INFORMATION STATE

We now come to the main concept of this paper. Assume that we have an uncertain dynamical system Σ with system function F controlled via an admissible feedback control law G. The solution of the implicit equations thus obtained leads to maps $F_u:\Omega \mapsto U$ and $F_y:\Omega \mapsto Y$. These satisfy $F_u = GF_y$ and $F(\cdot,F_u(\cdot)) = F_y(\cdot)$. Now, for every ω, an observation $F_y(\omega)$ appears and the decision maker will process these observations through the system G and obtain the control $GF_y(\omega)$. As time progresses the decision maker is hence presented with an ever increasing series of measurements and thus faces the problem of what information about these measurements to store for use in his future decisions, and how he should update this stored information at each step. This is the problem of determining the *information state*. Although the concept is obviously related to that of sufficient statistic there seem to be no treatments in the statistics literature to cover our problem. The solution given here is very much in the spirit of [3].

The problem of determining the information state is very much like the problems treated in deterministic realization theory [8]. However there is one crucial difference. In realization theory one would construct a state space realization of the map $G:Y \to U$, whereas in the case of information state it suffices to realize the map G on the "effective" input space $Y' := F_y(\Omega)$. Since there is no reason to expect Y' to be closed under concatenation it is not possible to straightforwardly apply realization theory ideas. In fact a suitable theory requires a somewhat more general concept of dynamical system in state space form: system over bundles. This will now be introduced. For simplicity, we will only consider time-invariant discrete time systems:

A (set theoretic version of a) *bundle* is defined by a surjection $\pi: B \mapsto X$ with B the *bundle*, X the *base space*, and π the *projection* of B onto the base space X. A bundle is denoted by $\pi: B \to X$. The set $\pi^{-1}(x)$ is called the *fibre* above x. A (*discrete time*) *state space dynamical system over a bundle* is defined by a bundle $\pi: B \mapsto X$, and maps $f: B \mapsto X$ and $r: B \mapsto Y$, with X the *state space*, f the *next state map*, Y the *output alphabet*, and r the *read-out map*. The fibre $\pi^{-1}(x)$ plays the role of the input alphabet at x. However, since there is in general no natural way of identifying fibres one cannot assume that $B = X \times U$ (with U then the input alphabet in the usual sense). Bundles of the type $\pi: B = X \times U \mapsto X$ with $\pi: (x,u) \mapsto x$ are called *trivial*. The usual input/state/output dynamical systems thus correspond to the case that the bundle is trivial. There are many situations (e.g., in mechanics or, for that matter, the system $\dot{x} = u$, $|x| \leq 1$) where bundles offer a natural approach to system modelling. In fact some recent work [9,10] suggests that these ideas are gaining ground. Furthermore, the bundle approach is quite akin to the situation treated in automata theory [11].

Consider now the problem of processing the inputs (to the controller) in $Y = F_y(\Omega)$ to outputs (to the controller) according to $y \mapsto u = Gy$, taking into consideration memory and updating. Assume for simplicity that $T = \{0,1,2,\ldots\}$. A *state space realization* of G on Y' is defined by (i) a bundle $\pi: B \mapsto X$, a *next state map* $f: B \mapsto X$ and a *read-out map* $r: B \mapsto U$, (ii) an *input injection* which is a family of partial maps $\iota_x: Y \mapsto \pi^{-1}(x)$ (a *partial map* is simply a map defined on a subset of the domain space only), and (iii) an *initial state* $x_0 \in X$, such that for all $y \in Y'$ there holds $(Gy)(t) = r(\iota_{x(t)}(y(t)))$ with $x(t)$ recursively defined by: $x(0) = x_0, x(t+1) = f(\iota_{x(t)}(y(t)))$. Note that the effect of the input injection is simply to identify each fibre $\pi^{-1}(x)$ with a subset of the input alphabet. We will call the element $x(t)$ defined recursively above the *information state* corresponding to this realization.

It is of interest to generalize the usual "Nerode equivalence" construction of ordinary realization theory to the situation at hand. The problem then is to declare when, for the given map $G: \mathcal{Y}' \mapsto \mathcal{U}$ two past inputs in $\mathcal{Y}'$ are equivalent. Let us consider two inputs $y_1, y_2 \in \mathcal{Y}'$ and assume for simplicity that we want to find out if in some sense they lead to the same information state at time $t' \in T$ (by shifting the time axis in one of the inputs it suffices to consider only this case). Let $\mathcal{Y}_+$, $\mathcal{Y}'_+$ and $\mathcal{U}_+$ denote the restriction of elements of $\mathcal{Y}$, $\mathcal{Y}'$ and $\mathcal{U}$ to $[t', \infty) \cap T$. Now the past $y: (-\infty, t') \cap T \mapsto Y$ of an element of $\mathcal{Y}'$ defines in an obvious way a partial map from $\mathcal{Y}'_+$ into $\mathcal{U}_+$. This map tells us how future inputs produce future outputs. However, since $\mathcal{Y}'$ need not in general be closed under concatenation it is only a partial map. It would be logical to define two pasts equivalent if the induced operators agree on the intersection of their domains but this does *not* lead to an equivalence relation. However if two pasts are such that the induced operators do not agree on the intersection of their domains then they are *incompatible* in the sense that they will certainly not lead to the same information state. The idea of an abstract construction of information state comes down to constructing a partition of the space of pasts of $\mathcal{Y}'$ such that no two states in the same piece of the partition are incompatible and to do this in a minimal way (this may be defined formally but can be thought of as a partition having a minimal number of pieces). Unfortunately, no effective construction for such a minimal partition is known. Moreover it can be shown through examples that it will not follow that any two minimal realizations are equivalent (which again may be defined formally in a rather obvious manner).

4 APPLICATIONS OF INFORMATION STATE

In applications to stochastic control one can show a number of interesting theorems, for example, a generalization of a result in [12] which essentially states that if a controlled partially observed Markov process is regulated by minimizing an additive cost functional of the form $\{\sum_{t_0}^{t_1} L(x_t, u_t, t)\}$ then the conditional density of the state given past observations and past controls will be an information state for the resulting optimal feedback control law. In general however this information state cannot be expected to be minimal in any sense. We will not dwell upon this application but rather develop another one: namely that the state of the plant together with the information state of the controller forms a state of the closed loop system.

We will now define an uncertain dynamical system in (recursive) state space form. For simplicity we will only consider the discrete time case $T = \{0,1,2,...\}$ with input space $\mathcal{U} = U^T$ and an output space $\mathcal{Y} = Y^T$. Let X be a set called the state space. The system is then defined by the next state map $f: \Omega \times X \times U \times T \mapsto X$, the read out map $r: \Omega \times X \times U \times T \mapsto Y$ and an initial state map $x_0: \Omega \to X$. Now for every $\omega \in \Omega$ and $u \in \mathcal{U}$ this yields a state trajectory x and an output trajectory y obtained in the obvious way. As far as their dependence on ω is concerned we postulate for all u that if ω_1 yields the trajectory pair (x_1,y_1) and ω_2 yields the trajectory pair (x_2,y_2) and if at some time $t' \in T$ there holds $x_1(t') = x_2(t')$ then there should exist an $\omega_3 \in \Omega$ such that the ensuing trajectory pair (x_3,y_3) satisfies

$$(x_3,y_3)(t) = \begin{cases} (x_1,y_1)(t) & \text{for } t < t' \\ (x_2,y_2)(t) & \text{for } t \geq t' \end{cases}$$

It is easily seen that the following property then holds: Let Σ be the family of all possible input/state/output trajectories of an uncertain dynamical system in state space form as we let u and ω vary. Then for each time t, Σ may be viewed as a subset of $R^- \times R^0 \times R^+$ with $R^- := U^{[0,t)} \times X^{[0,t)} \times Y^{[0,t)}$, $R^0 := X$ and $R^+ := U^{[t,\infty)} \times X^{(t,\infty)} \times Y^{[t,\infty)}$. Consider now the subset Σ_a of Σ consisting of all those elements with $x(t) = a$. Obviously $\Sigma = \bigcup_{a \in X} \Sigma_a$. The property of state which seems crucial to us and which follows from the assumptions above is that Σ_a is a subset of the form $R_a^- \times \{a\} \times R_a^+$ which expresses the desired "Markov" property of x.

Assume now that Σ is a discrete time uncertain dynamical system in state space form. This induces an uncertain input/output dynamical system with system function $F: \Omega \times \mathcal{U} \mapsto \mathcal{Y}$

defined by $F(\omega,u)(t) := r(\omega,x(t), u(t),t)$ and x recursively defined by $x(t+1) = f(\omega,x(t), u(t),t), x(0) = x_0(\omega)$. Let $G: Y \to U$ be an admissible control law and assume that G has been realized by a bundle system as explained in section 3 with z(t) the ensuing information state at time t. Altogether we thus end up with maps of Ω into U^T, Y^T, X^T and Z^T and we would like to prove that the pair (x(t), z(t)) is a state for the closed loop system, which requires us to show that if $\omega_1 \mapsto v_1 := (u_1,y_1,x_1,z_1)$ and $\omega_2 \mapsto v_2 := (x_2,y_2,x_2,z_2)$ are such that, for some $t' \in T$, $x_1(t') = x_2(t')$ and $z_1(t') = z_2(t')$ then there should exist ω_3 such that $\omega_3 \mapsto v_3 := (u_3,y_3,x_3,z_3)$ with $v_3(t) = \begin{cases} v_1(t) & t < t' \\ v_2(t) & t \geq t' \end{cases}$. It is actually almost trivial to prove this at the level of generality used here. Indeed, as shown above the family of all (u,x,y)'s which could result in the plant from varying u and ω and which lead to a given state $x(t')$ at the time t' is a product set of past and futures. Now, from the definition of information state in section 3 it is easy to prove the analogous property for the family of (y,z,u)'s which could result in the feedback controller by varying ω. Now the closed loop system consists exactly of those (u',x',y')'s of the plant and those (y'',z'',u'')'s of the feedback controller for which $u' = u''$ and $y' = y''$. It is easily seen from this that (x,z) has indeed the required state property, i.e. that the pasts and futures compatible with a particular value of (x,z) at time t' forms a product set.

5 THE INFORMATION STATE IN KALMAN FILTERING

It is possible to generalize the above ideas to stochastic systems. An interesting specific question which one may ask in this context is to give conditions for the Kalman filter to be minimal. Consider the stochastic system:

$$dx = Ax\,dt + B\,dw_1$$

$$dy = Cx\,dt + \quad dw_2$$

on the time axis $\mathbb{R} = (-\infty,+\infty)$. Assume that $\mathrm{Re}\,\lambda < 0$ for $\lambda \in \sigma(A)$ and let w_1 be a Wiener process on $\mathbb{R}$. Then x is a well-defined zero mean stationary Guassian process. Moreover

$Q := E\{x(t)\,x^T(t)\} > 0 \iff (A,B)$ controllable. We will assume this to be the case. The observation y is corrupted by an additive Wiener process w_2 which, for simplicity and as usual, we will take to be independent of w_1. We will assume that (A,C) is observable.

The conditional mean of Cx(t) given $y(\tau)$ for $\tau \leq t$ is then governed by:

$$d\hat{x} = A\hat{x}\,dt + L(dy - C\hat{x}\,dt),\ \hat{z} = C\hat{x};$$

with $L := \Sigma C^T$ and Σ the positive definite solution of the Riccati equation $A\Sigma + \Sigma A^T - \Sigma C^T C\Sigma + BB^T = 0$.

It need not in general be true that the Kalman filter is minimal. Here we may simply interpret this minimality as the requirement that $\{A - LC, L, C\}$ is a minimal triple. Equivalent conditions for minimality are given next.

The following conditions are equivalent:

(i) The Kalman filter is minimal;
(ii) (A,L) is controllable and (A,C) is observable;
(iii) x is a "minimal" state space representation of y = Cx.

(This means the following: let z:= Cx. We call {x,C} a *Markov representation* of z. Now, if every matrix P for which Px is Markov and CP = C, is invertible, then the pair {x,C} is called a *minimal* Markov representation of z.)

(iv) $\{A, QC^T, C\}$ is a minimal triple (which actually implies (A,C) observable and Q > 0 and thus (A,B) controllable). This minimality condition may be interpreted as *stochastic observability* and *stochastic reconstructibility* in the sense that they require the functions $x(0) \mapsto f_+ : [0,\infty) \to \mathbb{R}^p$ and $x(0) \mapsto f_- : (-\infty,0] \to \mathbb{R}^p$, to be injective. Here $f_+(t), f_-(t) := E\{Cx(t)\,|\,x(0)\}$.

(v) The Mc-Millan degree [13] of $R_{zz}(t) := E\{z(t)z^T(t)\} = C\,e^{At}QC^T$ equals the dimension of x.

(vi) The Mc-Millan degree of the spectral density of z equals twice the dimension of x.

(vii) The transfer function $C(Is-A)^{-1}B$ contains no all-pass factors.

It is easy to prove that the Kalman filter will be minimal for generic A,B,C's. An unsolved problem is to give conditions analogous to those above for the minimality of the optimal LQG-controller. Again it may be shown that minimality is generic but to find the precise conditions in terms of system and cost-functional matrices is an interesting and challenging problem.

6 REFERENCES

1. Witsenhausen, H. S. "Minimax Control of Uncertain Systems", Report ESL-R-269, M.I.T., (1966).

2. Witsenhausen, H. S. "On Information Structures, Feedback and Causality", *SIAM J. Control,* **9**, 149-160, (1971).

3. Davis, M. H. A. and Varaiya, P. P., "Information States for Linear Stochastic Systems", *J. Math. An. and Appl.,* **37**, 384-402, (1972).

4. Willems, J. C. "Recursive Filtering", *Statistica Neerlandica,* **32**, 1-39, (1978).

5. Wonham, W. H. "Random Differential Equations in Control Theory", in *Probabilistic Methods in Applied Mathematics,* (Ed. A. T. Bharucha-Reid), **2**, 132-212, Academic Press, (1970).

6. Lindquist, A. "On Feedback Control of Linear Stochastic Systems", *SIAM J. Control,* **11**, 323-343, (1973).

7. Fleming, W. J. and Rishel, R. W. "Deterministic and Stochastic Optimal Control", Springer, (1975).

8. Kalman, R. E., Falb, P. L. and Arbib, M. A. "Topics in Mathematical Systems Theory", McGraw Hill, (1969).

9. Takens, F. "Variational and Conservative Systems I", Report ZW-7603, Un. of Groningen, (1976).

10. Brockett, R. W. "Control Theory and Analytical Mechanics", in Geometric Control Theory (Eds: C. Martin and R. Hermann), 1-48, Math. Sci. Press, (1977).

11. Eilenberg, S. "Automata, Languages and Machines", Academic Press, (1974).

12. Striebel, C. "Optimal Control of Discrete Time Stochastic Systems", Springer, (1975).

13. Brockett, R. W. "Finite Dimensional Linear Systems", Wiley, (1970).

CHAPTER 6

STATE ESTIMATION

6.1 THE GEOMETRY OF THE CONDITIONAL DENSITY EQUATION

R. W. Brockett

(Division of Applied Sciences, Harvard University, USA)

and

J. M. C. Clark

(Department of Computing and Control, Imperial College, UK)

1 INTRODUCTION

In this paper we sketch out a circles of ideas concerning the relationship between problems in estimation theory and the theory of representations of Lie groups. This approach gives a completely new vantage point on estimation problems and suggests a method of classification which accurately reflects the intrinsic complexity of propagating the conditional density. We consider mainly the finite state problem in this paper but this point of view also shows that linear filtering has an intrinsic relationship with the Heisenberg algebra and certain of its generalizations.

2 PRELIMINARIES

Let x be a finite state Markov process taking on values in $S = \{x_1, x_2, \ldots, x_n\}$, a subset of the real numbers, and let w be a standard Wiener process. Suppose that $\tilde{p}_i$ is the probability that $x = x_i$ and suppose that

$$\frac{d}{dt}\tilde{p}(t) = A\tilde{p}(t); \quad dy = xdt+dw$$

In 1965 Wonham [1] gave a derivation of the equation satisfied by the conditional probability $\rho(t,x|Y_t)$ where Y_t indicates the y process on the interval $[0,t]$. In terms of the notation

$$B = \begin{bmatrix} x_1 & 0 & \dots & 0 \\ 0 & x_2 & \dots & 0 \\ \cdot & \cdot & \cdot & \cdot \\ 0 & 0 & \dots & x_n \end{bmatrix}; \quad b = \begin{bmatrix} x_1 \\ x_2 \\ \vdots \\ x_n \end{bmatrix}$$

Wonham's equation is the Ito equation

$$d\rho = A\rho + (B-\langle b,\rho\rangle I)\rho(dy-\langle b,\rho\rangle dt)$$

It is a straight forward exercise in the Ito calculus to verify that if $\Sigma\rho_i(0) = 1$ then $\Sigma\rho_i(t) = 1$ for all $t > 0$ and that if p satisfies the Ito equation

$$dp = Apdt + Bpdy \qquad (*)$$

then the elements of p_i remain nonnegative if they are initially nonnegative and that the equality

$$\rho(t) = p(t)/\Sigma p_i(t)$$

holds for all $t > 0$ if it holds initially. We call (*) the unnormalized conditional density equation.

For our purposes it is more insightful to reexpress (*) in Fisk-Stratonovich form (see [2]). Indicating Stratonovich differentials by d^+ we have

$$dp = (A - \frac{1}{2}B^2)pdt + bpd^+y$$

It is intuitive that if we find that $S(\rho_0)$ is the reachable set for the control problem

$$\dot{p} = (A - \frac{1}{2}B^2)p + uBp; \quad p(0) = p_0$$

then the support for the measure associated with p is confined to the closure of $S(p_0)$. This is, in fact, a theorem of Stroock and Varadhan [3]. Moreover it is known from the controllability results based on Chow's theorem that the reachable set has non empty interior in the set

$$\hat{S}(p_0) = \{\exp\{A-\frac{1}{2}B^2,B\}_{LA}\}_G p_0$$

where $\{A - \frac{1}{2}B^2,B\}_{LA}$ indicates the matrix Lie algebra generated by $A-\frac{1}{2}B^2$ and B and $\{\exp M\}_G$ indicates the subgroup of the n by n invertible matrices consisting of elements which are expressible as products of exponentials of members of M. These and related facts are discussed in [4].

From this point of view we see that

$$L = \{A-\frac{1}{2}B^2,B\}_{LA}$$

reflects the complexity of the conditional density equation. To make this more precise we recall that if L_i is a basis for a matrix Lie algebra then the solution of the linear matrix equation

$$\dot{\Phi} = (\sum_{i=1}^{r} L_i\phi_i(t))\Phi$$

can be expressed as

$$\Phi = \exp(\sum_{i=1}^{r} L_i f_i(t))$$

or as

$$\Phi(t) = \exp L_1 g_1(t)\exp L_2 g_2(t)\ldots\exp L_r g_r(t)$$

on an interval of positive length $0 \leqslant t \leqslant \varepsilon$. In order to establish such results in a constructive way, i.e. in a way which gives differential equations for the parameters f_i or g_i we need to differentiate the expressions for Φ and systematically employ the identities

$$[L_i,L_j] = \Sigma\gamma_{ijk}L_k$$

which define the Lie algebra. Usually this is easier to do for the product form but in special cases the sum form may be easier. In the former case the key identity is

$$e^{L} L_0 e^{-L} = \sum_{k=0}^{\infty} \frac{1}{k!} \mathrm{ad}_L^k (L_0)$$

where

$$\mathrm{ad}_L^k (L_0) = \underbrace{[L,[L,\ldots[L,L_0]\ldots]]}_{k \text{ brackets}}$$

3 GENERALITIES ON THE LIE ALGEBRA $\{A-\frac{1}{2}B^2,B\}_{LA}$

The Lie algebras which can arise as the controllability algebras of finite dimensional unnormalized conditional density equations satisfy certain basic restrictions which we describe in lemmas 1-3 below. In the following section we give some examples which suggest that these restrictions are the only essential ones. Throughout this paper we assume that the values x takes on are unrepeated i.e. that B has unrepeated eigenvalues. This assumption is without loss of generality because we can merge the states corresponding to identical values of x. Recall that a finite state Markov process is called irreducible if the infinitesimal generator (intensity matrix) A does not take the form

$$A = \begin{bmatrix} A_{11} & A_{12} \\ 0 & A_{22} \end{bmatrix}$$

with A_{ii} square for any ordering of the basis.

If L is a Lie algebra then the set of elements of the form $[L,L] = L^1$ is called the derived algebra, $[L^1,L^1] = L^2$ is the second derived algebra, etc. The Lie algebra is said to be solvable if the chain of derived algebras $L^1, L^2, \ldots$ becomes trivial beyond a certain point. A subalgebra $H \subset L$ is said to be an ideal if $[H,L] \subset H$ and an algebra is said to be semisimple if it has no solvable ideas. By the centre of a Lie algebra we understand an ideal H such that $[H,L] = \{0\}$. A Lie algebra is simple if it has no nontrivial ideals and $[L,L] \neq \{0\}$. A Lie algebra of n by n matrices is said to act irreducibly on $\mathbb{R}^n$ if there is no proper subspace $S \subset \mathbb{R}^n$ such that S is an invariant subspace for all elements of the algebra.

The first observation is that in this problem the two notions of irreducibility, one referring to Markov processes and the other referring to representations of Lie algebras, coincide.

Lemma 1. If A is irreducible in the sense that $\dot{p} = Ap$ defines an irreducible Markov process then $\{A-\frac{1}{2}B^2,B\}_{LA}$ acts irreducibly on $\mathbb{R}^n$.

Proof. Because B is diagonal with distinct eigenvalues we see that the invariant subspaces for B are all of the form span $\{e_{i_1},e_{i_2},\ldots e_{i_\nu}\}$ where e_{i_k} are standard basic vectors. But by definition of irreducibility of a Markov process, A leaves invariant no subspace of this form.

A Lie algebra is said to be reductive if it admits a faithful representation which is fully reduced (block diagonal as opposed to block triangular). Using a result from the theory of reductive Lie algebras we can establish the following.

Lemma 2. If A is the infinitesimal generator of an irreducible Markov process then $L = \{A-\frac{1}{2}B^2,B\}_{LA} = L^1 \oplus \{\alpha I\}$ where L^1 is semisimple and $\{\alpha I\}$ is the one dimensional centre consisting of the real multiples of the identity.

Proof. From Lemma 1 we see that $\{A-\frac{1}{2}B^2,B\}_{LA}$ acts irreducibly on $\mathbb{R}^n$. From the theory of reductive Lie algebras see [5, p. 371] we know that any Lie algebra which admits a faithful irreducible representation is of the form $L^1 \oplus Z$ where Z is the centre. Because $\mathrm{tr}(A-\frac{1}{2}B^2) \neq 0$ and $\mathrm{tr}(MN-NM) = 0$ we see that $L^1 \neq L$. But the only matrices which commute with all elements of an irreducible representation are the multiples of the identity and so the ideal is in fact $\{\alpha I\}$.

Our final result establishes a further substantial restriction on the kind of Lie algebras which can arise in this context. Recall that representations are studied in part by determining if there is any full rank matrix Q such that

$$M'Q + QM = 0$$

for all M in L. If this happens for some $Q = Q'$ one says that L leaves invariant a symmetric form; if it happens for some

$Q = -Q'$ we say that L leaves invariant a skew symmetric form. By the signature of a symmetric form we mean the number of positive eigenvalues minus the number of negative ones.

Lemma 3: Let A, B, and L be as in Lemma 2, then L^1 does not leave invariant a symmetric form of signature greater than one. Furthermore a necessary condition for L^1 to leave invariant a nondegenerate form (symmetric or skew symmetric) is that for $\alpha = \Sigma b_{ii}$ we have the symmetry condition

$$\{b_{ii}-\alpha\}_{i=1}^{n} = \{-(b_{ii}-\alpha)\}_{i=1}^{n}$$

Proof: By Lemma 2 we see that $I \in L$ and hence the zero trace diagonal matrix $B-\alpha I$ belongs to L^1. If we have $Q(B-\alpha I)+(B-\alpha I)'Q = 0$ then

$$q_{ij}(b_{ii}+b_{jj}-2\alpha) = 0$$

Now since the b_{ii} are distinct $(b_{ii}+b_{jj}-2\alpha) = 0$ can have at most one solution j for each value of i and hence there is at most one nonzero q_{ij} in each column and row of Q. For a suitable ordering of the basis (the ordering $b_{11} > b_{22} > \ldots > b_{nn}$) we see that $Q = \pm Q'$ takes the form

$$Q = \begin{bmatrix} \bigcirc & & & & q_{1,n} \\ & & & q_{2,n-1} & \\ & & \cdot^{\cdot^{\cdot}} & & \\ & q_{n-2,2} & & & \\ q_{n-1,1} & & & & \bigcirc \end{bmatrix}$$

where some of the anti-diagonals are possibly zero. Any symmetric form of this type has signature -1, 0 or 1. In fact if Q is nondegenerate it has signature zero for n even and signature ±1 for n odd. For $b_{ii}+b_{jj}-2\alpha=0$ to have n solutions we obviously need the symmetry condition cited.

4 A CLASS OF EXAMPLES

So far we have not given any examples. What we show here is that the Lie algebra can be of much lower dimension than the dimension of the space on which the conditional density equation evolves. This is significant because of its implication for the number of "sufficient statistics" needed to summarize the past observations. That is, it implies the existence of low-dimensional nonlinear filters. A large class of such examples exist, in fact, with some additional effort one may expect to show that any real irreducible representation of a simple Lie algebra subject to the conditions of Lemma 3 can arise as the L^1 corresponding to $L = \{A - \frac{1}{2}B^2, B\}_{LA}$. The problem is far from a standard one in Lie theory because of the particular requirements imposed on A by the fact that it must be the generator of a finite state Markov process. This means that $a_{ij} \geq 0$ for $i \neq j$ and that the columns of A should sum to zero.

We recall some facts about real, simple, three dimensional Lie algebras. There are two. One is isomorphic to the set of two by two matrices of zero trace, hereafter denoted by $s\ell(2,R)$, and one is isomorphic to the set of 3 by 3 skew symmetric matrices, so(3). In view of Lemma 3, skew symmetric matrices cannot be expected to play a role in our theory. What we want to do in this section is to show that, to within an obvious sort of equivalence, every irreducible representation of $s\ell(2,R)$ arises as the L^1 of $\{A - \frac{1}{2}B^2, B\}_{LA}$ for a suitable choice of A and B. The corresponding full Lie algebra $L = \{A - \frac{1}{2}B^2, B\}_{LA}$ will be isomorphic to $g\ell(2,R)$ the Lie algebra of real 2 by 2 matrices according to $g\ell(2,R) \simeq s\ell(2,R) \oplus \{\alpha I\}$. This means that there exists, for each positive integer n, n-state processes with 4 dimensional conditional mean generators.

Here are some facts we need from the theory of the representations of $s\ell(2,R)$. Define T_n^+, T_n^- and H_n as follows:

$$T_n^- = \begin{bmatrix} 0 & 0 & 0 & \cdots & 0 \\ n-1 & 0 & 0 & \cdots & 0 \\ 0 & n-2 & 0 & \cdots & 0 \\ & & & & 0 \\ 0 & 0 & 0 & 1 & 0 \end{bmatrix}; \quad T_n^+ = \begin{bmatrix} 0 & 1 & 0 & & 0 \\ 0 & 0 & 2 & & 0 \\ 0 & 0 & 0 & & 0 \\ & & & & n-1 \\ 0 & 0 & & 0 & 0 \end{bmatrix}$$

$$H_n = \begin{bmatrix} n-1 & 0 & \cdots & 0 \\ 0 & n-3 & \cdots & 0 \\ & & & \\ 0 & 0 & & -n+1 \end{bmatrix}$$

One sees that these three matrices form representations of $s\ell(2,R)$. If we add the multiples of the identity, we have a four dimensional Lie algebra isomorphic to $g\ell(2,R)$ and whose derived algebra L^1 is isomorphic to $s\ell(2,R)$. Now for any value of n notice that

$$T_n^+ + T_n^- - (n-1)I = A = \begin{bmatrix} -n-1 & 1 & 0 & \\ n-1 & -n-1 & 2 & \ddots \\ 0 & n-2 & n-1 & \\ & \ddots & \ddots & \end{bmatrix}$$

is the infinitesimal generator of a Markov process. For this choice of A, and for the choice $B = H_n$ we see that $\{A,B\}_{LA}$ is a four dimensional algebra isomorphic to $g\ell(2,R)$. However we must work with $A - \frac{1}{2}B^2$ and B, not just A and B. Introduce a diagonal change of basis sending T_n^+ and T_n^- into $RT_n^+R^{-1}$ and $RT_n^-R^{-1}$ respectively where

$$R = \mathrm{diag}(r_1, r_2, \ldots, r_n)$$

This change of basis leaves H_n and I alone and hence $RT_n^+R^{-1}$, $RT_n^-R^{-1}$, H_n and I also define a Lie algebra representation that is equivalent to $g\ell(2,R)$. Now choose B to be mH_n and A to be

$$A = R(T_n^+ + T_n^-)R^{-1} + \tfrac{1}{2}m^2H_n^2 + \alpha I$$

If the following equations are satisfied A is the infinitesimal generator of Markov process

$$(n-1)\frac{r_2}{r_1} + \tfrac{1}{2}m^2(n-1)^2 + \alpha = 0$$

$$\frac{r_1}{r_2} + (n-1)\frac{r_3}{r_3} + \tfrac{1}{2}m^2(n-3)^2 + \alpha = 0$$

$$\cdot\ \cdot\ \cdot\ \cdots = \cdots$$

$$(n-1)\frac{r_{n-1}}{r_n} + \tfrac{1}{2}m^2 + \alpha = 0.$$

There is a solution at $m = 0$, $r_i = 1$ and $\alpha = -(n-1)$. If we regard m and r_n as fixed, the linearization of the left-hand side in the remaining variables is

$$\begin{bmatrix} -(n-1), & n-1, & 0 & \cdots & & 1 \\ 1, & -(n-1), & n-2 & \cdots & & 1 \\ 0, & 2, & & & & \\ \cdots & \cdots & \cdots & \cdots & \cdots & \\ & & & n-2, & -(n-1), & 1 \\ 0 & 0 & 0 & 0, & n-1, & 1 \end{bmatrix} \begin{bmatrix} \delta r_1 \\ \delta r_2 \\ \\ \\ \delta r_{n-1} \\ \delta\alpha \end{bmatrix}$$

Since this does define an onto map, for $m \neq 0$ but small, the implicit function theorem implies that we can solve for the r_i and α so as to make $\{A - \frac{1}{2}B^2, B\}_{LA}$ isomorphic to $g\ell(2,R)$.

Incidentally, these representations illustrate Lemma 3 in the following way. For n odd these representations leave invariant a symmetric form of signature 1 and for n even they leave invariant a skew form. Both are nondegenerate.

5 REMARKS ON THE INFINITE DIMENSIONAL CASE

If the state space has the cardinality of the integers or the real line then most of the specific results on representation theory no longer apply. In particular, the connections between irreducibility and solvability become meaningless or false. This situation is already exemplified by one dimensional linear estimation theory. In this case the Lie algebra generated by $A-\frac{1}{2}B^2$ and B is solvable but the relevant representation is irreducible. More concretely, for the model

$$dx = d\nu; \quad dy = xdt+d\omega$$

with V and W standard Wiener processes are as follows.

$$A = \frac{1}{2}\frac{\partial^2}{\partial x^2} \qquad B = \text{multiplication by } x$$

We have

$$\{(A-\frac{1}{2}B^2),B\}_{LA} =\{\frac{1}{2}\frac{\partial 2}{\partial x^2} - \frac{1}{2}x^2,x\}_{LA}$$

A basis for this algebra is

$$L_1 = \frac{1}{2}\frac{\partial^2}{\partial x^2} - \frac{1}{2}x^2$$

$$L_2 = x$$

$$L_3 = \frac{\partial}{\partial x}$$

$$L_4 = 1$$

This algebra is solvable and L^1 is the Heisenberg algebra. Although the equations for the Kalman-Bucy filter follow easily from the commutation relation of the Lie algebra we do not give the details here.

6 ACKNOWLEDGEMENT

Partially supposrted by the Army Research Office under Grant DAAG29-76-C-0139, the National Aeronautics and Space Administration under Grant NSG-2265, and the U.S. Office of Naval Research under the Joint Services Electronics Program Contract N00014-75-C-0648.

7 REFERENCES

1. Wonham, W. M. "Some Applications of Stochastic Differential Equations to Optimal Nonlinear Filtering", *J. SIAM*, Series A: Control, **2**, 3, (1965).

2. Clark, J. M. C. "An Introduction to Stochastic Differential Equations on Manifolds", in Geometric Methods in System Theory, Reide, Dordrecht, 131-149, (1973).

3. Stroock. D. W. and Varadhan, S. R. S. "On Degenerate Elliptic-Parabolic Operators of Second Order and Their Associated Diffusions", *Comm. Pure and Appl. Math.*, **25**, 651-713, (1972).

4. Brockett, R. W. "Lie Algebras and Lie Groups in Control Theory", in Geometric Methods in System Theory, Reidel, Dordrecht, 43-82, (1973).

5. Wolf, J. "Spaces of Constant Curvature", McGraw-Hill, N.Y., (1967).

6.2 FINITE-TIME LINEAR FILTERING, PREDICTION AND CONTROL

M. J. Grimble

(Department of Electrical and Electronic Engineering, Sheffield Polytechnic, UK)

1 INTRODUCTION

The finite-time linear filtering [1], prediction [2] and control [3,4] problems have recently been solved in both the s and z domains and in the process several new problems have been considered and new results have been obtained. The most interesting results have been concerned with the solution of the linear estimation problem and this is considered in detail in this paper. It is well known that the solution to the optimal finite-time filtering problem is a Kalman filter [5-7] with time-varying gain matrix. However, in practice the constant gain (infinite-time) Kalman filter or Wiener filter [8] is often used in finite-time filtering situations because the constant gain filter is easier to implement. The solution to this problem which is obtained in the complex frequency domain involves a time-invariant filter which combines some of the properties of both the time-varying Kalman and Wiener filters. At the end of some chosen filtering interval this filter will give a state estimate which is identical with that obtained from a full time-varying Kalman filter. The filter also has a fixed memory length when used outside the given interval. Limited memory filtering has previously been used [9-11] to overcome filter divergence problems due to modelling errors.

The time-varying Kalman filter gain matrix may be calculated directly from the expression for the time-invariant filter transfer function matrix. This result also enables a sub-optimal Kalman filter to be defined which is easier to implement in continuous-time systems than the optimal filter. The sub-optimal filter has a gain matrix which is identical with the optimal Kalman gain matrix at times zero and infinity. The new time-invariant and sub-optimal filters are of greater practical significance than the s-domain method for calculating the Kalman gain matrix [1] which was the initial objective of this work.

To solve the finite-time estimation problem in the s-domain this problem must be embedded within an equivalent infinite-time problem. This simplifies the calculation of the Laplace transforms. The infinite-time estimation problem is posed so that the solution will yield the desired solution to the finite-time problem under consideration [2]. The problem is thus one of determining a time-invariant linear estimator, with initial time $-\infty$, which minimizes the mean-square estimation error at some chosen time T. The projection theorem [12] is used to derive the necessary and sufficient condition for optimality in the form of a Wiener-Hopf equation. This is transformed into the s-domain and the solution for the estimator transfer function matrix is obtained.

2 THE SYSTEM AND NOISE DESCRIPTION

It is assumed that estimates of the state variables are required for a linear time-invariant system which has been in operation from time $t = -\infty$. The following system description is more general than was used previously. The observations signal is assumed to be delayed, the noise is assumed to be coloured and the plant has both stochastic and deterministic inputs. The plant is defined by the following state equations:

$$\dot{x}(t) = Ax(t) + D\omega(t) + u_1(t), \quad x(t) \in R^n, \; \omega(t) \in R^q \tag{1}$$

$$z(t) = Cx(t - \tau_0) + v(t - \tau_0), \quad z(t) \in R^m \tag{2}$$

The plant disturbance $\omega(t)$ and the measurement noise $v(t)$ are assumed to be zero-mean, stationary coloured noise processes with correlation functions $\phi_{\omega\omega}(\tau)$, $\phi_{\omega v}(\tau)$, $\phi_{vv}(\tau)$ and known rational spectral densities. The plant is assumed to be completely controllable and observable so that the infinite time solution exists and is unique. The transition matrix for the plant is denoted by $\Phi(t)$ and the impulse response matrices are denoted by $W(t) = C\Phi(t)D$, $W_0(t) = C\Phi(t)B$ and $W_0'(t) = \Phi(t)B$ where, to simplify the notation $\Phi(t) \triangleq 0$ for $t < 0$.

The plant input $u_1(t)$ includes a deterministic control signal $u(t)$ which is assumed to be zero for $t \leq 0$ and an impulse $x_0'\delta(t)$ which initializes the state at time zero. The random vector x_0' is assumed to have a known mean m_0 and covariance Σ_0, and to be uncorrelated with the noise processes. Thus,

$$u_1(t) \triangleq Bu(t) + x_0'\delta(t) \tag{3}$$

and

$$x(t) = \int_{-\infty}^{t} \Phi(t - \tau)(Bu(\tau) + D\omega(\tau) + x_0'\delta(\tau))\, d\tau, \text{ for all } t \tag{4}$$

At the limiting time instants before and after time zero,

$$x(0^-) \triangleq \int_{\infty}^{0} \Phi(-\tau)D\omega(\tau)\, d\tau \tag{5}$$

$$x(0^+) \triangleq x(0^-) + x_0' \tag{6}$$

where $E\{x(0^-)\} = 0$, $E\{x(0^+)\} = m_0$. The impulse therefore merely establishes the given mean value for the plant state at time zero and thus

$$x(t) - \bar{x}(t) = \int_{-\infty}^{t} \Phi(t - \tau)(D\omega(\tau) + (x_0' - m_0)\delta(\tau))\, d\tau \tag{7}$$

3 THE STATE ESTIMATION PROBLEM

The estimation problem may now be defined and the time-invariant estimator be introduced. The estimator may be represented in the form:

$$\hat{x}(t_1|t) = d_0(t,\tau_1) + \int_{-\infty}^{t} h_1(t - \sigma, \tau_1)z(\sigma)\, d\sigma \tag{8}$$

where the zero-input response is defined by:

$$d_0(t,\tau_1) = \Psi(t - \tau_0,\tau_1)\hat{x}_0' + \int_0^t h_2(t - \sigma,\tau_1)u(\sigma)\, d\sigma \quad \text{for } t \geqslant 0 \tag{9}$$

and $d_0(t,\tau_1) \triangleq 0$ for $t < 0$. The prediction interval $\tau_1' \triangleq t_1 - t$

and τ_1 is the total equivalent prediction interval $\tau_1 \triangleq \tau_0 + \tau_1'$. The estimator output (8) may be written more concisely in operator form as:

$$\hat{x}(t_1|t) = d_0(t,\tau_1) + (H_1 z)(t). \tag{10}$$

The operator H_1 in this equation should be replaced by H_{10} when the intergration interval in (8) begins at time zero.

The optimal linear time-invariant estimator is required which will minimize the estimation error criterion:

$$J(\tilde{x}(t_1|t)) = E\{\tilde{x}(t_1|t)^T\tilde{x}(t_1|t)\} \tag{11}$$

$$\tilde{x}(t_1|t) \triangleq x_i(t_1) - \hat{x}(t_1|t) \tag{12}$$

at some chosen fixed time $t = T > 0$. The state estimates are required to be unbiased for all time and the system is assumed to be in operation from time $-\infty$. The ideal output of the estimator at time $t > 0$ ($t_0 \triangleq t - \tau_0$) is defined as:

$$x_i(t_1) = x(t_1) - \Psi(t_0,\tau_1)x(0^-). \tag{13}$$

The solution for this infinite-time problem will yield the desired solution to the finite-time problem of interest [1]. This follows because the estimation error equation and in turn the cost function is the same for the two problems. This equation may be expanded by noting that, from (4) and (9), for the estimator to be unbiased then:

$$\Psi(t_0,\tau_1) \triangleq \Phi(t_1) - (H_{10}C\Phi)(t_0). \tag{14}$$

The estimation error equation follows from equations (4), (10), (12) and (13):

$$\tilde{x}(t_1|t) = x'(t_1) - d_0(t,\tau_1) - (H_{10}z')(t - \tau_0) \tag{15}$$

where x' and z' denote the system state and observations respectively corresponding to a plant with initial condition vector x_0'. This vector is assigned the mean m_0 and covariance

Σ_O of the initial state vector in the finite-time filtering problem. It follows that the estimation error equation which is used to calculate the optimal estimator is the same as that found in the finite-time situation. This arises as a consequence of the second term in equation (13) which removes the effect of plant inputs before time zero. Since the estimator depends upon the inputs over the positive-time interval only, the estimator has a finite memory length, that is, the impulse response is zero for $\tau > T$.

4 THE WIENER-HOPF INTEGRAL EQUATION

The following theorem may be derived using the projection theorem [12] and the decomposition theorem:

Theorem: A necessary and sufficient condition for $\hat{x}(T_1|T)$ to be a minimum variance unbiased estimator for $x_i(T_i)$ is that the estimator impulse response $\hat{h}_1(T - \sigma, \tau_1)$ must satisfy the following Wiener-Hopf equation:

$$\text{cov}[x_i(T_1), z(\lambda)] = \int_{-\infty}^{T} \hat{h}_1(T - \sigma, \tau_1)\,\text{cov}[z(\sigma), z(\lambda)]\; d\sigma \qquad (16)$$

for all $\lambda \in (-\infty, T]$ and the matrix $\hat{d}(T, \tau_1)$ must satisfy

$$\hat{d}_O(T, \tau_1) = \bar{x}_i(T_1) - \int_{-\infty}^{T} \hat{h}_1(T - \sigma, \tau_1)\bar{z}(\sigma)\; d\sigma \qquad (17)$$

Note that the optimal estimate is required at the fixed time $t = T$ of the predicted state at time $T_1 = T + \tau_1'$ and that $T_O \triangleq T - \tau_O$. The Wiener-Hopf equation may now be expanded using the results in appendix 1. The left-hand side of this equation becomes:

$$\int_{-\infty}^{\infty}\int_{-\infty}^{\infty} \Phi(T_1-\sigma_1)D\phi_{\omega\omega}(\sigma_1-\sigma_2)W^T(\lambda-\tau_O-\sigma_2)\; d\sigma_2\, d\sigma_1 + \Phi(T_1)\Sigma_O\Phi^T(\lambda-\tau_O)C^T$$

$$-\Psi(T_O,\tau_1)\phi_{x_Ox}(\tau_O-\lambda)C^T + \int_{-\infty}^{\infty}(\Phi(T_1-\sigma)-\Psi(T_O,\tau_1)\Phi(-\sigma)D\phi_{\omega v}(\sigma-\lambda+\tau_O)\; d\sigma$$

and the right-hand side of (16) becomes:

$$\int_{-\infty}^{T} \hat{h}_1(T - \sigma,\tau_1)\Big[\int_{-\infty}^{\infty}\int_{-\infty}^{\infty} W(\sigma - \sigma_1)\phi_{\omega\omega}(\sigma_1 - \sigma_2)W^T(\lambda - \sigma_2)d\sigma_2\, d\sigma_1$$

$$+ C\Phi(\sigma - \tau_0)\Sigma_0\Phi^T(\lambda - \tau_0)C^T + \phi_{vv}(\sigma - \lambda)$$

$$+ \int_{-\infty}^{\infty} W(\sigma - m)\phi_{\omega v}(m - \lambda) + \phi_{v\omega}(\sigma - m)W^T(\lambda - m)\ dm\Big]\ d\sigma$$

Collecting the terms involving Σ_0 in the Wiener-Hopf equation:

$$(\Phi(T_1) - \int_{-\infty}^{T} \hat{h}_1(T - \sigma,\tau_1)C\Phi(\sigma - \tau_0)\ d\sigma)\Sigma_0\Phi^T(\lambda - \tau_0)C^T$$

$$= \Psi(T_0,\tau_1)\Sigma_0\Phi^T(\lambda - \tau_0)C^T \tag{18}$$

The Wiener-Hopf equation may therefore be expressed in the form:

$$\Psi(T_0,\tau_1)\Big[\Sigma_0\Phi^T(\lambda - \tau_0)C^T - \phi_{x_0x}(\tau_0 - \lambda)C^T - \int_{-\infty}^{\infty}\Phi(-\sigma)D\phi_{\omega v}(\sigma - \lambda + \tau_0)d\sigma\Big]$$

$$- \int_{-\infty}^{T} \hat{h}_1(t - \sigma,\tau_1)\Big[\int_{-\infty}^{\infty}\int_{-\infty}^{\infty} W(\sigma - \sigma_1)\phi_{\omega\omega}(\sigma_1 - \sigma_2)W^T(\lambda - \sigma_2)\ d\sigma_2\ d\sigma_1$$

$$+ \phi_{vv}(\sigma - \lambda) + \int_{-\infty}^{\infty} W(\sigma - m)\phi_{\omega v}(m - \lambda) + \phi_{v\omega}(\sigma - m)W^T(\lambda - m)\ dm\Big]d\sigma$$

$$+ \int_{-\infty}^{\infty}\int_{-\infty}^{\infty}\Phi(T_1 - \sigma_1)D\phi_{\omega\omega}(\sigma_1 - \sigma_2)W^T(\lambda - \tau_0 - \sigma_2)\ d\sigma_2\ d\sigma_1$$

$$+ \int_{-\infty}^{\infty}\Phi(T_1 - \sigma)D\phi_{\omega v}(\sigma - \lambda + \tau_0)\ d\sigma = 0 \text{ for all } \lambda \in (-\infty,T] \tag{19}$$

The matrix on the left side of this equation may be denoted by $G_T(\tau)$ where $\tau \triangleq T_1 - \lambda + \tau_0 = T + \tau_1 - \lambda$. The necessary and sufficient condition for optimality may now be expressed in the form:

$$G_T(\tau) = 0 \text{ for all } \tau \in [\tau_1, \infty) \tag{20}$$

The advantage of the s-domain approach becomes apparent in the next section since the above equation is simplified considerably by transformation into the s-domain.

5 SOLUTION IN THE s-DOMAIN

The solution for the transfer function matrix of the optimal time-invariant estimator is obtained below. The Wiener-Hopf equation is transformed (with respect to τ) into the s-domain using the two-sided Laplace transform. The transforms of each of the terms in this equation are given in appendix 2. By using these results (16) becomes:

$$G_T(s) = \psi(T_0,\tau_1)\left[\Sigma_0 \Phi^T(-s)C^T - \Phi(s)D\Phi_{\omega\omega}(s)W^T(-s) - \Phi(s)D\Phi_{\omega v}(s)\right]e^{-sT_1}$$

$$- \hat{H}_1(s,T,\tau_1)Y(s)Y^T(-s)e^{-s\tau_1} + \Phi(s)D(\Phi_{\omega\omega}(s)W^T(-s) + \Phi_{\omega v}(s)) \tag{21}$$

where $Y(s)Y^T(-s)$ represents the (generalized) spectrally factored matrix [14-16]:

$$Y(s)Y^T(-s) = W(s)\Phi_{\omega\omega}(s)W^T(-s) + \Phi_{vv}(s) + W(s)\Phi_{\omega v}(s) + \Phi_{v\omega}(s)W^T(-s) \tag{22}$$

By post-multiplying equation (21) by the matrix $(Y^T(-s))^{-1}e^{-s\tau_1}$, it can be seen that the resulting left-hand side will contain the transform of a term which is zero for positive τ (from the condition for optimality). The estimator is required to be causal which implies that $\hat{H}(s,T,\tau_1)Y(s)$ is the transform of a function which is zero for negative τ. The optimal time-invariant estimator follows from these results by equating the transforms of positive τ terms:

$$\hat{H}_1(s,T_0,\tau_1) = (N_{11}(s,\tau_1) + \Psi(T_0,\tau_1)(N_{12}(s,T_0) + N_2(s,T_0))Y(s)^{-1} \tag{23}$$

where

$$N_{11}(s,\tau_1) \triangleq \{\Phi(s)D(\Phi_{\omega v}(s) + \Phi_{\omega\omega}(s)W^T(-s))(Y^T(-s))^{-1}e^{s\tau_1}\}_+ \quad (24)$$

$$N_{12}(s,T_0) \triangleq -\{\Phi(s)D(\Phi_{\omega v}(s) + \Phi_{\omega\omega}(s)W^T(-s))(Y^T(-s))^{-1}e^{-sT_0}\}_+ \quad (25)$$

$$N_2(s,T_0) \triangleq \Sigma_0\{M^T(-s)e^{-sT_0}\}_+ \quad (26)$$

and

$$M(s) \triangleq (Y(s))^{-1}C\Phi(s) \quad (27)$$

The first term in equation (23) represents either the Wiener predictor ($\tau_1 > 0$) or Wiener filter ($\tau_1 = 0$), that is

$$\hat{H}_1(s,\infty,\tau_1) = N_{11}(s,\tau_1)Y(s))^{-1} \quad (28)$$

The remaining terms in (23) are introduced because the estimation interval is finite and they ensure that the impulse response of the estimator is zero for $\tau > T_0$. The matrix $\psi(T_0,\tau_1)$ is the only unknown in equation (23) and an expression is derived in the following section from which this may be calculated.

6 ZERO-INPUT RESPONSE OF THE ESTIMATOR

The second equation (section 4) in the necessary and sufficient conditions for optimality requires that the estimate at time T be unbiased. This condition is achieved by defining the zero input response of the estimator as:

$$d_0(t,\tau_1) = \bar{x}(t_1) - (H_1\bar{z})(t)$$
$$= \Phi(t_1)m_0 + (W_0'u)(t_1) - (H_{10}C\Phi)(t - \tau_0) - H_{10}w_0u)(t - \tau_0) \quad (29)$$

By comparison with equation (9) define $\hat{x}_0' = m_0$,

$$\psi(t - \tau_0,\tau_1) = \Phi(t_1) - (H_{10}C\Phi)(t - \tau_0) \quad (30)$$

and

$$(\hat{H}_2 u)(t) = (W_O' u)(t_1) - (H_{1O} W_O u)(t - \tau_O) \tag{31}$$

The unknown matrix $\psi(T_O, \tau_1)$ may now be determined using (30) and the following results:

$$\hat{H}_1(s, T_O, \tau_1) C\Phi(s) = (N_{11}(s, \tau_1) + \psi(T_O, \tau_1)(N_{12}(s, T_O) + N_2(s, T_O))M(s) \tag{32}$$

$$\{M^T(-s) e^{-sT_O}\}_+ = L_1(M^T(T_O - \tau) H(T - \tau)) \tag{33}$$

$$N_{11}(s, \tau_1) = \Phi(\tau_1) N_{11}(s, O) \tag{34}$$

thus

$$\psi(T_O, \tau_1) = \Phi(\tau_1)(\Phi(T_O) - \int_O^{T_O} N_{11}(\tau, O) M(T_O - \tau) d\tau)$$

$$- \psi(T_O, \tau_1) \left[\int_O^{T_O} (N_{12}(\tau, T_O) + \Sigma_O M^T(T_O - \tau)) M(T_O - \tau) \right] d\tau$$

The matrix $\psi(T_O, \tau_1)$ may therefore be calculated using,

$$\psi(T_O, \tau_1) = \Phi(\tau_1) \psi(T_O, O) \tag{35}$$

where

$$\psi(T_O, O) = (\Phi(T_O) - I_{11}(T_O))(I + I_{12}(T_O) + \Sigma_O S(T_O))^{-1} \tag{36}$$

and

$$I_{11}(T_O) = \int_O^{T_O} N_{11}(\tau, O) M(T_O - \tau) d\tau \tag{37}$$

$$I_{12}(T_O) = \int_O^{T_O} N_{12}(\tau, T_O) M(T_O - \tau) d\tau \tag{38}$$

$$S(T_0) = \int_0^{T_0} M^T(T_0 - \tau)M(T_0 - \tau)d\tau \tag{39}$$

It follows that, $\psi(0,0) = I_n$ and $\psi(0,\tau_1) = \Phi(\tau_1)$. From (23) and (35) the optimal predictor is obtained by multiplying the filter output by the matrix $\Phi(\tau_1)$:

$$\hat{H}(s,T_0,\tau_1) = \Phi(\tau_1)\hat{H}(s,T_0,0) \tag{40}$$

If the initial state of the plant is known exactly $\Sigma_0 = 0$, $N_2(s,T_0) = 0$ and from (23) the estimator becomes:

$$\hat{H}(s,T_0,\tau_1) = (N_{11}(s,\tau_1) + \psi(T_0,\tau_1)N_{12}(s,T_0))Y(s)^{-1}$$

If the initial state is completely unknown $\Sigma_0 \to \infty$ and from (23) and (36) the estimator is given by:

$$\hat{H}(s,T_0,\tau_1) = (N_{11}(s,\tau_1) + \psi(T_0,\tau_1)N_2(s,T_0))Y(s)^{-1}$$

7 CALCULATION OF THE KALMAN GAIN MATRIX

The Kalman filter gain matrix may be calculated directly from the transfer function matrix for the optimal time-invariant filter. The relationship between these matrices is derived below. The optimal time-invariant filter corresponding to the usual Kalman filtering problem is obtained from (23) by setting $\Phi_{\omega\omega}(s) = Q$, $\Phi_{vv}(s) = R$, $\Phi_{\omega v}(s) = G$, $\Phi_{v\omega}(s) = G^T$ (white noise spectral density matrices), and $\tau_0 = \tau_1 = 0$. The filter impulse response matrix can be shown (appendix 3) to satisfy a finite-time version of the Wiener-Hopf equation:

$$\text{cov}[x'(T),z'(\lambda)] = \int_0^T \hat{h}_1(T - \sigma,0)\ \text{cov}[z'(\sigma),z'(\lambda)]d\sigma \tag{41}$$

where $x'(T)$ and $z'(\lambda)$ are the state and observations vectors for a plant with initial state x'_0. The time-varying impulse response matrix $\hat{h}(t,\sigma)$ for the corresponding Kalman filter must satisfy, at time $t = T$, the equation:

$$\mathrm{cov}\left[x'(T),\ z'(\lambda)\right] = \int_0^T \hat{k}(T,\sigma)\ \mathrm{cov}\left[z'(\sigma),z'(\lambda)\right]\ d\sigma \tag{42}$$

for all $\lambda \in [0,T]$. The solution for these equations is unique (the Wiener-Hopf equation is a necessary and sufficient condition for optimality [12]) and thus:

$$\hat{k}(T,\sigma) = \hat{h}_1(T - \sigma,0) \quad \text{for all } \sigma \in [0,T] \tag{43}$$

The Kalman filter gain matrix [5] is defined as $K(T) = \hat{k}(T,T)$ and thus from (43) and by use of the Laplace transform initial value theorem:

$$K(t) = \lim_{s\to\infty} s\hat{H}(s,t,0) \tag{44}$$

From (14) and (43) the transition matrix $\psi_k(t,0)$ for the Kalman filter is related to $\psi(T)$ for the time-invariant filter by:

$$\psi_k(T,0) = \psi(T) \tag{45}$$

Recall that the initial state for the Kalman filter is defined as $\hat{x}_0 = m_0$. Thus the initial condition responses and the filter impulse response matrices are identically equal at time $t = T$. It follows that for the same observations signal the time-invariant and time-varying Kalman filters will give the same state estimate at time T. The theoretical filtering error covariance at time T is the same for the two filters and may be calculated using the usual Riccati equation [17].

It is interesting that, in general, a constant gain Kalman filter which is equivalent to the proposed time invariant filter does not exist. To show that this is the case assume that such a constant gain filter with gain K_T does exist. The impulse response must satisfy equation (43) at time $t = T$:

$$\hat{k}(T,\sigma) = \psi_k(T,\sigma)K(\sigma) = \phi_k(T - \sigma)K_T \quad \text{for all } \sigma \in [0,T] \tag{46}$$

where $\phi_k(t)$ is the transition matrix for this constant gain filter. However, for the case when $\Sigma_0 = 0$, then $K(0) = 0$ and the equation cannot be satisfied for all σ.

A sub-optimal Kalman filter [1] may also be defined from these results. The calculation of the time-varying gain matrix is simplified by approximating the matrix $\psi(T_0, \tau_1) = \psi(T,0)$ (for $\tau_0 = \tau_1 = 0$) in equation (23) by the identity matrix. The gain matrix then becomes:

$$\bar{K}(t) = \lim_{s\to\infty} s(N_{11}(s,0) + N_{12}(s,t) + N_2(s,t))Y(s))^{-1} \tag{47}$$

The matrix is equal to the optimal gain matrix at times zero and infinity. The difference between these gain matrices at other times depends upon the system and noise descriptions. The analogue circuitry necessary to generate the time-varying gain matrix is simplified considerably by the above approximation. The sub-optimal filter and the corresponding sub-optimal controller [4], have promising characteristics and require further investigation.

8 PERFORMANCE OF THE OPTIMAL TIME-INVARIANT ESTIMATOR

The time-invariant estimator gives the same state estimate at time T as that given by the Kalman estimator. The estimate at this time therefore has a lower mean square error than could be obtained with a Wiener estimator. The state estimates within the interval $[0,T]$ may also be an improvement on those available from a Wiener estimator. To compare the Wiener, Kalman and time-invariant estimators let $g(s) \triangleq (Y(s))^{-1}z(s)$, $N_{22}(s,T) = N_{12}(s,T) + N_2(s,T)$ and $m_0 = 0$, then the state estimates from each of these filters are respectively:

$$\hat{x}_w(t|t) = \int_0^t N_{11}(\tau)g(t-\tau)\,d\tau \tag{48}$$

$$\hat{x}_k(t|t) = \int_0^t N_{11}(\tau)g(t-\tau)\,d\tau + \psi(t)\int_0^t N_{22}(\tau,t)g(t-\tau)\,d\tau \tag{49}$$

$$\hat{x}_g(t|t) = \int_0^t N_{11}(\tau)g(t-\tau)\,d\tau + \psi(T)\int_0^t N_{22}(\tau,T)g(t-\tau)\,d\tau \tag{50}$$

Equation (49) for the Kalman estimate follows from the equivalence of the time-invariant and Kalman estimators at time T (in this case T is replaced by t).

The time-invariant estimator has several advantages over the Kalman estimator when the minimum variance estimate is required at one particular point in time. Such an application arises in fixed interval prediction [19] where the state estimate is required at the end of an interval of specified length T. In this case, the Kalman and time-invariant predictors give identical estimates of the future plant states, at time $t = T$. The implementation of the continuous-time Kalman estimator is complicated by the analogue system which generates the time-varying gain matrix and the multipliers which are necessary. The time invariant estimator, however, is in transfer function matrix form and may be implemented using standard techniques.

It has been noted that the impulse response of the time-invariant estimator is zero for $\tau > T$. This is a consequence of the e^{-sT} terms in the expression for the estimator transfer function matrix. These operator terms represent time delays which cannot easily be implemented in continuous-time systems. To overcome this problem the continuous-time estimator is only used to give state estimates within the interval $[0,T]$. Any e^{-sT} terms in the transfer function solution for the estimator may therefore be neglected since they do not affect the state estimates within this interval. This implementation problem does not arise in discrete-time systems [18] and the finite memory property can be exploited. The state estimate at any time $t > T$ will then depend upon the observations in a moving window $[t - T,t]$. This type of limited memory estimator has well known advantages in reducing divergence problems [10]. The discrete-time estimator equations are similar to those of section 5 but are expressed in z-transfer function matrix form. The calculation of a state estimate using this estimator involves either fewer operations than for the Kalman estimator or less computer storage if the gain matrix is not calculated sequentially, since the gains are not time varying.

9 THE DETERMINISTIC FINITE-TIME OPTIMAL REGULATOR PROBLEM

The fixed-point estimation problem is the dual of the optimal deterministic (open loop) state regulator problem. The s-domain solution for the finite-time problem is given below [4]. Let Q_1, R_1 and Σ_1 respectively, denote the performance criterion state, control and end-state weighting matrices and let T represent the optimization interval. Assume also that the noise inputs, for the system described in section

2, are zero. The optimal control is given by $\hat{u}(s) = -K_1(s,T)x_o$ and the open loop controller is given by:

$$K_1(s,T) = -Y(s)^{-1}(N_{11}(s) + (N_{12}(s,T) + N_2(s,T))G(T)) \tag{51}$$

where

$$N_{11}(s) \triangleq \{M^T(-s)Q_1\Phi(s)\}_+ \tag{52}$$

$$N_{12}(s,T) \triangleq -\{M^T(-s)Q_1\Phi(s)e^{-sT}\}_+ \tag{53}$$

$$N_2(s,T) \triangleq \{M^T(-s)e^{-sT}\}_+ \Sigma_1 \tag{54}$$

$$M(s) \triangleq \Phi(s)BY(s)^{-1} \tag{55}$$

and where the matrix G(T) is defined as:

$$G(T) = (I_n + I_{12}(T) + S\Sigma_1)^{-1}(\Phi(T) - I_{11}(T)) \tag{56}$$

and

$$I_{11}(T) \triangleq \int_o^T M(T - \tau)N_{11}(\tau)\, d\tau \tag{57}$$

$$I_{12}(T) \triangleq \int_o^T M(T - \tau)N_{12}(\tau,T)\, d\tau \tag{58}$$

$$S(T) \triangleq \int_o^T M(T - \tau)M^T(T - \tau)\, d\tau \tag{59}$$

The optimal state feedback solution for this problem may be calculated using the above results since the feedback gain matrix [4] is given by:

$$K(t) = \lim_{s\to\infty} sK_1(s,T-t) \tag{60}$$

A similar z-domain solution may be obtained for the discrete-time state regulator problem.

10 THE STOCHASTIC OPTIMAL REGULATOR PROBLEM

The infinite-time stochastic control problems, for constant systems and stationary noise, have been solved [20-23] in the s-domain but finite-time problems have not been considered. A solution to the LQG optimal regulator problem may, however, be obtained using the previous s-domain solutions for the optimal gain matrices (equations (44) and (60)). That is, by invoking the separation principle [24,25] and by using the s-domain solutions for the Kalman filtering and deterministic optimal control problems, the solution may be obtained for the stochastic finite-time optimal regulator problem.

The time-invariant estimator can be used in a state feed-back stochastic regulator system. The properties of such a system are now being investigated. It should also be possible to solve certain finite-time stochastic optimal control problems in the s-domain, without the aid of the separation principle. This is also an area of current research by the author.

11 CONCLUSIONS

The proposed time-invariant estimator has some interesting and useful properties. It is surprising that a time-invariant estimator can give the same state estimate as that resulting from the time-varying Kalman estimator, at some chosen time. It is also interesting that, in general, this estimator cannot be implemented using a constant gain Kalman estimator. The time-invariant estimator is obtained in transfer function matrix form and thus standard frequency domain techniques may be used for its implementation. The finite memory property (for $t > T$) is particularly important in discrete systems since the full estimator may easily be implemented. This property cannot be achieved in the continuous time case without introducing delay elements. It is therefore more convenient to neglect the delay terms and to restrict operation to the interval of most interest $t \in [0,T]$.

The time-invariant estimator is related closely to the Wiener and Kalman estimators. If the fixed time T tends to infinity the estimator reduces to the Wiener estimator. The Kalman filter gain and transition matrices can be calculated directly from the s-domain results for the time-invariant filter. A sub-optimal Kalman filter may also be defined which is based upon an approximation in the s-domain equations.

The s-domain solution procedure for the time-invariant estimator and the Kalman gain matrix is more direct and is more flexible than the usual time-domain methods. For example, the plant noise signals were assumed to be coloured rather than white but this makes little difference to the solution procedure. The plant was also assumed to contain a measuring system delay and this can be incorporated easily. The s-domain solutions for both the filtering and control problems also exist when the equivalent time-domain solutions fail, for example, due to singular noise R or control weighting R_1 matrices. The s-domain approach has the advantage over the state-space methods that a significant number of engineers prefer transfer function techniques.

There are several areas worthy of future research. The solution for the finite-time stochastic optimal control problems with and without system transport delays requires further study. The calculation of the s-domain (or z-domain) controllers and filters also needs to be considered. Simple problems can be solved by hand but high order systems require computer algorithms. Finally, these systems must be implemented in order to determine any practical disadvantages which may not be apparent at present.

12 ACKNOWLEDGEMENTS

This work is continuing under the support of the United Kingdom Science Research Council and GEC Electrical Projects Ltd., Rugby, as part of a joint paper on the design of filters and controllers for dynamic ship positioning systems. I am grateful for the helpful comments of Mr. R. J. Patton of Sheffield City Polytechnic and Dr. F. Boland of the University of Sheffield.

13 REFERENCES

1. Grimble, M. J. "Solution of the Kalman filtering problem for stationary noise and finite data records", *Int. J. Systems Science,* **10**, 2, 177-196, (1979).

2. Grimble, M. J. "Solution of the linear estimation problem in the s-domain", *Proc. IEE,* **125**, 6, 541-549, June, (1976).

3. Grimble, M. J. "S-domain solution for the fixed end point optimal control problem", *Proc. IEE,* **124**, 9, 802-808, (1977).

4. Grimble, M. J. "The design of finite-time optimal multivariable systems", *Int. J. Systems Sci.*, **9**, 3, 311-334, (1978).

5. Kalman, R. E. and Bucy, R. S. "New results in linear filtering and prediction theory", *Trans of the ASME*, **83**, 95-108, March, (1961).

6. Kalman, R. E. "A new approach to linear filtering and prediction problems", *Trans of the ASME*, 35-45, March, (1960).

7. Kalman, R. E. "New methods in Wiener Filtering theory", Proc. of the Symposium on Engineering Applications of Random Function Theory and Probability, 270-388, (1961).

8. Sage, A. P. "Optimum systems control", Prentice-Hall, (1968).

9. Jazwinski, A. H. "Stochastic processes and filtering theory", Academic Press, (1970).

10. Jazwinski, A. H. "Limited memory optimal filtering", *IEEE Trans. on Auto. Contr.*, 558-563, October (1968).

11. Price, C. F. "An analysis of the divergence problem in the Kalman filter", *IEEE Trans. on Auto. Contr.*, 699-702, December, (1968).

12. Deutsch, R. "Estimation theory", Chapter 3, Prentice-Hall, (1965).

13. Shaked, U. "A general transfer function approach to linear stationary filtering and steady-state optimal control problems", Research Report, Department of Engineering, University of Cambridge, No. TR126, (1976).

14. Grimble, M. J. "Factorization procedure for a class of rational matrices", *Int. J. Control.*, July, (1978).

15. Youla, D. C. "On the factorization of rational matrices", IRE Trans. on Information Theory, 172-189, July, (1961).

16. Davis, M. C. "Factoring the spectral matrix", *IEEE Trans.*, **AC-8**, 296-305, (1963).

17. Davis, M. H. A. "Linear estimation and stochastic control", Chapman and Hall, (1977).

18. Grimble, M. J. "A finite-time linear filter for discrete-time systems", Sheffield City Polytechnic Research Report, No. EEE/19/July 1978, to be published *Int. J. Systems Science*.

19. Sage, A. P. and Melsa, J. L. "Estimation theory with applications to communications and control", 349, McGraw Hill, (1971).

20. Shaked, U. "A general transfer function approach to the steady state linear quadratic Gaussian stochastic control problem", *Int. J. Control,* **24**, 6, 771-800, (1976).

21. Gawthrop, P. J. "Developments in optimal and self-tuning control theory", University of Oxford, Department of Engineering Science Report 1239/78, (1978).

22. Grimble, M. J. "The design of stochastic optimal feedback control systems", Proceedings IEE, **125**,11, 1275-1284, November, (1978).

23. Weston, J. E. and Bongiorno, J. J. "Extension of analytical design techniques to multivariable feedback control systems", *IEEE Trans. on Automatic Control,* **AC–17**, 5, 613-620, October, (1972).

24. Kwakernaak, H. and Sivan, R. "Linear optimal control systems", Wiley Interscience, (1972).

25. Athans, M. "The role and use of the stochastic linear quadratic Gaussian problem in control system design", *IEEE Trans. on Auto. Control,* **AC–16**, 6, 529-551, December, (1971).

APPENDIX I

Covariance Terms in the Wiener-Hopf Equation

The Wiener-Hopf equation may be expanded and simplified using the following results.

$$\operatorname{cov}\left[\underline{x}_i(T_1), z(\lambda)\right] = \operatorname{cov}\left[\underline{x}(T_1), x(\lambda - \tau_0)\right]C^T - \psi(T,\tau_1)\phi_{x_0 x}(\tau_0 - \lambda)C^T$$
$$+ \int_{\infty}^{\infty} (\Phi(T_1 - \sigma) - \psi(T,\tau_1)\Phi(-\sigma)D\phi_{\omega v}(\sigma - \lambda + \tau_0)\, d\sigma \tag{61}$$

where

$$\phi_{x_O x}(\tau_O - \lambda) \triangleq E\{x(O^-)x^T(\lambda - \tau_O)\} \tag{62}$$

$$\operatorname{cov}\left[x(t), x(\lambda)\right] = \int_{-\infty}^{\infty}\int_{-\infty}^{\infty} \Phi(t - \sigma_1) D\phi_{\omega\omega}(\sigma_1 - \sigma_2) D^T \Phi^T(\lambda - \sigma_2)\, d\sigma_2\, d\sigma_1 + \Phi(t)\Sigma_O \Phi^T(\lambda) \tag{63}$$

$$\operatorname{cov}\left[z(\sigma), z(\lambda)\right] = C \operatorname{cov}\left[x(\sigma - \tau_O), x(\lambda - \tau_O)\right] C^T + \phi_{vv}(\sigma - \lambda) + \int_{-\infty}^{\infty} W(\sigma - m)\phi_{\omega v}(m - \lambda) + \phi_{v\omega}(\sigma - m) W^T(\lambda - m)\, dm \tag{64}$$

$$C \operatorname{cov}\left[x(\sigma - \tau_O), x(\lambda - \tau_O)\right] C^T = \int_{-\infty}^{\infty}\int_{-\infty}^{\infty} W(\sigma - \sigma_1)\phi_{\omega\omega}(\sigma_1 - \sigma_2) W^T(\lambda - \sigma_2)\, d\sigma_2\, d\sigma_1 + C\Phi(\sigma - \tau_O)\Sigma_O \Phi^T(\lambda - \tau_O) C^T \tag{65}$$

APPENDIX 2

Two-sided Laplace Transforms

The two-sided Laplace transforms of each of the terms in the Wiener-Hopf equation are given below. The transforms are with respect to the variable $\tau = T - \lambda + \tau_1$.

$$L_2(\Phi^T(\lambda - \tau_O)) = \Phi^T(-s)e^{-sT_1} \tag{66}$$

$$L_2(\Phi_{x_O x}(\tau - \lambda)) = \Phi(s) D\Phi_{\omega\omega}(s) D^T \Phi^T(-s) e^{-sT_1} \tag{67}$$

$$L_2\left(\int_{-\infty}^{\infty} \Phi(-\sigma) D\phi_{\omega v}(\sigma - \lambda + \tau_O)\, d\sigma\right) = \Phi(s) D\Phi_{\omega v}(s) e^{-sT_1} \tag{68}$$

$$L_2\left(\int_{-\infty}^{\infty}\int_{-\infty}^{\infty} \Phi(T_1 - \sigma_1) D\phi_{\omega\omega}(\sigma_1 - \sigma_2) W^T(\lambda - \tau_0 - \sigma_2)\, d\sigma_2\, d\sigma_1\right)$$

$$= \Phi(s) D\Phi_{\omega\omega}(s) W^T(-s) \tag{69}$$

$$L_2\left(\int_{-\infty}^{\infty} \Phi(T_1 - \sigma) D\phi_{\omega v}(\sigma - \lambda + \tau_0)\, d\sigma\right) = \Phi(s) D\Phi_{\omega v}(s) \tag{70}$$

$$L_2\left(\int_{-\infty}^{\infty} \hat{h}(T - \sigma, \tau_1) \int_{-\infty}^{\infty}\int_{-\infty}^{\infty} W(\sigma - \sigma_1) \phi_{\omega\omega}(\sigma_1 - \sigma_2) W^T(\lambda - \sigma_2)\, d\sigma_2\, d\sigma_1\, d\sigma\right)$$

$$= \hat{H}(s, \tau_1) W(s) \Phi_{\omega\omega}(s) W^T(-s) e^{-s\tau_1} \tag{71}$$

$$L_2\left(\int_{-\infty}^{\infty} \hat{h}(T - \sigma, \tau_1) \phi_{vv}(\sigma - \lambda)\, d\sigma\right) = \hat{H}(s, \tau_1) \Phi_{vv}(s) e^{-s\tau_1} \tag{72}$$

$$L_2\left(\int_{-\infty}^{\infty} \hat{h}(T - \sigma, \tau_1) \int_{-\infty}^{\infty} W(\sigma - m) \phi_{\omega v}(m - \lambda)\, dm\, d\sigma\right)$$

$$= \hat{H}(s, \tau_1) W(s) \Phi_{\omega v}(s) e^{-s\tau_1} \tag{73}$$

$$L_2\left(\int_{-\infty}^{\infty} \hat{h}(T - \sigma, \tau_1) \int_{-\infty}^{\infty} \phi_{v\omega}(\sigma - m) W^T(\lambda - m)\, dm\, d\sigma\right)$$

$$= \hat{H}(s, \tau_1) \Phi_{v\omega}(s) W^T(-s) e^{-s\tau_1} \tag{74}$$

APPENDIX 3

Finite-Time Wiener-Hopf Equation

The optimal time-invariant estimator is shown below to satisfy a finite-time version of the Wiener-Hopf equation (16). Recall that $\hat{h}_1(\tau) = 0$ for all $\tau > T$, and let $\tau_0 = \tau_1 = 0$, then (16) becomes

$$\text{cov}[x_i(T), z(\lambda)] = \int_0^T \hat{h}_1(T - \sigma, 0) \,\text{cov}[z(\sigma), z(\lambda)]\, d\sigma \tag{75}$$

Let $x'(t)$ denote the state of the plant at time t with initial state x'_0, then from (30)

$$\operatorname{cov}\left[x_i(T), z(\lambda)\right] = E\{(x'(T) + (H_{10}C\Phi)(T)x(0^-))z^T(\lambda)\} + \bar{x}'(T)(\bar{z}'(\lambda))^T \tag{76}$$

and note that for σ, $\lambda > 0$

$$E\{z'(\sigma)z^T(\lambda)\} = E\{z'(\sigma)(v(\lambda) + Cx(\lambda))^T\} = E\{z'(\sigma)(z'(\lambda))^T\}$$

since

$$E\{z'(\sigma)(C\Phi(\lambda)x(0^-))^T\} = 0$$

Substituting into the Wiener-Hopf equation (75):

$$\operatorname{cov}\left[x'(T), z'(\lambda)\right] = \int_0^T \hat{h}_1(T - \sigma, 0) \operatorname{cov}\left[z'(\sigma), z'(\lambda)\right] d\sigma . \tag{77}$$

6.3 BAYESIAN DETECTION AND ESTIMATION OF JUMPS IN LINEAR SYSTEMS

A. F. M. Smith
(University of Nottingham, UK)

and

U. E. Makov
(Chelsea College London, UK)

1 INTRODUCTION

In recent years, there has been a great deal of interest in the problems of detecting and estimating jumps in systems modelled by the linear Kalman-Bucy filter as well as in the problem of ensuring satisfactory state estimation in the possible presence of such jumps. See [1]-[7], and the numerous references contained therein, for a reasonably complete coverage of the literature.

Our purposes in this paper are threefold. First, to note that a unified overview of the problems and methods proposed in the literature can be achieved by representing the system, together with possible jumps, as a set of alternative possible Kalman-Bucy filters. Secondly, to note the computational problems involved in the optimal (Bayesian) updating procedures applied to this set of filters, and to examine approximations to the optimal solution. Thirdly, we report on various simulation studies we have made of the performance of such approximate solutions, and we attempt to draw some tentative conclusions regarding their value for jump detection and estimation as well as for state estimation.

2 A LINEAR SYSTEM WITH POSSIBLE JUMPS

We shall consider the discrete-time linear system

$$x(k+1) = A(k)x(k) + B(k)u(k+1) + \Gamma(k)w(k) \tag{1}$$

$$z(k+1) = H(k+1)x(k+1) + v(k+1) \tag{2}$$

where all vectors and matrices are assumed appropriately dimensioned, $w(k)$ and $v(k+1)$ are independent, zero-mean Gaussian sequences with known covariance matrices Q and R, respectively, $x(0)$ is the initial state condition, assumed to have a Gaussian distribution with mean $\hat{x}(0)$ and covariance matrix $P^x(0|0)$, and $u(k+1)$ represents possible jumps in the states of the system.

The interpretation, and hence the representation, of the jumps will, of course, depend on the context being modelled. We shall here consider two of the many possible assumptions that might be made:

(i) $B(k) = I$ and the $u(k+1)$ take values in a finite set of known vectors $\{u_0, u_1, \ldots, u_m\}$, where $u_0 = 0$ corresponds to "no jump";

(ii) the $u(k+1)$ take values in the set $\{u_0, u_1\}$, where $u_0 = 0$ and u_1 is unknown, but a priori assigned a Gaussian distribution with mean $\hat{u}_1(1)$ and covariance matrix $P^u(0|0)$.

The first assumption might be appropriate, for example, in applications to tracking a manoeuvering target, where the jumps correspond to pilot choices. It is often possible to choose a set of values $u_1, \ldots, u_m$ in such situations to reflect adequately the range of options open to the pilot. The second assumption might well be reasonable in situations where a system is subject to an impulse having the same, but unknown, effect whenever it occurs.

We shall now show that both (i) and (ii), in conjunction with (1) and (2), lead to a model which consists of a set of alternative Kalman-Bucy discrete-time filters.

Let us first consider (i). By defining

$$u^* = \left[u_0^T, u_1^T, \ldots, u_m^T\right]^T , \tag{3}$$

$$x^* = \begin{bmatrix} x(k) \\ u^* \end{bmatrix}, \tag{4}$$

$$A_i^*(k) \begin{bmatrix} A(k) & \delta_{i,0}B(k) \ldots \delta_{i,m}B(k) \\ 0 & I \end{bmatrix} \tag{5}$$

and

$$H^{*}(k+1) = \left[H(k+1) \quad 0\right], \tag{6}$$

where $\delta_{i,j}$ is the usual Kroenecker delta, we see that the possible system jumps $u(k+1) = u_i$, $i = 0,\ldots,m$, correspond to the set of alternative Kalman-Bucy discrete-time filters, M_i, $i = 0,\ldots,m$, given by

$$x^{*}(k+1) = A_i^{*}(k)x^{*}(k) + \Gamma^{*}(k)w(k) \tag{7}$$

$$z(k+1) = H^{*}(k+1)x^{*}(k+1) + v(k+1) \tag{8}$$

where $\Gamma^{*}(k) \triangleq \begin{bmatrix} \Gamma(k) \\ 0 \end{bmatrix}$.

The a priori Gaussian distribution for $x^{*}(0)$ has mean equal to $\left[\hat{x}(0) \;\; u^{*}\right]^T$ and covariance matrix $P(0|0)$ with leading block diagonal $P^{x}(0|0)$ and zeros elsewhere.

In the case of assumption (ii), we again arrive at equations (3) - (8), but with $m = 1$, and an a priori Gaussian distribution for $x^{*}(0)$ having mean $\left[\hat{x}(0) \;\; 0 \;\; \hat{u}_1(1)\right]^T$ and covariance matrix $P(0|0)$ of the form

$$P(0|0) = \begin{bmatrix} P^{x}(0|0) & 0 & 0 \\ 0 & 0 & 0 \\ 0 & 0 & p^{u}(0|0) \end{bmatrix} \tag{9}$$

We now examine what happens when the Bayesian algorithm is applied, successively, to equations (7) and (8). Let us suppose that we have observed $Z(n) = \left[z(1), z(2), \ldots, z(n)\right]$, and we wish to compute $p\left[x^{*}(n) \,|\, Z(n)\right]$, the current joint probability density for the components of $x^{*}(n)$, given $Z(n)$.

If the true sequence of jump values (or equivalently $A_i^{*}(k)$ matrices) were known, this (Gaussian) density would have a known form, corresponding to successive application of the Kalman-Bucy updating equations using the appropriate matrices. (See, for example, [8].) In general, there are $(m+1)^n$ possible sequences of the type $(M_{i_1}, M_{i_2}, \ldots, M_{i_n})$. We shall call S(n) the set of all possible sequences, and denote by s(n) a generic element of this set. It then follows that

$$p[x^*(n)|Z(n)] = \sum_{s(n)\varepsilon S(n)} p[x^*(n)|Z(n),s(n)]p[s(n)|Z(n)]. \quad (10)$$

As we have already remarked, the first factor within the summation is, for given s(n), a Gaussian density with known mean and covariance matrix, obtained from successive application of the standard updating equations.

So far as the second factor is concerned, by Bayes theorem we have

$$p[s(n)|Z(n)] \propto p[Z(n)|s(n)]\ p[s(n)], \quad (11)$$

where

$$p[Z(n)|s(n)] = p[z(n)|Z(n-1),s(n)]..p[z(2)|Z(1),s(2)]p[z(1)|s(1)]. \quad (12)$$

and, here, $s(j) = [M_{i_1}, M_{i_2}, \ldots, M_{i_j}]$, and $p[s(n)]$ is the a priori probability attached to the model sequence s(n). For example, in an obvious notation, we might make the Markov assumption

$$p[s(n)] = \pi(i_1)\ \pi_{i_1 i_2} \ldots \pi_{i_{n-1} i_n}. \quad (13)$$

In general, we can choose the $p[s(n)]$ to reflect our knowledge about the frequency and type of jumps in any particular application.

From (12), we see that we require terms of the form

$$p[z(j)|Z(j-1),s(j)] = \int p[z(j)|x^*(j)]p[x^*(j)|Z(j-1),s(j)]dx^*(j). \quad (14)$$

By standard Kalman-Bucy theory, this is easily seen to be of Gaussian form with mean

$$H^*(j)\hat{x}^*[j|j-1,\ s(j)] \tag{15}$$

and covariance matrix

$$R + H^*(j)P[j|j-1,\ s(j)]\,[H^*(j)]^T, \tag{16}$$

where the new terms appearing in (15) and (16) are defined by the standard results

$$p[z(j)|x^*(j)] \equiv N\{H^*(j)x^*(j),\ R\} \tag{17}$$

$$p[x^*(j)|Z(j-1),s(j)] \equiv N\{\hat{x}^*[j|j-1,s(j)],\ P[j|j-1,s(j)]\}. \tag{18}$$

<u>Inference about the states x(k) of the original system</u> can be based upon the appropriate components of (10). Current estimates of the states have the form of weighted averages of state estimates based upon particular jump histories. The weights are the posterior probabilities of these particular jump histories.

<u>Detection of jumps</u> can be based upon these latter probabilities, given by equation (11). For example, in both cases (i) and (ii), $p[u(k+1) = u_i|Z(n)]$ is obtained by summing (11) over all s(n) such that $u(k+1) = u_i$ $(k \leq n-1)$. These probabilities might be combined with a suitable loss structure if formal decisions about the occurrence of jumps are required.

<u>Inference about jump size</u>. In case (i), the components of the joint density in (10) corresponding to u^* have a degenerate distribution since the u_i are, a priori, assumed known, and inferences about jump size at particular time points are again based on equation (11). In case (ii), the components of (10) corresponding to u_1 have a marginal distribution which is a weighted average of marginal Gaussian distributions each conditioned on a particular s(n).

Although this analysis provides, in a sense, a complete solution to the problem, (given the model we have chosen), we note that there are a great many difficulties in its implementation.

We see from (10) that, for n observations, the complete updating procedure requires the processing (and associated computer storage) of up to $(m+1)^n$ possible system histories, and the carrying out of the corresponding Kalman-Bucy filter calculations. Of course, in practice this number may be considerably smaller if we adopt a prior distribution over S(n) which assigns zero to many possible histories, but even with constraints of this type, there will be some kind of computational and storage explosion as n increases.

For this reason, we consider in the next section some possible alternatives and approximations to the full analysis of section 2.

3 APPROXIMATE PROCEDURES

There appear to be two general approaches to cutting through the computational explosion of the optimal solution.

(i) Selection. As n increases, the possible histories of the system build up in a tree-like fashion. One way of preventing this is to select a branch (or a small number of branches) at each stage, thus acting as if there is a single history (or only a small number). This procedure is akin to Decision Directed procedures as encountered in unsupervised learning contexts. Indeed, the problem under consideration is precisely an example of an unsupervised learning procedure. A general study of approximations based on selection would need to consider how often to select, how many branches to select and how to select them. We know of no proper study of this problem. Particular approaches have been considered in [1], [4] and [2]. The former two papers deal with selections based upon generalized likelihood ratio tests, the latter considers Bayes decision rules for observations considered three at a time assuming that at most one jump has occurred. We shall consider later a scheme (which we call (DD)) based on choosing the most likely branch after each observation and then assuming this to be the correct decision. Thus equation (10), for n = 1, reduces to $p[x^*(1)|Z(1), \tilde{s}(1)]$, where $\tilde{s}(1)$ maximizes $p[s(1)|Z(1)]$ over S(1). This acts now as an a priori (Gaussian) distribution for z(2), and the process is repeated.

(ii) Averaging. Instead of selecting branches and basing inferences on the assumption that the chosen branch is the true one, we could try to approximate (10) by a smaller mixture of Gaussian densities, in order to keep storage and computational problems within bounds. A general study of approximations based on averaging would need to consider how often to average, how much collapsing is required in forming a smaller mixture, how should the process of collapsing be carried out. An important reference here (not much referred to in the Engineering literature) is [9]. Given a set of (m+1) Kalman-Bucy discrete-time filters as in (7) and (8), the proposal is to collapse the resulting mixture of $(m+1)^2$ densities, after each observation, to a mixture of (m+1). The mixture of (m+1) component densities arising from branches from each model M_i are replaced by a single Gaussian density having the same mean and covariance matrix.

A similar proposal, but collapsing from (m+1) to a single density at each stage is presented in [10]. It proceeds as follows. Writing

$$p[x^*(1)|Z(1),s(1)] \equiv N\{\hat{x}^*[1|1,s(1)],\ P[1|1,\ s(1)]\}$$

we replace (10), after one observation, by a Gaussian distribution with mean and covariance matrix given by

$$\hat{x}^*(1) = \sum_{s(1)\varepsilon S(1)} P[s(1)|Z(1)]\hat{x}^*[1|1,\ s(1)]$$

$$P(1|1) = \sum_{s(1)\varepsilon S(1)} P[s(1)|Z(1)]\{P[1|1,s(1)] + [\hat{x}^*[1|1,s(1)] - \hat{x}^*(1)]$$

$$[\hat{x}^*[1|1,s(1)] - \hat{x}^*(1)]^T\},$$

respectively. The procedure is then repeated for the second observation using this as an a priori distribution, and so on. We shall refer to this Probabilistic Editing procedure as the (PE) method.

An alternative, even simpler, form of approximation is the following. Considering equation (5), we note that supervised learning for the system given by (7) and (8) would correspond to knowing, for each $A_i(k)$, whether the corresponding delta was 0 or 1. Instead, equation (11) enables us to find the expected value of each delta, given the actual, unsupervised learning situation.

Given (11), we can calculate $\pi_i(k)$, the probability that model M_i applied at time k, by summing (11) over all sequences which have M_i at time k. We then form

$$A^*(k) = \sum_i \pi_i(k)\ A_i^*(k)$$

and reprocess Z(n) using (7) and (8) with $A^*(k)$ in place of $A_i^*(k)$ in (7). This is equivalent to replacing all the delta's by their expected values, given Z(n), as a method of approximating the supervised solution. We shall call this the Quasi-Bayes (QB) approach. Studies of this type of approximation in other contexts are given in [11], [12] and [13]. Although somewhat further removed from the optimal solution than the (PE) method, (QB) requires much less computer storage, and thus its performance is worth studying if storage constraints are present.

4 SIMULATION STUDIES

The studies we shall report here all relate to a situation where the state vector x(k) has four dimensions, positions and velocities with respect to rectangular coordinates, and the jump structure is that of assumption (ii) of section 2. In fact, we shall always take $B(k) = [1\ 1\ 0\ 0]^T$ and $u_1 = 1$, with prior variance unity. In other words, when jumps occur, they are unit additions to the current position coordinates. $P^x(0|0)$ is always taken to be the unit matrix, and

$$A(k) = \begin{bmatrix} 1 & 0 & 1 & 0 \\ 0 & 1 & 0 & 1 \\ 0 & 0 & 1 & 0 \\ 0 & 0 & 0 & 1 \end{bmatrix}, \quad \Gamma(k) = \begin{bmatrix} 1 & 0 \\ 0 & 1 \\ 1 & 0 \\ 0 & 1 \end{bmatrix}, \quad H(k+1) = \begin{bmatrix} 1 & 0 \\ 0 & 1 \\ 0 & 0 \\ 0 & 0 \end{bmatrix}^T .$$

Our first study compared the (DD), (PE) and (QB) approximations based on selection or averaging after each single observation. A number of simulations were made with varying run conditions. In particular, various choices were made of the true initial state, x(0), the estimated initial state and jump, $\hat{x}(0)$ and $\hat{u}_1(1)$, state noise and measurement noise, Q and R (both assumed to be diagonal matrices), and true jump probability π (assumed constant, with independence of u(k) for different k).

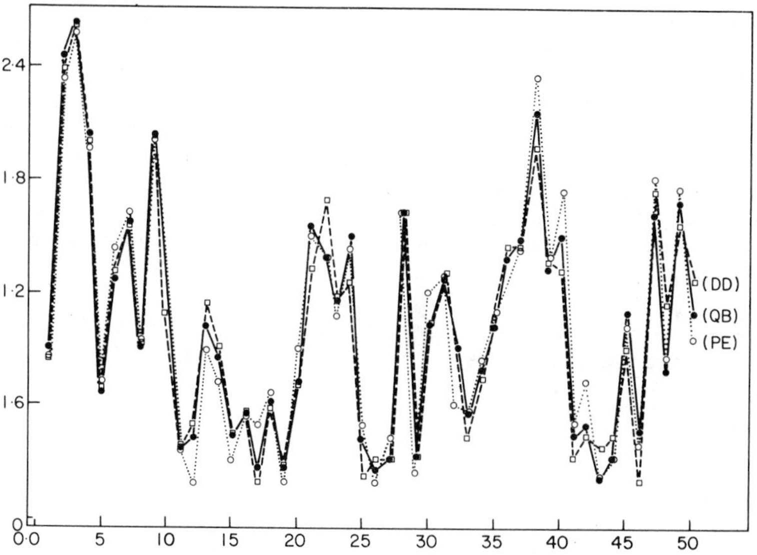

Fig. 1 Tracking error versus number of observations

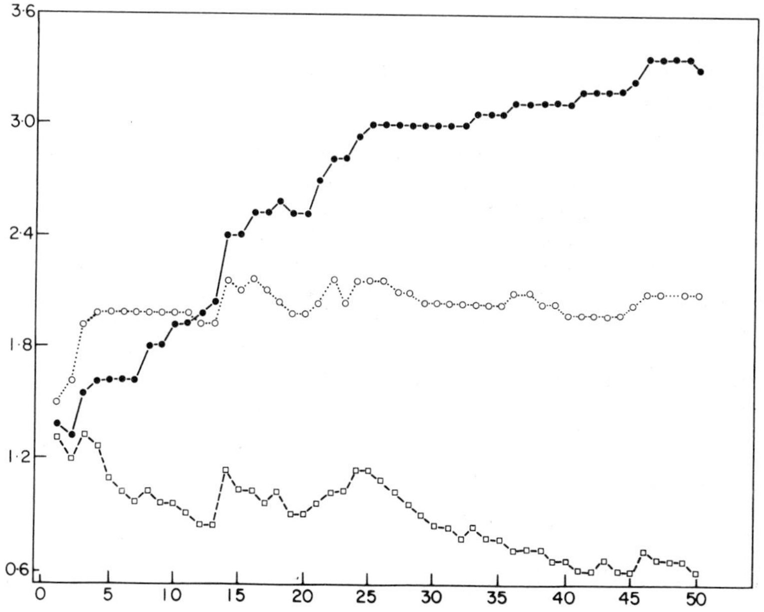

Fig. 2 Jump estimates versus number of observations

The conclusions are as follows. For a wide range of assumptions, the three approximations all display a comparable and satisfactory performance in tracking (as measured by the Euclidean distance of the estimated position from the true position). If we "decide" on a jump occurrence if the estimated jump probability is greater than one-half, again all three methods perform similarly, though not very satisfactorily. So far as estimation of jump size is concerned, their behaviour is very different. The (PE) method is almost always the most satisfactory, and (QB) the least.

Figs. 1 and 2 show, respectively, the tracking and jump size estimation performances for the first 50 observations simulated under the following conditions:

$x(0)$: 0.0 0.0 1.0 0.0

$\hat{x}(0)$: -5.0 -5.0 3.0 3.0

$u_1 = 1,\ \hat{u}_1(1) = 0.5,\ Q_{ii} = 0.25,\ R_{ii} = 0.75,\ \pi = 0.5$

The total numbers of detection errors (false positives plus false negatives) were:

(DD) = 21, (QB) = 19, (PE) = 19.

These plots are very typical of those arising from a wide range of conditions.

The (PE) method is consistently either comparable with (QB), or better, whereas (QB) is somewhat easier to calculate. To obtain some insight into the improvements attainable by allowing n to increase from 1 before collapsing the tree, we have studied the changing performance of (QB) for different n. We conclude that there can be striking improvements in performance in detecting jumps, and some degree of improvement in tracking and jump estimation. Figs. 3 and 4 show the absolute errors in tracking and estimation, respectively, for the first 48 observations simulated from the following situation:

$x(0)$: 0.0 0.0 1.0 0.0

$\hat{x}(0)$: -5.0 -5.0 3.0 3.0

$u_1 = 1,\ \hat{u}_1(1) = 0.5,\ Q_{ii} = 0.25,\ R_{ii} = 0.75,\ \pi = 0.25.$

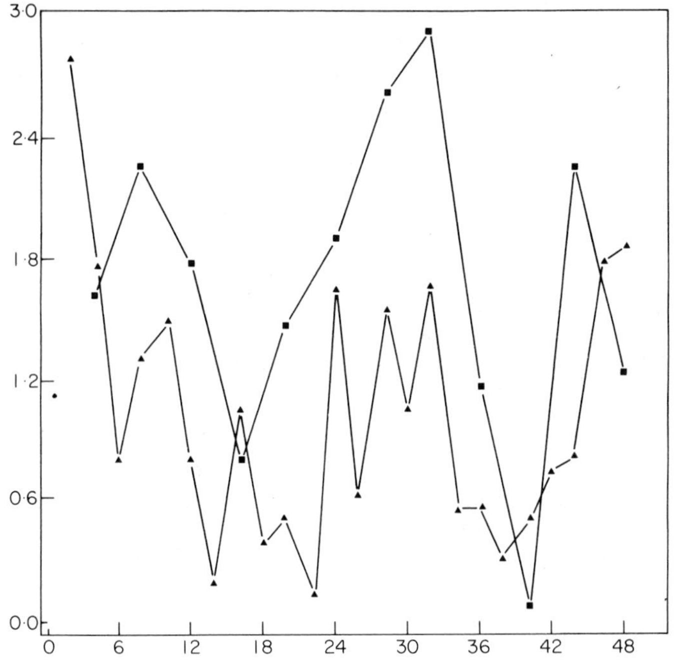

Fig. 3 Absolute tracking error versus number of observations

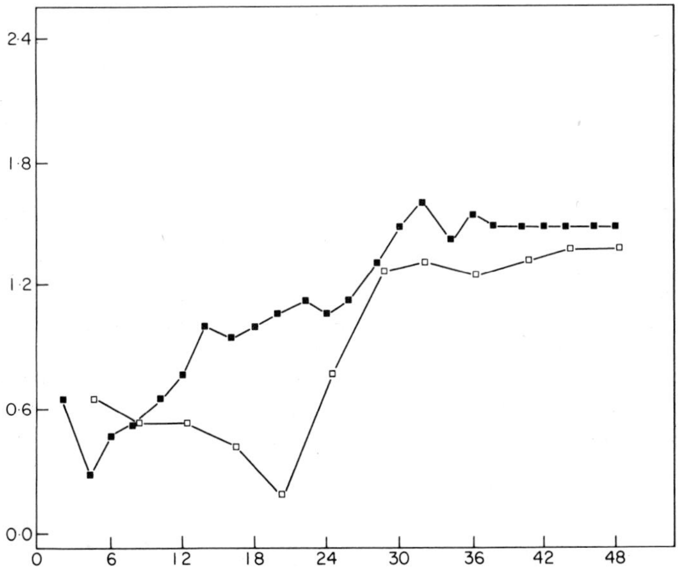

Fig. 4 Absolute jump estimation error versus number of observations

Comparison is made with collapsing after n = 2 and after n = 4. The total number of detection errors were

n = 2 33 errors

n = 4 16 errors.

A reduction to about 50% is typical of the improvements obtained for a wide range of conditions.

We have further studied the effects of strong prior knowledge regarding the number, but not location, of jumps. For example, if after n = 5 we are willing to state that we know there to have been 3 jumps, we can reduce from considering $2^5 = 32$ possible histories to ${}^5C_3 = 10$ histories. (QB) and, hence, (PE) perform very well on studies of collapsing after 5 observations using this smaller number of possible histories. If the number of jumps is unknown, another possibility is to use (11) to estimate the most likely number of jumps and then to reprocess the observations as if this were the true number. (QB) and, hence, (PE) seem to perform quite well using this approach.

5 CONCLUSIONS

On the basis of the limited studies undertaken thus far, we believe that the (PE) approximation gives a satisfactory performance for all aspects of estimation and jump detection. Naturally, it is best to allow n to increase as far as computational constraints allow before approximating the optimal solution given by (10) and (11).

If computational constraints (particularly storage) make (PE) unfeasible, we believe that (QB) offers a good alternative for tracking, but performs badly for jump estimation if n is small. Jump detection errors can be cut dramatically by allowing the tree to develop at least as far as n = 4.

If prior information about the number of jumps occurring within a particular sampling window is available, this again improves dramatically the performance of all the approximations.

6 REFERENCE

1. Willsky, A. S. and Jones, H. L. "A Generalized Likelihood Ratio approach to State Estimation in Linear Systems subject to Abrupt Changes, Proc. IEEE Conf. on Dec. and Contr., (1974).

2. Sanyal, P. "Bayes' Detection Rule for Rapid Detection and Adaptive Estimation Scheme with Space Applications", *IEEE Trans. Automat. Contr.*, **AC-19**, 228-231, (1974).

3. Moose, R. L. "An Adaptive State Estimation Solution to the Manoeuvering Target Problem", *IEEE Trans. Automat. Contr.*, **AC-20**, 359-362, (1975).

4. Willsky, A. S. and Jones, H. L. "A Generalized Likelihood Ratio Approach to the Detection and Estimation of Jumps in Linear Systems", *IEEE Trans. on Automat. Contr.*, **AC-21**, 108-112, (1976).

5. Murphy, D. J. "Batch Estimation of a Jump in the State of a Stochastic Linear System", *IEEE Trans. on Automat. Contr.*, **AC-22**, 275-276, (1977).

6. Gholson, N. H. and Moose, R. L. "Manoeuvering Target Tracking using Adaptive State Estimation", *IEEE Trans. Aerosp. Electron. Syst.*, **AES-13**, 310-317, (1977).

7. Vaca, M. V. and Tretter, S. A. "Optimal Estimation for Discrete Time Jump Processes", *IEEE Trans. on Inf. Th.*, **IT-24**, 289-296, (1978).

8. Jazwinski, A. H. "Stochastic Processes and Filtering Theory", Academic Press, New York, (1970).

9. Harrison, P. J. and Stevens, C. F. "Bayesian Forecasting", *J. R. Statist. Soc. B.*, **38**, 205-247, (1976).

10. Athans, M., Whiting, R. H. and Gruber, M. "A Suboptimal Estimation Algorithm with Probabilistic Editing for False Measurements with Applications to Target Tracking with Wake Phenomena", *IEEE Trans. on Automat. Contr.*, **AC-22**, 372-384, (1977).

11. Makov, U. E. and Smith, A. F. M. "A Quasi-Bayes Approximation to Unsupervised Filters", Proc. Conf. on Measurement and Contr., (Meco. 77), 102-106, (1977).

12. Makov, U. E. and Smith, A. F. M. "A Quasi-Bayes Unsupervised Learning Procedure for Priors", *IEEE Trans. on Inf. Th.*, **IT-23**, 761-763, (1977).

13. Smith, A. F. M. and Makov, U. E. "A Quasi-Bayes Sequential Procedure for Mixtures", *J. R. Statist. Soc. B*, **40**, 106-112, (1978).

6.4 A STOCHASTIC VARIATIONAL APPROACH TO THE DUALITY BETWEEN ESTIMATION AND CONTROL: DISCRETE-TIME

M. Pavon and R. J. B. Wets

(University of Kentucky, Lexington, USA)

1 INTRODUCTION

In this paper we gave a new derivation of a duality relation between estimation problems and certain associated control problems, first exhibited by Kalman in 1960. The novelty of this approach resides in the fact that we do not rely on formal arguments. We first show that the estimation problem can be viewed as a stochastic optimization problem that can be embedded in a class of variational problems of the Bolza type. A natural construction leads us to a stochastic dual problem. We then show how various assumptions are needed, in order to reduce this stochastic dual problem to a deterministic problem. We restrict ourselves to the discrete time case.

2 THE ONE-STEP ESTIMATION PROBLEM

By $(w_t, t = 0,1,\dots,T)$ we denote a vector-valued discrete-time Gaussian process defined on $\{0,1,\dots,T\}\times\Omega$ with values in R^p where (Ω,A,P) is the underlying probability space. We write tl for t + 1 and thus in particular Tl = T + 1. The state dynamics are described by the system of finite difference equations: for $t = 1,\dots,T$

$$x_{t1}(\omega) = A_t x_t(\omega) + B_t w_t(\omega) \quad \text{a.s.},$$

or equivalently

$$\Delta x_t(\omega) = (A_t - I)x_t(\omega) + B_t w_t(\omega) \quad \text{a.s.},$$

with initial conditions

$$x_1(\omega) = B_0 w_0(\omega) \quad \text{a.s.}$$

where

$$\Delta x_t(\omega) = x_{t1}(\omega) - x_t(\omega).$$

The n-vector x_t represents the *state* of the system. The matrices A_t and B_t are n×n and n×p respectively and I is the n×n identity.

Note that $(x_t, t = 1,\ldots,T,T1)$ is a discrete-time vector valued Gaussian process, since for t = 0,...,T we have that

$$x_{t1}(\omega) = \Sigma_{\tau=0}^{t} (\Pi_{\sigma=t}^{\tau 1} A_\sigma) B_\tau w_\tau(\omega) \quad \text{a.s.,}$$

with the convention that $\Pi_{\sigma=t}^{t1} A_\sigma = I$.

Instead of the actual state of the system we observe a vector $y_t \in R^m$ which depends linearly on x_t, itself disturbed by an additive "noise", namely

$$y_t(\omega) = C_t x_t(\omega) + D_t w_t(\omega) \quad \text{a.s.}$$

where C_t and D_t are respectively m×n and m×p matrices. The information process $(y_t, t = 1,\ldots,T)$ is also a discrete-time Gaussian process; for t = 1,...,T we have that

$$y_{t1}(\omega) = C_{t1}\left[\Sigma_{\tau=0}^{t} (\Pi_{\sigma=t}^{\tau 1} A_\sigma) B_\tau w_\tau(\omega)\right] + D_{t1} w_{t1}(\omega) \quad \text{a.s.}$$

Let $\mathcal{Y}$ be the sigma-field induced on Ω by the random vectors $(y_t, t = 1,\ldots,T)$, i.e. $\mathcal{Y}$ is the smallest sigma-field with respect to which all random variables y_t are measurable. The $\mathcal{Y}$-measurability of a function means that it depends only on the information.

The *one-step estimation (or prediction) problem* consists in finding a function γ yielding the "best" estimate of x_{T1} on the basis of the information collected. More precisely, we need to find a $\mathcal{Y}$-measurable function γ from Ω into R^n which minimizes

$$J(\gamma) = E\{\tfrac{1}{2}\| x_{T1}(\omega) - \gamma(\omega)\|^2\}$$

where $\|\cdot\|$ denotes the Euclidean norm. Since $x_{T1} \in L^2(\Omega, \mathcal{A}, P; R^n) = L_n^2(\mathcal{A})$ it is natural to restrict γ to the same class of functions. Also the criterion might not be well-defined otherwise. This means that γ must belong to $L_n^2(\mathcal{Y})$, a closed linear subspace of $L_n^2(\mathcal{A})$. We can thus express the one-step estimation problem as follows:

EP Find $\gamma \in L_n^2(\mathcal{Y}) \subset L_n^2(\mathcal{A})$ such that $J(\gamma)$ is minimized.

This is an optimal recourse problem [4], [5]. The recourse function γ must be nonanticipative, i.e. $\mathcal{Y}$-rather than $\mathcal{A}$-measurable. The objective is convex, in fact strictly convex. Thus the problem admits a unique solution γ^* which must satisfy the following conditions [4]: for almost all ω,
$\gamma^*(\omega) = \arg\min_\gamma [\tfrac{1}{2}\| x_{T1}(\omega) - \gamma\|^2 - \rho(\omega)\cdot\gamma]$ where $\rho \in L_n^2(\mathcal{A})$ with $(E^{\mathcal{Y}}\rho)(\omega) = 0$ a.s., it then follows that,

$$\gamma^*(\omega) =_{a.s.} (E^{\mathcal{Y}} x_{T1})(\omega).$$

This is neither a surprising nor a new result, but this derivation is uncommonly simple and provides an alternative to the standard projection-type arguments [1].

In the following sections we obtain a dual problem associated with EP. We do this in the framework developed for stochastic variational problems. The "duality" between EP and a certain deterministic linear-quadratic regulator problem, already observed by R. Kalman [2] acquires in this context a precise interpretation, which until now was believed to be of a purely formal nature. This approach provides new insight into the important relationship between estimation and control. It shows how some of the assumptions can be relaxed without impairing the basic results. Finally it opens some new avenues for the study of "non-standard" estimation problems.

3 A VARIATIONAL FORMULATION OF THE ESTIMATION PROBLEM

Under the assumptions of section 2, since $(y_t, t = 1,\dots,T)$ is a Gaussian process and $\mathcal{Y}$ is the sigma-field generated by $(y_t,\ 1 \leq t \leq T)$, it follows that if $z \in L^2(\mathcal{Y})$ and z is a minimum point there exist $u_{ij} \in R$, $i = 1,\dots,m, j = 1,\dots,T$ such that

$$z(\omega) = \Sigma_{ij} u_{ij} y_{ij}(\omega)$$

In particular, if $\gamma \in L_n^2(\mathcal{Y})$, then

$$\gamma(\omega) = -\Sigma_{t=1}^{T} U_t y_t(\omega)$$

where the U_t are (fixed) n by m matrices, i.e. γ is a linear combination of the observations. The minus sign is introduced for symmetry purposes and will be justified by the further development. One way to interpret the above equation is as follows: each observation y_t contributes incrementally to the construction of the estimator. Define

$$\Delta\gamma_t(\omega) = -U_t(C_t x_t(\omega) + D_t w_t(\omega)) \qquad t = 1,\dots,T$$

$$\gamma_1(\omega) = 0.$$

where $\Delta\gamma_t = \gamma_{t1} - \gamma_t$. These equations provide us with a description of the *estimation* process. A collection $U = (U_s,\ 1 \leq s \leq T)$ is a sequence of *estimator* weights. They provide a way to process the information to construct an estimator of x_{T1}. Finding the best estimator is thus equivalent to finding optimal weights. The original problem can be given the following equivalent formulation:

WP Find U that minimizes f(U)

where $f(U) = \text{Min } E\{\Phi_{\ell,L}(\omega, x(\omega), \gamma(\omega); U) \mid x \in L_N^2(A),\ \gamma \in L_N^2(A)\}$

$$\Phi_{\ell,L}(\omega,x,\gamma) = \ell(\omega, x_1, \gamma_1, x_{T1}, \gamma_{T1}) + \Sigma_{t=1}^{T} L_t(\omega, x_t, \gamma_t, \Delta x_t, \Delta\gamma_t; U_t)$$

$$\ell(\omega,x_1,\gamma_1,x_{T1},\gamma_{T1}) = \begin{cases} \tfrac{1}{2}\|x_{T1} - \gamma_{T1}\|^2 & \text{if } \begin{cases} x_1 = B_O w_O(\omega) \\ \gamma_1 = O \end{cases} \\ +\infty & \text{otherwise} \end{cases}$$

$$L_t(\omega,x_t,\gamma_t,\Delta x_t,\Delta\gamma_t;U_t) = \begin{cases} O & \text{if } \begin{cases} \Delta x_t = (A_t - I)x_t + B_t w_t(\omega) \\ \Delta\gamma_t = -U_t(C_t x_t + D_t w_t(\omega)) \end{cases} \\ +\infty & \text{otherwise.} \end{cases}$$

and $N = n\cdot T1$.

The problem is thus to find optimal weights; to weights U correspond an (optimal) estimator obtained by satisfying the dynamical equations. In other words, the value of f(U) is determined by solving a discrete-time stochastic Bolza problem.

The dual of this variational problem yields a new representation of f(U). It is in this form that we are able to exploit the specific properties of this problem, in particular to reduce WP to a deterministic control problem.

4 THE DUAL REPRESENTATION OF f

For fixed U, we consider the Bolza problem.

$$\text{VP} \quad \text{Min } E\{\Phi_{\ell,L}(\omega,x(\omega),\gamma(\omega);U) \mid x \in L^2_N(A),\ \gamma \in L^2_N(A)\}.$$

We embed VP in a class of Bolza problems, by submitting the state variables to (global) variations. Let ψ express the dependence of the minimum of VP on these variations, specifically.

$$\psi(r,\eta;U) = \inf_{(x,\gamma)} E\{\Phi_{\ell,L}(\omega,x(\omega)\frown r(\omega),\gamma(\omega)\frown\eta(\omega);U)\}$$

where both r and η are in $L^2_N(A)$; more precisely

$$\Phi_{\ell,L}(\omega,x\frown r,\gamma\frown\eta) = \ell(\omega,x_1 + r_0,\gamma_1 + \eta_0,r_{T1},\gamma_{T1})$$

$$+ \Sigma_{t=1}^{T} L_t(\omega,x_t,\gamma_t,\Delta x_t + r_t,\Delta\gamma_t + \eta_t).$$

Costate variables are paired with the variations through the bilinear form:

$$<(q,\alpha), (r,\eta)> = E\{\Sigma_{t=0}^{T}(q_t(\omega)\cdot r_t(\omega) + \alpha_t(\omega)\cdot\eta_t(\omega))\}.$$

The problem VD, dual to VP, is obtained as the conjugate of ψ, namely

$$\text{VD} \quad \text{Min } E\{\Phi_{m,M}(\omega,q(\omega),\alpha(\omega);U) \,|\, q \in L_N^2(A),\ \alpha \in L_N^2(A)\}$$

where

$$E\{\Phi_{m,M}(\omega,q(\omega),\alpha(\omega);U)\} = \sup_{r,\eta}\left[<(q,\alpha),(r,\eta)> - \psi(r,\eta;U)\right].$$

The notation $\Phi_{m,M}$ appearing in VD is consistent, as we shall see, with the fact that

$$\Phi_{m,M}(\omega,q,\alpha;U) = m(\omega,q_0,\alpha_0,q_T,\alpha_T) + \Sigma_{t=1}^{T} M_t(\omega,q_t,\alpha_t,\Delta q_t,\Delta\alpha_t)$$

where

$$m(\omega,q_0,\alpha_0,q_T,\alpha_T) = \ell^*(\omega,q_0,\alpha_0,-q_T,-\alpha_T)$$

$$= \sup_{(r_1,\eta_1,r_{T1},\eta_{T1})}\left[q_0\cdot x_1+\alpha_0\cdot\gamma_1-q_T\cdot x_{T1}-\alpha_T\cdot\gamma_{T1}+\ell(\omega,x_1,\gamma_1,x_{T1},\gamma_{T1})\right]$$

$$= \begin{cases} q_0 B_0 w_0(\omega)+\tfrac{1}{2}\|\alpha_T\|^2 & \text{if } q_T = -\alpha_T \\ +\infty & \text{otherwise} \end{cases}$$

and

$$M_t(\omega,q_t,\alpha_t,\Delta q_t,\Delta\alpha_t;U_t) = L_t^*(\omega,\Delta q_t,\Delta\alpha_t,q_t,\alpha_t)$$

$$= \sup_{(x_t,\gamma_t,\Delta x_t,\Delta\gamma_t)} \Delta q_t \cdot x_t + \Delta\alpha_t \cdot \gamma_t + q_t \cdot \Delta x_t + \alpha_t \cdot \Delta\gamma_t - L_t(\omega,x_t,\gamma_t,\Delta x_t,\Delta\gamma_t;U)$$

$$= \begin{cases} (q_t B_t - \alpha_t U_t D_t) w_t(\omega) & \text{if } \begin{cases} -\Delta q_t = q_t(A_t - I) - \alpha_t U_t C_t \\ -\Delta\alpha_t = 0 \end{cases} \\ +\infty & \text{otherwise.} \end{cases}$$

with

$$\Delta q_t = q_t - q_{t-1} \quad \text{and} \quad \Delta\alpha_t = \alpha_t - \alpha_{t-1}\ ^{\dagger}.$$

THEOREM. Problems VP and VD are solvable. Moreover if (x^*,γ^*) is an optimal solution of VP and (q^*,α^*) is an optimal solution of VD, then

$$\Phi_{\ell,L}(\omega,x^*(\omega),\gamma^*(\omega);U) = -\Phi_{m,M}(\omega,q^*(\omega),\alpha^*(\omega);U) \quad \text{a.s.}$$

Proof. (Sketch) Problem VP admits only one feasible solution (a.s.), the optimal one. As for VD note that $\Delta\alpha_t(\omega) =_{\text{a.s.}} 0$ for all t. Thus $\alpha_t(\omega) =_{\text{a.s.}} \alpha_T(\omega)$. On the other hand from the definitions of M_t and Δq_t we have that

$$q_{t-1}(\omega) =_{\text{a.s.}} q_t(\omega)A_t - \alpha_T(\omega)U_t C_t.$$

---

† For matrix multiplication, we adopt the convention that a vector appearing to the left of a matrix is a row vector, it is a column vector if it appears to the right. Thus if a and b are two vectors, by ab or a•b we mean the inner product of two vectors. Otherwise we rely on the notation a □ b, which denotes the matrix obtained by multiplying the column vector a by the row vector b.

Combining this with the fact that a feasible solution to VD must satisfy $q_T(\omega) = -\alpha_T(\omega)$, we obtain a recursive definition for each q_{t-1} by substitution, in particular.

$$q_{T-\ell}(\omega) =_{a.s.} -\alpha_T(\omega)\left[\Sigma_{s=0}^{\ell} U_{T1-s}C_{T1-s}(\Pi_{\tau=T}^{T1-\ell+s} A_\tau)\right]$$

where by definition

$$U_{T1}C_{T1} = I$$

and we use the convention

$$\Pi_{\tau=T}^{T1-\ell+s} A_\tau = I \quad \text{when } T1 - \ell + s > T, \text{ i.e. } s + 1 > \ell.$$

Let

$$Q_{T-\ell} = \Sigma_{s=0}^{\ell} U_{T1-s}C_{T1-s}(\Pi_{\tau=T}^{T1-\ell+s} A_\tau),$$

with this notation we have that

$$q_t(\omega) = -\alpha_T(\omega)Q_t.$$

Thus the dual variational problem is equivalent to the following problem:

VD$'$ Find $\alpha_T \in L_n^2(A)$ such that

$$E\{\tfrac{1}{2}\|\alpha_T(\omega)\|^2 - \alpha_T(\omega)\cdot\left[\Sigma_{t=0}^{T} Q_tB_t + U_tD_t\right]w_t(\omega)]\} \text{ is minimized,}$$

with $U_0D_0 = 0$, i.e. we seek to minimize a coercive (strictly convex) functional on a Hilbert space. Thus VD$'$, and consequently VD, has an almost surely unique optimal solution, namely

$$\alpha_T(\omega) =_{a.s.} \Sigma_{t=0}^{T} (Q_tB_t + U_tD_t)w_t(\omega)$$

The optimal value of VD$'$, and thus of VD, is given by

$$E\left[-\tfrac{1}{2}\left\| \Sigma_{t=0}^{T} (Q_t B_t + U_t D_t) w_t(\omega) \right\|^2\right].$$

This completes the proof of the first statement of the theorem. The second statement is derived from the duality between VP and VD. Ignoring the functional dependence on U, we have that

$$E\{\Phi_{\ell,L}(\omega, x^*(\omega), \gamma^*(\omega))\} = \inf E\{\Phi_{\ell,L}(\omega, x(\omega), \gamma(\omega))\}$$

$$= \psi(0,0) = -\inf E\{\Phi_{m,M}(\omega, q(\omega), \alpha(\omega))\}$$

$$= E\{\Phi_{m,M}(\omega, q^*(\omega), \alpha^*(\omega))\}$$

The middle equality follows from the lower semicontinuity of ψ and thus

$$\psi(0,0) = \psi^{**}(0,0) = -\inf \psi^*.$$

It remains only to show that in fact

$$\psi^*(q,\alpha) = E\{\Phi_{m,M}(\omega, q(\omega), \alpha(\omega))\},$$

where m and M_t are as defined earlier. Note that

$$\psi^*(q,\alpha) = \mathrm{Sup}_{(r,\eta)} \Big[E(q_0(\omega)\cdot r_0(\omega) + \alpha_0(\omega)\cdot \eta_0(\omega) + \Sigma_{t=1}^{T} (q_t(\omega)\cdot r_t(\omega)$$

$$+ \alpha_t(\omega)\cdot \eta_t(\omega))) - \inf_{(x,\gamma)} E\{\Phi_{\ell,L}(\omega, x(\omega)\frown r(\omega), \gamma(\omega)\frown \eta(\omega))\}\Big]$$

$$= \operatorname{Sup}_{(r,\eta,x,\gamma)} E\Big[q_O(\omega)(x_1(\omega)+r_O(\omega))+\alpha_O(\omega)(\gamma_1(\omega)+\eta_O(\omega))$$

$$-q_O(\omega)x_1(\omega)-\alpha_O(\omega)\gamma_1(\omega)$$

$$-\ell(\omega,x_1(\omega)+r_O(\omega),\gamma_1(\omega)+\eta_O(\omega),x_{T1}(\omega),\gamma_{T1}(\omega))$$

$$+\Sigma_{t=1}^{T}(q_t(\omega)(\Delta x_t(\omega)+r_t(\omega))+\alpha_t(\omega)\cdot(\Delta\gamma_t(\omega)+\eta_t(\omega))$$

$$-L_t(\omega,x_t(\omega)+r_t(\omega),\gamma_t(\omega)+\eta_t(\omega),\Delta x_t(\omega),\Delta\gamma_t(\omega)))$$

$$-\Sigma_{t=1}^{T}(q_t(\omega)\cdot\Delta x_t(\omega)+\alpha_t(\omega)\cdot\Delta\gamma_t(\omega))\Big]$$

The "integration by parts" formula for summands yields

$$\Sigma_{t=1}^{T}q_t\cdot\Delta x_t=\Sigma_{t=1}^{T}q_tx_{t1}-\Sigma_{t=1}^{T}q_tx_t+q_Ox_1-q_Ox_1=-q_Ox_1-\Sigma_{t=1}^{T}\Delta q_t\cdot x_t+q_Tx_{T1}$$

with $\Delta q_t=q_t-q_{t-1}$. Similarly

$$\Sigma_{t=1}^{T}\alpha_t\cdot\Delta\gamma_t=-\alpha_O\gamma_1-\Sigma_{t=1}^{T}\Delta\alpha_t\cdot\gamma_t+\alpha_T\gamma_{T1}.$$

Substituting in the above expression for ψ^*, cancelling the appropriate terms, writing $(\bar{x}_1,\bar{\gamma}_1)$ for $(x_1+r_O,\gamma_1+\eta_O)$ and $(\Delta\bar{x}_t,\Delta\bar{\gamma}_t)$ for $(\Delta x_t+r_t,\Delta\gamma_t+\eta_t)$, we see that we can interchange the sup and expectation operators. We get the following form of ψ^*:

$$\psi^*(q,\alpha)=E\{\operatorname{Sup}_{(\bar{x}_1,\bar{\gamma}_1,x_{T1},\gamma_{T1})}q_O(\omega)\bar{x}_1+\alpha_O(\omega)\bar{\gamma}_1-q_T(\omega)x_{T1}-\alpha_T(\omega)\gamma_{T1}$$

$$-\ell(\omega,\bar{x}_1,\bar{\gamma}_1,x_{T1},\gamma_{T1})\}$$

$$+\Sigma_{t=1}^{T}E\{\operatorname{Sup}_{(x_t,\gamma_t,\Delta\bar{x}_t,\Delta\bar{\gamma}_t)}(\Delta q_t(\omega)x_t+\Delta\alpha_t(\omega)\gamma_t+q_t\Delta\bar{x}_t+\alpha_t\Delta\bar{\gamma}_t)$$

$$-L_t(\omega,x_t,\gamma_t,\Delta\bar{x}_t,\Delta\bar{\gamma}_t)\}$$

or in conjugate functional notation

$$\psi^*(q,\alpha)=E\Big[\ell^*(\omega,q_0(\omega),\alpha_0(\omega),-q_T(\omega),-\alpha_T(\omega))+$$

$$+\Sigma_{t=1}^{T}L_t^*(\omega,\Delta q_t(\omega),\Delta\alpha_t(\omega),q_t(\omega),\alpha_t(\omega))\Big]$$

$$=E\Big[\Phi_{m,M}(\omega,q(\omega),\alpha(\omega))\Big]$$

The above and the duality relations yield

$$E\{\Phi_{\ell,L}(\omega,x^*(\omega),\gamma^*(\omega))+\Phi_{m,M}(\omega,q^*(\omega),\alpha^*(\omega))\} = 0.$$

The Young-Fenchel inequality between a convex function and its conjugate gives the following

$$\Phi_{\ell,L}(\omega,x,\gamma)+\Phi_{m,M}(\omega,q,\alpha) \geq 0.$$

This is obtained from the conjugacy between ℓ and m, L_t and M_t, and the integration by parts formula for (discrete) summands. The combination of the two preceding relations yield the second statement of the theorem. This completes the proof.

COROLLARY. The function f(U) has the following representation

$$f(U) = E\{\tfrac{1}{2}\Sigma_{t=0}^{T}\|Z_t w_t(\omega)\|^2\}$$

where

$$\alpha_T(\omega)(Q_T-I)=_{a.s.}0,$$

$$\alpha_T(\omega)(Q_{t-1}-Q_tA_t-U_tC_t)=_{a.s.}0 \qquad t=1,\dots,T$$

$$Z_t-Q_tB_t-U_tD_t=0 \qquad t=0,\dots,T$$

and U_0D_0 is defined to be 0.

Proof. The form of this representation of f(U) essentially appears in the proof of the theorem. The rest naturally follows from the duality between VP and VD.

5 THE CONTROL PROBLEM ASSOCIATED WITH THE WP PROBLEM

In what follows we shall assume that the covariance matrix $E[\alpha_T^*(\omega)\square\alpha_T^*(\omega)]>0$. This can be done without any loss of generality. It turns out that with the optimal weights U, the quantity $\alpha_T^*(\omega)=x_T(\omega)-(E^y x_T)(\omega)$ and this is the error term associated with the optimal estimate. To see this, one needs to write down the transversality and Euler-Lagrange conditions associated with the variational problem VP. This would lead us too far astray. If the covariance of the estimation error is singular there exist nontrivial predictable directions, cf. [3] for example. In such directions the matrix equations which we obtain eliminating $\alpha_T(\omega)$ are trivially satisfied by optimal quantities. Hence we could safely ignore them.

The results of the preceding section and the covariance condition yield a new formulation of the problem WP:

WP′ Find $U = (U_t, 1\leq t\leq T)$, $Q=(Q_t, 0\leq t\leq T)$ and $Z=(Z_t, 1\leq t\leq T)$

such that $Q_T=I$

$$Q_{t-1}=Q_tA_t+U_tC_t \qquad t=1,\dots,T$$

$$Z_t=Q_tB_t+U_tD_t \qquad t=0,\dots,T$$

and $$E\{\tfrac{1}{2}\Sigma_{t=0}^{T}\|Z_tw_t(\omega)\|^2\} \text{ is minimized.}$$

This is a *deterministic* problem. Note that actually the selection of weights U automatically determines the values of other variables Q and Z. If the random variables w_t are correlated, the correlation matrices will appear in the objective and it would be very simple indeed to solve the corresponding optimization problem.

However, if we also assume that the $(w_t; t=1,\dots,T)$ are uncorrelated normalized Gaussian variables and also uncorrelated with w_0, then problem WP′ takes on the form of a linear-quadrati regulator problem, namely

LQR Min $\frac{1}{2}$ trace $\{Q_0 P_0 Q_0' + \Sigma_{t=1}^{T} Z_t Z_t'\}$

$$Q_T = I$$

$$Q_{t-1} = Q_t A_t + U_t C_t \qquad t=1,\ldots,T$$

$$Z_t = Q_t B_t + U_t D_t$$

where P_0 is the covariance matrix of the random vector $B_0 w_0(\omega)$ and here the primes$'$ indicate that the two matrices involved in the multiplication are the transposed of each other. For this class of problems the optimal weights can be computed without recourse to the stochastics of the system.

6 CONCLUSION

The fact that solving WP and LQR both yield the optimal estimator is not new, this was the important contribution made by R. Kalman in [2]. However, as already pointed out in the introduction the derivation is new and sheds some light on the *principles* involved. In particular we gain new insight into the two system-theoretic concepts of controllability and observability. They appear here as the *same* property of two dual problems. In this article we have restricted our attention to the simplest case but from this development it should be clear which assumptions can be removed and still allow us to follow similar type arguments, e.g. the form of the loss function, the Gaussianess assumption can be completely ignored if we seek an optimal estimator in the linear span of the observations, i.e. a best wide sense estimator, etc... Finally, the framework clearly allows for the study of some nonlinear estimation problems.

7 ACKNOWLEDGEMENT

Supported in part by a grant from the National Science Foundation MCS 78-02864.

8 REFERENCES

1. Bucy, R. and Joseph, P. "Filtering for Stochastic Processes with Applications to Guidance", Interscience Publishers, New

York, (1968).

2. Kalman, R. "A new approach to linear filtering and prediction problems", *J. Basic. Engr.*, **82D**, 35-45, (1960).

3. Pavon, M. "Stochastic Realization and Invariant Directions of the Matrix Riccati Equation", *SIAM J. Cont. and Optimization*, (forthcoming).

4. Rockafellar, R. T. and Wets, R. "Nonanticipativity and L^1-Martingales in Stochastic Optimization Problems", *Mathematical Programming Studies*, **6**, 170-180, (1976); also in *Stochastic Systems: Modelling, Identification and Optimization*, **1**, ed. R. Wets, North-Holland Publishing Co., Amsterdam, (1976).

5. Rockafellar, R. T. and Wets, R. "The Optimal Recourse Problem in Discrete Time: L^1-Multipliers for Inequality Constraints, *SIAM J. Control and Optimization*, **16**, 16-36, (1978).

6.5 NON-LINEAR FILTERING: WHITE NOISE MODEL

A. V. Balakrishnan

(Department of Systems Science, School of Engineering and Applied Science, University of California, Los Angeles, California 90024, USA)

1 INTRODUCTION

In loose terms, any non-linear operation on observed data is referred to as "non-linear filtering". The basic model for observed data is that of a well-defined stochastic process ("signal" or "system response") with an additive "noise" process to account for the unavoidable error due to the sensor (measuring instrument).

Thus, letting y(t) represent the observed process, we have:

$$y(t) = s(t) + N(t), \; 0 < t \tag{1.1}$$

where the s(t) is the "signal" and N(t) the "noise". Here "t" represents time, and y(s), $0 \leq s \leq t$ is usually referred to as the observed time history up to time t, starting at some arbitrary point in time which is normalized to zero. A basic assertion is whether the time variable "t" should be treated as "continuous" or "discrete". In this paper we shall consider the continuous-time ("continuous parameter") model [for the reason that in the one engineering application of interest, viz., aircraft flight-test data reduction, although all processing is done by a digital computer, the sampling rate in the A-D converter is sufficiently high as to make the time-continuous model more attractive overall]. Of course, the continuous-time model increases the level of sophistication of the mathematical theory required. In particular the noise model used becomes more complex.

The noise process N(t) is, to begin with, not as well-defined as the signal. About all that one can say about the process is that since it accounts for the ("non-systematic") instrumentation error (and can be ascribed to thermodynamic origin), it can be modelled as a stationary Gaussian process with a spectral density which is constant over a range of frequencies wider than the frequency range of the signal (since the instrument, if well designed, is not supposed to "distort" the signal). In the absence of precise information about the noise bandwidth (which is characteristically the case in practice), it has been customary in the early engineering literature to translate the concept of "large" bandwidth to infinite bandwidth, and refer to it as "white noise". Of course, such a process must have infinite variance and leads to conceptual mathematical difficulties. Fortunately, so long as operations on the data are confined to "linear" operations, this does not create any real stumbling block, and "asymptotic" considerations can take care of matters, as in the Kalman Filter, for example.

However, serious difficulties at the "operations" level arise as soon as we wish to consider non-linear operations. Indeed, the simplest non-linear operation of squaring the instantaneous values (considering say $\|y(t)\|^2$), leads to ambiguous interpretations in the asymptotic sense of "infinite" bandwidth.

This was the situation about 1960, when a different noise model, purported to be more "rigorous", was introduced (see [1] and the many references therein). This proceeds as follows. First we rewrite (1.1) in the "integrated" version:

$$\begin{aligned} Y(t) &= \int_0^t y(s)\,ds \\ &= \int_0^t s(\sigma)\,d\sigma + \int_0^t N(\sigma)\,d\sigma \end{aligned} \tag{1.2}$$

If, formally, N(t) is "white noise" with spectral density matrix D, we see that

$$E\left(\int_0^t N(s)\,ds\right)\left(\int_0^t N(s)\,ds\right)^* = tD \tag{1.3}$$

But next these authors make the all-important step: make, in other words, the transition to:

$$Y(t) = \int_0^t s(\sigma)\,d\sigma + W(t) \tag{1.4}$$

where W(t) is now a Wiener process. Of course, $E[W(t)W(t)^*] = tD$ and in this sense the theory is "more rigorous", for now there is nothing ambiguous about the Wiener process. Moreover, there is the crucial advantage that for (1.4) one can lift bodily the well-developed apparatus of the Itô stochastic integral and concomitant mathematical theory. One result in particular (in many ways, one of the triumphs of the Itô theory) is that the (time-continuous) likelihood ratio, of signal-plus-noise to noise alone, ($0 \le t \le T < \infty$) for the Wiener process model (1.4) can be expressed in closed form:

$$\text{Exp} - \frac{1}{2}\left\{\int_0^T [\hat{s}(t), \hat{s}(t)]dt - 2\int_0^T [\hat{s}(t), dY(t)]\right\} \tag{1.5}$$

where $\hat{s}(t) = E[s(t) \mid B_Y(t)]$.

where $B_Y(t)$ is the sigma-algebra generated by Y(s), $0 \le s \le t$. The salient feature of this formula is the second term which is an Itô-integral.

This is indeed a remarkable formula, mathematically impeccable. But unfortunately it simply does not apply to our model neither in theory nor in practice. It does not apply in theory, because in the first place the Itô integral is defined only with probability one and all sample paths Y(t) under our model (1.1) have probability zero. Contrariwise, the sample paths Y(t) under model (1.2) are not absolutely continuous in t with respect to Lebesgue measure, with probability one. [This becomes of significance when we consider problems of stochastic control - see [2]]. But even more important, the formula (1.5) has no operational meaning for our model (1.2), for according to (1.2) we can only substitute for dY in (1.5) by y(t)dt, and this would be completely wrong (unless the operations involved on Y(•) are totally linear as in the trivial case where the signal is deterministic). It does not have an operational meaning because it was derived in the first place on the basis of an artificial model created only for mathematical convenience! Indeed, this situation is not atypical. One of the major advantages of the Itô integral is that any non-linear operation on Y(•) represented by the Wiener process model (1.4), can be represented as an Itô integral in Y(•). Thus, we can indeed express any non-linear operation - the only problem being, such a formula would have no operational meaning for our model (1.2)! It is interesting that the promulgators of both the Itô likelihood-ratio formula and non-linear filtering based on (1.4) (of which there is now an enormous "engineering" literature) apparently never bothered to use their formulas on any real data!

Fortunately, it turns out that the situation can be rectified by going back to the earlier asymptotic notion of white noise but now exploiting the notion of Gauss measure - of weak distributions on Hilbert Spaces. This is quite different from the "distribution" definition of Gelfand, etc. To make that distinction, we refer to it as "non-linear white noise". Of course, we cannot present an adequate exposition of the theory here. We shall only include the main concepts (in section 2) and some of the more important results in sections 3 and 4.

2 BASIC NOTIONS

Let H denote a real separable Hilbert Space. Let C denote the class of cylinder sets with Borel bases in finite-dimensional sub-spaces. By "Gauss measure", μ_G on H, we mean the weak distribution (or finitely additive measure) on C defined by the characteristic function:

$$C(h) = \int_H \text{Exp}\, i[x,h]\; d\mu_G(x) = \text{Exp}-\frac{1}{2}[h,h] \tag{2.1}$$

See [3] for more on these elementary notions. Let μ be any weak distribution (not necessarily Gauss measure). Let $f(\cdot)$ denote any Borel measurable function mapping H into H_r, another real separable Hilbert Space, of the form $f(x) = f(Px)$, where P is a finite dimensional projection on H (into H). Such a function is called a "tame" function and defines a random variable, since inverse images are in C. A random variable is defined as any Cauchy sequence in probability of tame functions. We note in this connection that any weak distribution on H can be extended to be countable additive (via the Kolmogorov extension theorem, essentially) on H', the algebraic dual. (For Gauss measure this is equivalent to going from white noise to Wiener process). But the extension is at the expense of introducing new "events" (sets) which do not necessarily have physical interpretation and also at the expense of making the outer measure of H zero. On the other hand, our ("Cauchy sequence") definition of random variables on H includes all random variables definable on H' (with countably additive extension probability measure).

Given any Borel measurable function f(•) mapping H into H_r, it need not be associatable with a random variable. We shall say that f(•) is a physical random variable if for any sequence $\{P_n\}$ finite-dimensional projections converging strangely to the identity, $\{f(P_n x)\}$ is a Cauchy sequence in probability, and

$$C(h) = \operatorname*{Lim}_n \int_H \operatorname{Expi}\left[f(P_n x)h\right] d\mu(x) \tag{2.2}$$

$h \varepsilon H_r$, is independent of the particular sequence $\{P_n\}$ chosen. The remarkable fact is that the functions occurring in likelihood ratios and optimal non-linear filtering are of this class, under the condition that the signal has finite energy.

3 BASIC RESULTS

Let, for each t, $0<t<\infty$, $W(t) = L_2\left[(0,t);R_n\right]$. By "white noise" in $\left[0,t\right]$, we shall mean the triple $(W(t), C(t), \mu_G)$ where μ_G denotes Gauss measure on C(t) the algebra of cylinder sets in Hilbert Space W(t). In our model of observed data

$$y(t) = s(t) + N(t), \quad 0<t<T<\infty \tag{3.1}$$

we assume that N(•) is "white noise" in $\left[0,T\right]$ in the above precise sense. The sample paths of N(•) are elements of W(T). The signal is of course smoother; we assume it is a well-measurable stochastic process, $s(t,\omega)$, ω denoting sample points in Ω, (Ω,β,p) being the probability triple in the usual notation. We assume further that

$$\int_0^T E\left[\| s(t,\omega)\|^2\right] dt <\infty$$

In particular $s(\cdot,\omega)$ yields a Borel measurable map into W(T). Hence, we can write in abstract version for (3.1): $y = s(\cdot,\omega) + N$. Under our assumption that signal and noise are independent, y induces a weak distribution on W(T) with characteristic function:

$$C_y(h) = \left(\int_\Omega \text{Expi} \int_0^T [s(t,\omega),h(t)]dt\ dp \right) . \text{ Exp}- \frac{1}{2}[h,h] \tag{3.2}$$

Our first result concerns Radon-Nikodym derivatives. Let μ_y denote the weak distribution induced by y on W(T) and μ_G the Gauss measure. Then the functional

$$q(h) = \int_\Omega \exp- \frac{1}{2} \{\| s(t,\omega) \|^2 - 2[s(\cdot,\omega),h]\} \, dp \tag{3.3}$$

has the property that for any cylinder set C in ω(T),

$$\mu_y(C) = \lim_{n\to\infty} \int_C q(P_n x) d\mu_G$$

for any sequence $\{P_n\}$ of finite dimensional projections converging strongly to the identity. We call q(•) the Radon-Nikodym derivative of μ_y with respect to μ_G and note that it is a physical random variable with respect to μ_y and μ_G. See [4].

We can cast (3.3) in a slightly different form in which it is readily recognized as the likelihood ratio. We follow [5]. Thus, we have: (see [5] for the derivation):

$$q(h) = \text{Exp} - \frac{1}{2} \{ \int_0^T \| \hat{s}(t,h) \|^2 dt - 2 \int_0^T [\hat{s}(t,h),h(t)]dt + \int_0^T P(t,h)dt\} \tag{3.4}$$

where

$$\hat{s}(t;h) = \int_\Omega s(t,\omega) \ B(s(\cdot,\omega);h;t)dp \tag{3.5}$$

$$P(t,h) = \int_{\Omega} \| s(t,\omega) \|^2 \, B(s(\cdot\ \omega);h;t)dp - \| \hat{s}(t,h) \|^2 \tag{3.6}$$

where

$$B(s(\cdot;\omega);h;t) =$$

$$\frac{\text{Exp} - \frac{1}{2}\int_0^t \{\| s(\sigma,\omega)\|^2 - 2[s(\sigma,\omega),h(\sigma)]\}d\sigma}{\int_{\Omega}\{\text{Exp} - \frac{1}{2}\int_0^t \{\| s(\sigma,\omega)\|^2 - 2[s(\sigma,\omega),h(\sigma)]\}d\sigma\}\cdot dp} \tag{3.7}$$

The main point is that $\hat{s}(t;h)$ is a function which is a physical random variable with respect to μ_y and moreover, yields the conditional expectation of $s(t,\omega)$ given $y(s)$, $0\le s\le t$, in the sense that $s(t;P_nP(t)y)$ and $E[s(t,\omega)\,|\,P_nP(t)y]$, where $P(t)$ denotes the operator mapping $W(T)$ into itself:

$$P(t)f = g;\ g(s) = f(s) \quad 0<s<t$$
$$= 0 \quad T>s>t,$$

are equivalent Cauchy sequences in probability. Similarly, the first term in (3.6) defines the conditional expectation of $\| s(t,\omega)\|^2$ given $y(s)$, $0\le s\le t$, and $P(t,h)$ can be identified as the conditional mean square error, also a physical random variable with respect to μ_y.

Moreover, we show (see [6]) that if we define ϕ mapping $W(t)$ into itself by $z = \phi(y)$; $z(t) = \hat{s}(t;y)$; then $\nu = y - \phi(y)$, the "innovation" also defines white noise in $[0,T]$.

In the Gaussian case: i.e. when the signal is a Gaussian process, $\phi(\cdot)$ is linear, and further, $P(t;h)$ is independent of $h(\cdot)$; and in that case we can generalize R_n to be any real separable Hilbert Space as well. See [4]. The formula (3.4), specialized to the Gaussian case, has been used for the problem of identifying aircraft stability and control derivatives from flight data in wind gust, on real (as opposed to simulated) data, in what may be considered as a test case for evaluating the non-linear white noise model theory. See [7].

4 NON-LINEAR FILTERING: CONDITIONAL DENSITY

Much of non-linear filtering theory is concerned with the calculation of $\hat{s}(t;y)$, particularly in the case where the signal s(t) is characterised by a stochastic differential equation. As potentially the most useful case, we shall consider a time-invariant system where the signal is given by:

$$s(t) = C(x(t)) \tag{4.1}$$

$$d\,x(t) = F(x(t))dt + \sigma(x(t))\,dW(t) \tag{4.2}$$

where the differential is taken in the Itô sense, $x(.) \varepsilon R_m$, and x(0) is independent of the Wiener process W(t), and possesses a probability density. We remark that we could have also expressed the stochastic equation in the white noise sense (as in [8]), but this is immaterial, since what is needed is a mechanism for specifying uniquely the probability distributions of the signal process. Here we actually prefer the Itô version, since the infinitesimal generator (denote it L) of the process x(t) is more directly deduced from the equation. Of course we assume the usual sufficient conditions on F(.) and σ(.) to guarantee a unique strong solution to (4.2), as can be found for example in [9]. We assume furthermore that the process has a transition density, and denote the first order density by p(t,x).

As is customary in non-linear filtering theory, we shall actually seek to characterise the conditional density of x(t) given the data y(s), $s < t$. In principle this is more general since $\hat{s}(t;y)$ is only the first moment of C(x) and can be computed, from the conditional density, as indeed can any desired moment, the second moment being particularly desirable. Things are not quite so straightforward, since, as is well known, the evolution equation satisfied by the conditional density contains implicitly in it the first moment. Nevertheless, the evolution equation for the conditional density is usually what is derived in non-linear filtering theory based on the Wiener process noise model. Our objective here is to derive the similar equation based on our non-linear white noise model, with special attention to the noticeable differences. In this vein, we shall make the same assumptions as in the Wiener process theory, as may be found for example in [1,10]. These conditions do not depend on the noise model used. Of course we do assume that the signal has finite energy:

$$\int_0^T E(\|C\, x(t,.)\|^2)dt < \infty \tag{4.3}$$

Leaving the details of the derivation to [6], let f(.) denote any infinitely differentiable function, mapping R_m into R_I, and vanishing outside a compact subset. Define (the conditional expectation):

$$\hat{f}(x(t,.);P(t)y)$$

$$= \frac{\int_\Omega f(x(t,\omega))\ (\exp -\tfrac{1}{2}\int_0^t \{\|s(\sigma,\omega)\|^2 - [s(\sigma,\omega),y(\sigma)]\}d\sigma)\ dp}{\int_\Omega (\exp -\tfrac{1}{2}\int_0^t \{\|s(\sigma,\omega)\|^2 - [s(\sigma,\omega),y(\sigma)]\}\ d\sigma)\ dp} \tag{4.4}$$

Fix y(.) in W(T). Let r(t;x;P(t)y) denote for each t, the Baire function over R_m defined by:

$$r(t;x(t;\omega);P(t)y)$$

$$= \frac{E(\exp -\tfrac{1}{2}\int_0^t \{\|s(\sigma,\omega)\|^2 - 2\,[s(\sigma,\omega),\ y(\sigma)]\}\ d\sigma / x(t\ \omega))}{E(\exp -\tfrac{1}{2}\int_0^t \{\|s(\sigma,\omega)\|^2 - 2\,[s(\sigma,\omega),\ y(\sigma)]\}\ d\sigma)}$$

Next define:

$$p(t;x;P(t)y) = r(t;x;P(t)y)\ p(t,x)$$

Then, since:

$$f(x(t,.);P(t)y) = \int_{R_m} f(x)\ p(t;x;P(t)y)\ d|x|$$

it is natural to call p(t;x;P(t)y) the conditional density of x(t,ω) given y(s), s < t. By direct differentiation of (4.4) we obtain that: (see [6]) the conditional density will satisfy the evolution equation:

$$p(t;x;P(t)y) = L^*(p(t;\cdot;P(t)y))(x)$$
$$+ [C(x) - \hat{s}(t;y),\ y(t) - \hat{s}(t;y)]\ p(t;x;P(t)y)$$
$$+ \tfrac{1}{2}(P(t;y) - \|C(x) - \hat{s}(t;y)\|^2)\ p(t;x;P(t)y) \tag{4.5}$$

where we have assumed that $p(t;\cdot;P(t)y)$ belongs to the domain of L^*, the distributional adjoint of L. The main thing is to note the appearance of the third term in parenthesis which does not appear in the Wiener process version of the equation; see [1] for the Wiener process version. If the conditional density is to be calculated from observed $y(.)$ according to model (1.1), then the Wiener process version has no operational meaning, while (4.5) does. The equation now explicitly contains the second conditional moment. It is a stochastic partial differential equation in that it contains the forcing term involving $y(.)$; it is actually "bilinear" in the innovation, and the latter being white noise, it is of the kind treated in [8].

5 ACKNOWLEDGEMENT

Research supported in part under AFOSR grant no. 732942, Applied Math. Divn. USAF.

6 REFERENCES

1. Liptser, R. S. and Shiryayev, A. N. "Statistics of Random Processes", Springer-Verlag, New York, (1978).

2. Balakrishnan, A. V. "On Stochastic Bang-bang Control", Int. Symp. on Stochastic Differential Systems, Vilnius, August, 1978, Proceedings (to be published by Springer-Verlag, 1979).

3. Balakrishnan, A. V. "Applied Functional Analysis", Springer-Verlag, (1976).

4. Balakrishnan, A. V. "Radon-Nikodym Derivatives of a Class of Weak Distributions of Hilbert Spaces", *Journal of Appl. Maths. and Opt.*, **3**, 209-225, (1977).

5. Balakrishnan, A. V. "Likelihood Ratios for Signals in Additive White Gaussian Noise", *Journal of Appl. Math. and Opt.*, **3**, 341-356, (1977).

6. Balakrishnan, A. V. "Non-linear White Noise Theory", in Multivariate Analysis V, edited by P. R. Krishnaiah, Academic Press, (1979).

7. Iliff, K. W. "Identification and Stochastic Control with Application to Flight Control in Turbulence", Dissertation UCLA, 1973, (UCLA-ENG-7340, 1973).

8. Balakrishnan, A. V. "Stochastic Bilinear Partial Differential Equations", in "Variable Structure Systems", edited by R. Mohler and A. Ruberti, Springer-Verlag, (1975).

9. Gikhman, I. and Skhorokhod, A. V. "Stochastic Differential Equations", Springer-Verlag, (1970).

10. Rozovsky, B. L. "Stochastic Partial Differential Equations", Mat. Sbornik, 96 (138): 2, 314-341, (1975).

6.6 SOME NEW ALGORITHMS FOR NONLINEAR STATE ESTIMATION AND FILTERING

A. Halme

(Division of Control Engineering, University of Oulu, Finland)

1 INTRODUCTION

Nonlinear state estimation and filtering problems with Gaussian statistics are often treated by the aid of Bayesian maximum likelihood or maximum *a posteriori* (MAP) technique (see e.g. [1 - 3]) because it leads to a numerically tractable optimization problem, for which different recursive finite state algorithms can be developed. If no *a priori* knowledge of the process and observation noise is available, the use of the least-squares criterion when fitting the system trajectory to observed data leads also to an analog problem (e.g. [4]). Compared to such heuristic techniques as linearized or extended Kalman filter this approach gives a more exact way to treat the estimation problem. The mathematical problem encountered is deterministic fixed-interval optimal control problem, which can be transformed to an equivalent two-point boundary value problem (TPBVP). The solution of TPBVP gives smoothing estimates for the system state at inner time points and the filtering estimate at the end point of the interval. The smoothing estimates can be solved with standard techniques, such as the sweep method, quasilinearization or successive approximations.

The filtering solution in a recursive form has turned out more difficult and only partial success has been attained thus far. The invariant imbedding method mostly used (e.g. [2 - 5]) has required quite rough approximations to lead to a solution. Recently this approach has been improved [11] in the case where the nonlinearities are of polynomial type. However, the result is computationally rather difficult because the filter is approximated by a series of filters connected sequentially in quite a complicated way. Stationary properties of the filters as the interval length increases to infinity have not been studied. In the following another approach to treat the TPBVP will be tried. The Green's function method and corresponding integral equation has been chosen instead of invariant imbedding. It will be shown that for time invariant system an exact static filter can be found in conditions corresponding to the well known linear case. In the non-static case a successive approximation technique is considered. Continuous time formulation has been chosen to make the presentation easier. Results concerning the discrete counterpart may be found along the same lines.

2 PROBLEM STATEMENT

The system equations are supposed to be of the form

$$dx(t) = f(x(t),u(t))dt + G(x(t),u(t))dw \tag{1}$$

$$z(t) = h(x(t),u(t)) + v(t), \tag{2}$$

where x is n-dimensional state vector, u m-dimensional control vector and z r-dimensional observation vector. The functions f, G and h are supposed to be properly well-behaved and have linear parts so that they can be written in the forms

$$f(x,u) = Ax + Bu + f_1(x,u) \tag{3}$$

$$h(x,u) = cx + Du + h_1(x,u) \tag{4}$$

$$G(x,u) = G + G_1(x,u). \tag{5}$$

The process disturbance w and observation noise v are supposed to be independent zero mean Gaussian white processes such that

$$E(w(t)w(\tau)') = Q\delta(t-\tau) \tag{6}$$

$$E(v(t)v(\tau)') = R\delta(t-\tau) \tag{7}$$

$$E(w(t)v(t)') = 0 \tag{8}$$

where Q and R are covariance matrices.

The system is considered on the interval $[0,T]$ and the initial state x(0) is supposed to be a Gaussian stochastic variable with mean x_0 and covariance matrix P_0. Denote by X(T) and Z(T) the functions x(.) and z(.) on the interval

$[0,T]$. It is a well known result (see e.g. [7]) that the conditional probability $p(X(T)|Z(T))$ may be written in the form

$$p(X(T)|Z(T)) = \frac{p(Z(T)|X(T))\, p(X(T))}{p(Z(T))}$$

$$= a \exp \{-\frac{1}{2}\| x(0) - x_0 \|^2_{P_0^{-1}}$$

$$- \frac{1}{2}\int_0^T [\| z(t) - h(x(t),u(t))\|^2_{R^{-1}} \tag{9}$$

$$+ \| (G(x(t),u(t))w(t)\|^2_{[G(x,u)QG(x,u)']^{-1}}]dt\},$$

where a does not depend on $X(T)$. The maximum likelihood estimate is obtained by maximizing $P(X(T)|Z(T))$, which is equivalent to minimizing the quadratic criterion

$$J = \frac{1}{2}\| x(0) - x_0 \|^2_{P_0^{-1}} + \frac{1}{2}\int_0^T [\| z(t) - h(x(t),u(t))\|_{R^{-1}}$$
$$+ \| w(t)\|^2_{Q^{-1}}]dt. \tag{10}$$

Standard use of minimum principle or Euler-Lagrange technique (w as the minimising variable) leads to the following TPBVP

$$\dot{x}(t) = f(x(t),u(t)) - G(x(t),u(t))QG(x(t),u(t))'\, p(t) \tag{11}$$

$$\dot{p}(t) = \frac{\partial}{\partial x} h(x(t),u(t))' R^{-1}(z(t) - h(x(t),u(t)))$$
$$- \frac{\partial}{\partial x} f(x(t),u(t))'\, p(t) \tag{12}$$
$$+ [\frac{\partial}{\partial x} G(x(t),u(t)) \otimes QG(x(t),u(t))'\, p(t)]\, p(t)$$

$$p(0) = -P_0^{-1}(x(0) - x_0) \tag{13}$$

$$p(T) = 0, \tag{14}$$

where $\otimes$ denotes a suitable product operation relating to the derivation of $p' G(x,u)w$ in the Hamiltonian. By taking into account Eq. (3) - (5) and using a usual abbreviation by dropping t the TPBVP can be further written in the form

$$\dot{x} = Ax + Bu - GQG' p + F_1(x,p,u) \tag{15}$$

$$\dot{p} = -C' R^{-1} CX + C' R^{-1} Du - A' p + C' R^{-1} z + F_2(x,p,u,z), \tag{16}$$

$$\begin{bmatrix} I & P_0 \\ 0 & 0 \end{bmatrix} \begin{bmatrix} x(0) \\ p(0) \end{bmatrix} + \begin{bmatrix} 0 & 0 \\ 0 & I \end{bmatrix} \begin{bmatrix} x(T) \\ p(T) \end{bmatrix} = \begin{bmatrix} x_0 \\ 0 \end{bmatrix} \tag{17}$$

where F_1 and F_2 mean certain functions essentially nonlinear with respect to (x,p). Further simplification in notation is obtained by denoting

$$y = \begin{bmatrix} x \\ p \end{bmatrix}, \tag{18}$$

$$F(y,t) = \begin{bmatrix} F_1(z,p,u(t)) \\ F_2(x,p,u(t),z(t)) \end{bmatrix} \tag{19}$$

$$V = \begin{bmatrix} A & -GQG' \\ -C' R^{-1} C & -A' \end{bmatrix} \tag{20}$$

$$f(t) = \begin{bmatrix} Bu(t) \\ -C' R^{-1} Du(t) + C' R^{-1} z(t) \end{bmatrix} \tag{21}$$

$$M = \begin{bmatrix} I & P_O \\ O & O \end{bmatrix} \quad (22)$$

$$N = \begin{bmatrix} O & O \\ O & I \end{bmatrix} \quad (23)$$

$$c = \begin{bmatrix} x_O \\ O \end{bmatrix} \quad (24)$$

after which the TPBVP can be simply written as

$$\dot{y}(t) = Vy(t) + f(t) + F(y(t),t) \quad (25)$$

$$My(O) + Ny(T) = c. \quad (26)$$

The solution of Eq. (25) - (26), which will be denoted also by $y(\cdot|T)$, gives the smoothing estimate for the trajectory $x(\cdot)$ and the value $y(T|T)$ the filtering estimate for $x(T)$.

3 FILTERING IN THE STATIONARY CASE $T\to\infty$

It is not likely that one can find a simple finite dimensional state form recursive presentation for the filtering $x(T|T)$. In fact it has been proved (e.g. [8]) that in the nonlinear case exact solution of the filtering cannot be generally expressed by a finite dimensional state model. However, it will turn out that such a representation can be found in this case when $T\to\infty$, which corresponds to the situation where in the linear case the Kalman filter gain approaches a stationary value.

Provided that the linear part of the TPBVP (25) - (26) (i.e. the problem with $F\equiv O$) has a unique solution and $f(s)+F(y,s)$ satisfies certain general conditions, the problem can be represented as an equivalent integral equation [6],

$$y(t|T) = H(t;O,T)c + \int_O^T G(t,s;O,T)\left[f(s) + F(y(s|T),s)\right]ds \quad (27)$$

where $H(.;O,T)$ and $G(.,.;O,T)$ are the Green's functions

defined by

$$H(t;0,T) = \Phi(t,0)\left[M+N\Phi(T,0)\right]^{-1} \tag{28}$$

$$G(t,s;0,T) = \begin{cases} G^{-}(t,s;0,T) = \Phi(t,0)\left[M+N\Phi(T,0)\right]^{-1}M\Phi(0,s) & 0\le s<t \\ G^{+}(t,s;0,T) = -\Phi(t,0)\left[M+N\Phi(T,0)\right]^{-1}N\Phi(T,s) & t\le s\le T \end{cases} \tag{29}$$

$$\Phi(t,0) = \exp tV = \text{the state transition matrix of the linear part of (25).} \tag{30}$$

Proceed now by setting $t=T$ and considering the derivative of $y(T|T)$ with respect to T. After some manipulation the following equation can be written

$$\begin{aligned} \frac{d}{dT}\, y(T|T) &= \frac{d}{dT}\, H(T;0,T)c + (G^{+}(T,T;0,T)+I)(f(T)+F(y(T|T),T)) \\ &+ \int_0^T \frac{d}{dT}\, G^{-}(T,s;0,T)(f(s)+F(y(s|T),s))ds \\ &+ \int_0^T G(T,s;0,T)\, \frac{\partial}{\partial y} F(y(s|T),s)\, \frac{\partial}{\partial T}\, y(s|T)ds. \end{aligned} \tag{31}$$

For the derivatives one obtains

$$\frac{d}{dT}\, H(T;0,T) = (V-K(T))H(T;0,T) \tag{32}$$

$$\frac{d}{dT}\, G^{-}(T,s;0,T) = (V-K(T))G^{-}(T,s;0,T), \tag{33}$$

where

$$K(T) = \Phi(T,0)\left[M+N\Phi(T,0)\right]^{-1}NV \tag{34}$$

The equation (31) can further be written to the form

$$\frac{d}{dT} y(T|T) = (V-K(T))y(T|T) + (G^{+}(T,T;O,T)+I)(f(T)+F(y(T|T),T))$$
$$+ \int_{O}^{T} G^{-}(T,s;O,T) \frac{\partial}{\partial y} F(y(s|T),s) \frac{\partial}{\partial T} y(s|T)ds. \tag{35}$$

Equation (35) is interesting, because it represents $y(T|T)$ "almost" in a state equation form. The only problem is the integral term, which actually means that the filtering estimate $y(T|T)$ is dependent on the whole smoothing estimate $y(.|T)$. Note that the integral term exists only when the system is truly nonlinear. Consider more closely the integral term. It will be shown that it vanishes when $T\to\infty$. This is due to vanishing of the function $\frac{\partial}{\partial T} y(.|T)$ when $T\to\infty$. To prove this, it can be first noted that the "sensitivity function" $\frac{\partial}{\partial T} y(.|T)$ satisfies the equation

$$\frac{\partial}{\partial T} y(t|T) = \frac{\partial}{\partial T}H(t;O,T)c + G(t,T;O,T) \quad (f(T) + F(y(T|T),T))$$
$$+\int_{O}^{T} \frac{\partial}{\partial T} G(t,s;O,T) \quad (f(s) + F(y(s|T),s))ds \tag{36}$$
$$+\int_{O}^{T} G(t,s;O,T) \frac{\partial}{\partial y} F(y(s|T),s) \frac{\partial}{\partial T} y(s|T)ds.$$

It is a well known fact (e.g. [9]) that the eigenvalues of V are symmetric as to the imaginary axis. Except the trivial case where all eigenvalues are on the imaginary axis, this then implies that V necessarily has some eigenvalues with positive real parts. It can be shown that this implies for all t

$$\lim_{T\to\infty} \frac{\partial}{\partial T} H(t;O,T) \to O, \tag{37}$$

$$\lim_{T\to\infty} G(t,T;O,T) \to O,$$

For large values of T equation (36) becomes thus

$$\frac{\partial}{\partial T} y(t|T) \approx \int_0^T G(t,s;0,T) \frac{\partial}{\partial y} F(y(s|T),s) \frac{\partial}{\partial T} y(s|T)\,ds. \tag{38}$$

Further

$$\lim_{T\to\infty} G^-(t,s;0,T) = \Phi(t,0)\Phi(0,s) \tag{39}$$

$$\lim_{T\to\infty} G^+(t,s;0,T) = -\Phi(t,0)\Phi(0,s) \tag{40}$$

Thus the function $G(t,s;0,T)$ is unbounded when $T\to\infty$ as regards to integration variable s and time t. It follows easily that the equation (38) cannot have other finite solutions than $\frac{\partial}{\partial T} y(.|T) \equiv 0$ when $T\to\infty$. Because it must vanish also when $t\to T$ the integral term in (35) vanishes.

Go back to the equation (35).

When $T\to\infty$ it can be written in the form

$$\frac{d}{dT} y(T|T) = (V-K(T))y(T|T) + (G^+(T,T;0,T)+I)(f(I) + F(y(T|T),T)) \tag{41}$$

The filter equation for $x(T|T)$ can now be obtained from (41) by simply setting

$$y(T|T) = \begin{bmatrix} x(T|T) \\ 0 \end{bmatrix}$$

After some algebra it follows

$$\frac{d}{dT} x(T|T) = f(x(T|T),u(T)) + G^+_{12}(T,T;0,T) \frac{\partial}{\partial x} h(x(T|T),u(T))' R^{-1} \cdot \left[z(T) - h(x(T|T),u(T))\right] \tag{42}$$

where G_{12}^{+} is from the partition

$$G^{+} = \left[\begin{array}{c|c} G_{11}^{+} & G_{12}^{+} \\ \hline G_{21}^{+} & G_{22}^{+} \end{array}\right] \tag{43}$$

It can be shown that

$$G_{12}^{+}(T,T;O,T) = \left[\Phi_{12}(T,O)-\Phi_{11}(T,O)P_O\right]\left[\Phi_{22}(T,O)-\Phi_{21}(T,O)P_O\right]^{-1} \tag{44}$$

where Φ_{ij} are from the partition

$$\Phi = \left[\begin{array}{c|c} \Phi_{11} & \Phi_{12} \\ \hline \Phi_{21} & \Phi_{22} \end{array}\right] \tag{45}$$

But (44) is the well known solution of the matrix Riccati equation (see e.g. [10])

$$\frac{dP}{dT} = AP + PA' - PC'R^{-1}CP + GQG'$$
$$P(O) = P_O. \tag{46}$$

Thus $G_{12}^{+}(T,T;O,T)$ is the error covariance matrix related to the Kalman filter of the linear problem $f_1 = h_1 = G_1 = O$. The Kalman filter can be obtained also directly from (35) without considering $T\to\infty$. In the case P(T) has a steady solution, say $\bar{P}$, as $T\to\infty$ the filter equation can be written in the following form for practical calculations

$$\frac{d}{dt}\, x(t|t) = f(x(t|t),u(t)) + \bar{P}\,\frac{\partial}{\partial x}\, h(x(t|t),u(t))'\, R^{-1} \cdot \left[z(t) - h(x(t|t),u(t))\right] \tag{47}$$

The result may be compared to that obtained by invariant imbedding method (e.g. [7]). The filters are quite similar, however the correction terms differ slightly. A great advantage of the filter obtained here is that $\bar{P}$ can be precalculated and only one differential equation is needed for the estimate. The author has not succeeded to derive a useful equation for the covariance matrix of the estimation error, but equations like those given in [7] may be used as a first approximation.

4 APPROXIMATIONS IN THE NONSTATIONARY CASE

If the system is time-variant or the estimation time history is so short that the integral term in the equation (35) is meaningful, recursive approximation methods may be used to solve the filtering equation. Suppose that successive approximation of the fixed point method is used to solve the equation (35). Set the initial value

$$y^{(1)}(t|T) = H(t;O,T)c + \int_O^T G(t,s;O,T)f(s)ds, \tag{48}$$

and the iteration steps as follows

$$y^{(i+1)}(t|T) = H(t;O,T)c + \int_O^T G(t,s;O,T)(f(s) + F(y^{(i)}(s),s))ds. \tag{49}$$

Then

$$\frac{d}{dT} y^{(1)}(T|T) = (V-K(T))y^{(1)}(T|T) + (G^{+}(T,T;O,T)+I)f(T), \tag{50}$$

$$\frac{\partial}{\partial T} y^{(1)}(t|T) = \frac{\partial}{\partial T} H(t;O,T)c + G^{+}(t,T;O,T)f(T) + \int_O^T \frac{\partial}{\partial T} G(t,s;O,T)f(s)ds \tag{51}$$

$$\frac{d}{dT} y^{(i+1)}(T|T) = (V-K(T))y^{(i+1)}(T|T)$$
$$+ (G^{+}(T,T;O,T)+I)(f(T)+F(y^{(i)}(T|T),T)) \tag{52}$$
$$+ \int_{O}^{T} G(T,s;O,T)\frac{\partial}{\partial y}F(y^{(i)}(s|T),s)\frac{\partial}{\partial T}y^{(i)}(s|T)ds$$

$$\frac{\partial}{\partial T}y^{(i+1)}(t|T) = \frac{\partial}{\partial T}H(t;O,T)c+G^{+}(t,T;O,T)(f(T)+F(y^{(i)}(T|T),T)$$
$$+\int_{O}^{T}\frac{\partial}{\partial T}G(t,s;O,T)(f(s)+F(y^{(i)}(s|T),s))ds \tag{53}$$
$$+\int_{O}^{T} G(t,s;O,T)\frac{\partial}{\partial y}F(y^{(i)}(s|T),s)\frac{\partial}{\partial T}y^{(i)}(s|T)ds.$$

Looking more closely at the equations (49) - (52) it can first be noted that the first approximation gives the Kalman filter for $x^{(1)}(T|T)$ provided that assumption $p^{(1)}(T|T)=O$ is made. The "sensitivity function" $\frac{\partial}{\partial T}y^{(1)}(.|T)$ can be calculated directly from equation (51). However, the integral term must be recalculated for every new value of T. It can be quite easily shown that $\frac{\partial}{\partial T}H$, G^{+} and $\frac{\partial}{\partial T}G$ vanish when $T\to\infty$, so that $\frac{\partial}{\partial T}y^{(1)}$ goes to zero with T. In the general step the assumption $p^{(i)}(T) = p^{(i+1)}(T) = O$ may again be made. The equation (52) gives then

$$\frac{d}{dT}x^{(i+1)}(T|T) = Ax^{(i+1)}(T|T)+Bu(T)+f_1(x^{(i)}(T|T),u(T))$$
$$+ P(T)C'R^{-1}(z(T)-Cx^{(i+1)}(T|T)-Du(T)) \tag{54}$$
$$+ P(T)\frac{\partial}{\partial x}h_1(x^{(I)}(T|T),u(T))'R^{-1}(z(T)-h(x^{(i)}(T|T),u(T)))$$
$$- P(T)C'R^{-1}h_1(x^{(i)}(T|T),u(T))$$
$$+ \int_{O}^{T} G(T,s;O,T)\frac{\partial}{\partial y}F(y^{(i)}(s|T),s)\frac{\partial}{\partial T}y^{(i)}(s|T)ds$$

For the estimate $x^{(i+1)}(T|T)$ this is almost a linear Kalman filter. The additional terms come from the previous step. Again the integral must be recalculated for every new value of T. Also the smoothing $y^{(i)}(.|T)$ is now needed. In practice this is a tedious procedure at least if n is greater say 2 or 3. However, also in this case $\frac{\partial}{\partial T}y^{(i)}(.|T) \to 0$ when $T\to\infty$ and the value of the integral term in (54) decreases as T increases. It is interesting to note that for a large T the equation (54) actually means a series of linear Kalman filters connected sequentially through a nonlinear correction term.

5 CONCLUDING REMARKS

The Green's functions method used here is not generally a convenient method for numerical solution of TPBVP's, because of numerical instability and matrix inverses in the calculation of the Green's matrices (28) - (29). However, in this case it has turned out a useful tool for analytic treatment of the problem, substituting the invariant imbedding method usually used. In strict mathematical sense quite a lot of details have been ignored above. Such questions like conditions for the existence of solutions when $T\to\infty$ are, however, extremely difficult and from the practical point of view unnecessary to solve. When considering different limit properties of the Green's functions it should be noted that V has been supposed a constant matrix. If V is time-dependent similar results may be difficult to get and have not been considered by the author. Although no numerical study has been included it is quite clear that due to its proper form the filter (41) (or (47)) works. As to the successive approximation technique considered only numerical exercises show its feasibility. The discrete form presentation may improve its usefulness considerably. Discrete forms are available for the Green's functions [6], but their use seems to require a lot of algebra to obtain the corresponding results.

6 REFERENCES

1. Bryson, A. E. and Ho. Y. C. "Applied Optimal Control", Ginn (Blaisdell), Boston, (1969).

2. Sage, A. P. and Ewing, W. S. "On filtering and smoothing algorithms for non-linear state estimation", *Int. J. Control,* **11**, No. 1, (1970).

3. Sage, A. P. "Maximum a posteriori filtering and smoothing algorithms", *Int. J. Control,* **11**, No. 2, (1970).

4. Detchmendy, D. M. and Sridhar, R. "Sequential estimation of states and parameters in noisy nonlinear systems", I. Basic Eng., *Trans ASME,* Series D, **88**, No. 2, (1966).

5. Bellman, R. E., Kagiwada, H. H., Kalaba, R. E. and Sridhar, R. "Invariant imbedding and nonlinear filtering theory", *J. Astronaut. Sci.,* **13**, No. 3, (1966).

6. Falb, P. L. and DeJong, J. L. "Some successive approximation methods in control and oscillation theory", Academic Press, New York, (1969).

7. Sage, A. P. and Melsa, J. C. "Estimation theory with applications to communications and control", McGraw-Hill, New York, (1971).

8. Kushner, H. "Introduction to stochastic control", Holt Rinehart and Winston, New York, (1971).

9. Mårtensson, C. "New approaches to the numerical solution of optimal control problems", Lund Institute of Technology, (1972).

10. Athans, M. and Falb, P. L. "Optimal Control", McGraw-Hill, New York, (1966).

11. Orava, J. "State estimation of polynomial type non-linear differential systems", doctoral dissertation, Acata Polytechnica Scandinavica Ma 26, (1975).

6.7 FINITE-DIMENSIONAL NONLINEAR ESTIMATION FOR A CLASS OF SYSTEMS IN CONTINUOUS AND DISCRETE TIME

S. I. Marcus

(Department of Electrical Engineering, The University of Texas at Austin, USA)

S. K. Mitter

(Department of Electrical Engineering and Computer Science, Massachusetts Institute of Technology, USA)

D. Ocone

(Department of Mathematics, Massachusetts Institute of Technology, USA)

1 INTRODUCTION

In [1] - [3] we have shown that, for certain classes of nonlinear stochastic systems in both continuous and discrete time, the optimal conditional mean estimator of the system state given the past observations can be computed with a recursive filter of fixed finite dimension. The typical nonlinear system in these classes consists of a linear system with linear measurements and white Gaussian noise processes, which feeds forward into a nonlinear system described by a certain type of Volterra series expansion or by a bilinear or state-linear system satisfying certain algebraic conditions. It is our purpose in this paper to consider estimation problems similar to those in [1] - [3], to present simpler proofs that the estimators are indeed finite dimensional, to provide deeper insight into these problems by relating them to the homogeneous chaos of Wiener and to orthogonal polynomial expansions [4] - [8], [24], to explain the similarities and differences between the continuous and discrete time cases, and to prove some extensions of our previous results. The existence of polynomials in the innovations in the discrete time recursive estimator, in contrast to the continuous time estimator (as noted in [2]), is interpreted in terms of the homogeneous chaos. The existence of such polynomials in the innovations in the <u>optimal</u> filter suggests that suboptimal filter design in discrete time could be improved by incorporating such structure; this is in contrast to most discrete time estimator designs, such as the extended Kalman filter, in which the updated estimate is linear in the innovations (exceptions are the quasi-moment estimators of [9] and [10]) and the higher measurement space filter of [23].

2 PROBLEM STATEMENT

As in [1] - [3], the classes of systems considered in this paper are described as follows. It will be assumed that all random variables and processes are defined on a probability space $(\Omega, \mathcal{B}, P)$. In continuous time, we consider systems of the form, for $t \varepsilon [0,T]$,

$$dx(t) = A(t)x(t)dt + B(t)dw(t) \tag{1}$$

$$dy(t) = f(x(t),y(t),t)dt \tag{2}$$

$$dz_1(t) = C(t)x(t)dt + R^{\frac{1}{2}}dv(t) \tag{3}$$

where $x(t)\varepsilon \mathbb{R}^n$, $y(t)\varepsilon \mathbb{R}^m$, $z(t)\varepsilon \mathbb{R}^p$, w and v are standard vector Wiener processes, $R>0$, $x(0)$ is Gaussian, $\{x(0),y(0),w(t),v(s)\}$ are independent for all t and s, f is an analytic function of x and y, and $[A(t),B(t),C(t)]$ is completely controllable and observable.

The discrete time systems to be considered are of the form, for $t\varepsilon\{0;T\}$,

$$x(t+1) = A(t)x(t) + B(t)w(t) \tag{4}$$

$$y(t+1) = f(x(t),y(t),t) \tag{5}$$

$$z_2(t) = C(t)x(t) + R^{\frac{1}{2}}v(t) \tag{6}$$

where $T\varepsilon Z^+$, the set of positive integers, and $\{s;t\}$ is the set of integers $\{s,s+1,\dots,t\}$. The assumptions in (4) -(6) are the same as those in (1) - (3), except that w(t) and v(t) are zero-mean Gaussian white noise processes. Motivation for the study of systems of the form (1) - (3) and (4) - (6) is presented in [3].

The optimal estimate, with respect to a wide variety of criteria (including minimum mean square error), of $x(t)$ given the past observations $z_1^t \triangleq \{z_1(s),\ 0 \le s \le t\}$ or $z_2^t \triangleq \{z_2(s),\ s \varepsilon \{0;t\}\}$, is the conditional mean $\hat{x}(t|t)$ of $x(t)$ given the σ-field $F_t^{z_i}$ generated by z_i^t, also denoted $E[x(t)|z_i^t]$. It is assumed that all the relevant random variables are in $L_2(\Omega, \mathcal{B}, P)$, so the conditional expectation $\hat{x}(t|t)$ can also be interpreted as the orthogonal projection of $x(t)$ onto the subspace $L_2(\Omega, F_t^{z_i}, P)$ [14, App. A.]; this interpretation will be used in the sequel. Predicted and smoothed estimates will be used extensively, so we introduce the equivalent notations
$\hat{x}(s|t) \triangleq E[x(s)|z_i^t] \triangleq E^t[x(s)] \triangleq E[x(s)|F_t^{z_i}]$.

Thus our objective is the recursive computation of $\hat{x}(t|t)$ and $\hat{y}(t|t)$. The computation of $\hat{x}(t|t)$ can be performed by the recursive n-dimensional (linear) Kalman filter in continuous or discrete time. It is, in general, not possible to compute $\hat{y}(t|t)$ with a recursive estimator of fixed finite dimension. It has been proved in [1] - [3] that if the non-linear system (2) or (4) is characterized by a certain type of finite series expansion or by certain bilinear or state-affine equations, then $\hat{y}(t|t)$ can be computed by such a recursive finite dimensional estimator. Some of the major results can be summarized as follows.

Let the Volterra series expansions for the ith components of $y(t)$ in (2) and (4) be given by

$$y_i(t) = w_{0i}(t) + \sum_{k=1}^{\infty} \int_0^t \cdots \int_0^t \sum_{\alpha_1,\ldots,\alpha_k=1}^{n} w_{ki}^{(\alpha_1,\ldots,\alpha_k)}(t,\sigma_1,\ldots,\sigma_k) \cdot x_{\alpha_1}(\sigma_1)\ldots x_{\alpha_k}(\sigma_k)\, d\sigma_1 \ldots d\sigma_k \tag{7}$$

and

$$y_i(t) = w_{0i}(t-1) + \sum_{k=1}^{\infty} \sum_{\ell_1,\ldots,\ell_k=0}^{t-1} \sum_{\alpha_1,\ldots,\alpha_k=1}^{n} w_{ki}^{(\alpha_1,\ldots,\alpha_k)} \cdot (t,\ell_1,\ldots,\ell_k) \cdot x_{\alpha_1}(\ell_1)\ldots x_{\alpha_k}(\ell_k), \tag{8}$$

respectively. Here $w_{ki}^{(\alpha_1,\ldots,\alpha_k)}$ is called a kth order kernel, and a finite Volterra series expansion of order q is one such that all k^{th} order kernels are zero for $k>q$. In the continuous case (7), we consider, without loss of generality [11], only triangular kernels which satisfy $w_{ki}^{(\alpha_1,\ldots,\alpha_k)}(t,\sigma_1,\ldots,\sigma_k)=0$ unless $0 < \sigma_1 <\ldots < \sigma_k < t$. Such a kernel is separable if it can be expressed as a finite sum

$$w_{ki}(t,\sigma_1,\ldots,\sigma_k) = \sum_{j=1}^{m} \gamma_{j0}(t)\gamma_{j1}(\sigma_1)\ldots\gamma_{jk}(\sigma_k). \tag{9}$$

Similar definitions can be made in discrete time [2], but they are more complicated (this difficulty is related to the fact that the solution of a discrete time system may not be defined backward in time [12],[21]). Brockett [11] and Gilbert [13] have shown that the kernels in (7) are separable if f is analytic. Using variational expansions similar to those of Gilbert [13], it is straightforward to show that the kernels of the Volterra series (8) are also separable in the sense of [2],[12]; this is basically due to the fact that the kernels arise from the variational equations as products of pulse responses of linear systems. Brockett [11] has shown that a continuous time finite Volterra series has a bilinear realization if and only if it has separable kernels. The separability and realizability results are crucial in the proofs of the following two theorems.

<u>Theorem 1</u> [1]: Consider the system (1)-(3), and assume that (2) has a finite Volterra series expansion. Then $\hat{y}(t|t)$ can be computed with a finite dimensional recursive estimator--i.e., by a finite set of nonlinear stochastic differential equations driven by the innovations

$$\nu_1(t) \triangleq z_1(t) - \int_0^t C(s)\hat{x}(s|s)ds. \tag{10}$$

<u>Theorem 2</u> [2]: Consider the system (4)-(6), and assume that (5) has a finite Volterra series expansion. Then $\hat{y}(t|t)$ can be computed with a finite set of nonlinear difference equations driven by the innovations

$$\nu_2(t) = z_2(t) - C(t)A(t-1)\hat{x}(t-1|t-1). \tag{11}$$

The basic technique employed in [1] - [3] to prove these theorems is the augmentation of the state of the original system with additional states which arise as smoothed statistics of the original state. For the classes of systems considered here, it is shown that only a finite number of additional states (smoothed statistics) are required. We will see here, from a different point of view, how the additional filter states arise. In addition, we will prove results similar to Theorems 1 and 2 for some systems in which equations (2) and (4) for y(t) contain an additive noise term.

In this paper both the continuous and discrete time problems will be considered in a unified framework. It is useful first to contrast these problems with the estimation and prediction problems considered by Huang and Cambanis [8]. There the problem is that of estimating a nonlinear functional y of a Gaussian process $\{x(t),\ t\varepsilon S\}$, given observations of $\{x(t),\ t\varepsilon \tilde{S}\}$, where $\tilde{S}$ is a subset of S. In our problem the objective is to recursively estimate a nonlinear functional y(t) of x(·), given observations of linear functionals of x(·) plus noise. Although the elegant formulas of Huang and Cambanis cannot be applied here, the approach of utilizing the homogeneous chaos, or, equivalently, the Cameron-Martin orthogonal series decomposition of a Gaussian process [4] - [8], will prove to be quite useful in unifying and simplifying our results.

By employing the "innovations approach" [15], [16], the conditional expectations $\hat{y}(t|t)$ of Theorems 1 and 2 can equivalently be viewed as projections on Hilbert spaces generated by the innovations instead of the observations. For the discrete time problem (4)-(6) it can easily be shown recursively that $F_t^{z_2} = F_t^{\nu_2}$, so that $\hat{y}(t|t)$ is the projection of y(t) onto $L_2(\Omega, F_t^{\nu_2}, P)$; in fact $\nu_2(\cdot)$ is just obtained from the Gram-Schmidt orthogonalization of the sequence $z_2(\cdot)$. It has been shown [15], [16] for continuous time Gaussian processes (as in (1), (3)) that $F_t^{z_1} = F_t^{\nu_1}$; hence, $\hat{y}(t|t)$ is the projection of y(t) onto $L_2(\Omega, F_t^{\nu_1}, P)$. The innovations process $\nu_1(t)$ is a Wiener process with the same covariance as $R^{\frac{1}{2}}v(t)$ [14] - [16]; the innovations process $\nu_2(t)$ is a zero-mean Gaussian white noise sequence with $E[\nu_2(t)\nu_2(t)'] = C(t)P(t|t-1)C'(t) + R$, where $P(t|t-1)$ is the Kalman filter one-step error covariance matrix [17]. In both cases, the linear and nonlinear innovations are equal. Hence the estimation problem (1)-(3) or (4)-(6) can be reformulated as that of estimating $y(t)$, a nonlinear L_2-functional of the Gaussian process x^t; the estimate $\hat{y}(t|t)$ is the nonlinear L_2-functional of the innovations process (either ν_1^t or ν_2^t) which minimizes the mean square error. The expansion of such L_2-functionals of Gaussian processes is the subject of [4]-[8], and the application of these results to our recursive estimation problem is presented in the next section, where a new proof of Theorem 1 is presented and the corresponding proof of Theorem 2 is outlined.

3 L_2-FUNCTIONALS OF GAUSSIAN PROCESSES AND FINITE DIMENSIONAL ESTIMATION

Kallianpur [7] has generalized the earlier results of Cameron and Martin [5] and Itô [6] on the orthogonal decomposition of L_2-functionals of a Gaussian process. We will not require all of the isomorphisms presented in [7]; only the following decomposition in terms of Hermite polynomials will be utilized here [8]. Let $x(t)$, $t\varepsilon S$, be any zero-mean second order Gaussian process defined on $(\Omega,\mathcal{B},P)$; for our purposes S will be either an interval $[0,T]$ or the discrete time set $\{0;T\}$. Define the two Hilbert spaces associated with x: the nonlinear space $L_2(x)\underline{\underline{\Delta}}L_2(\Omega,F^x,P)$, where F^x is the σ-algebra generated by $x(t)$, $t\varepsilon S$; and the linear space $H(x)$, the closed subspace of $L_2(x)$ spanned by $x(t)$, $t\varepsilon S$.

<u>Lemma 1</u> [7],[8]: If $\{\xi_\gamma,\gamma\varepsilon\Gamma\}$ (Γ linearly ordered) is a complete orthonormal set (CONS) in $H(x)$, then the family

$$(p_{\gamma_1}!\ldots p_{\gamma_k}!)^{-\frac{1}{2}} H_{p_1}(\xi_{\gamma_1})\ldots H_{p_k}(\xi_{\gamma_k}),$$

$$p\geq 0,\ k\geq 1,\ p_1+\ldots+p_k = p,\ \gamma_1<\ldots<\gamma_k,$$

is a CONS in $L_2(x)$, where H_n is the nth normalized Hermite polynomial. That is, any L_2-functional θ of $x(\cdot)$ has the orthogonal series expansion

$$\theta = \sum_{p \geq 0} \sum_{\substack{p_1+\ldots+p_k=p \\ \gamma_1<\ldots<\gamma_k}} a^{p_1\ldots p_k}_{\gamma_1\ldots\gamma_k} H_{p_1}(\xi_{\gamma_1})\ldots H_{p_k}(\xi_{\gamma_k}). \quad (12)$$

Remark: If x has nonzero mean, the representation of Lemma 1 can be written with respect to a centred CONS, and the coefficients in (12) will depend on the mean of x.

Corollary 1 [6]: If $x(t)$, $t \in [0,T]$ is a standard Wiener process, then any $\theta \in L_2(x)$ has the orthogonal expansion

$$\theta = \sum_{p \geq 0} \int_0^T \int_0^{t_p} \cdots \int_0^{t_2} f_p(t_1,\ldots,t_p)dx(t_1)..dx(t_{p-1})dx(t_p) \quad (13)$$

$$\triangleq \sum_{p \geq 0} I_p(f_p)$$

where the integrals in (13) are iterated stochastic integrals; also, $I_p(f_p)$ and $I_q(f_q)$ are orthogonal for all $p \neq q$.

Now we consider the estimation problems of Theorems 1 and 2 in this framework. Assume throughout this section, for simplicity of notation, that x, y, z_1, and z_2 are all scalars; the following results also hold in the vector case. The state $y(t)$ (as given by (2) or (4)) is a nonlinear functional of x^t; assume that $y(t)$ has a finite Volterra series expansion (of the form (7) or (8)) of order q. It is then clear that $y(t)$ has a finite orthogonal series expansion (12) of order q -- i.e., with $a^{p_1\ldots p_k}_{\gamma_1\ldots\gamma_k} = 0$ for $p>q$. In the continuous case, the $\{\xi_i\}$ are centred versions of functionals of the form $\int_0^t \phi_i(s)x(s)ds$, while in discrete-time the $\{\xi_i\}$ are just centred linear combinations of the $x(s)$, $s\varepsilon\{1;t\}$. The estimate $\hat{y}(t|t)$ is a nonlinear L_2-functional of the Gaussian innovations process; thus it also has an orthogonal expansion of the form (12). In continuous time $\nu_1(t)$ is a Wiener process, so $\hat{y}(t|t)$ has the expansion (13) with $x(t) = R^{-\frac{1}{2}}\nu_1(t)$. In discrete time, $\nu_2(t)$ is an orthogonal sequence, so $\eta(t) \triangleq \left[C(t)^2 P(t|t-1)+R\right]^{-\frac{1}{2}}\nu_2(t)$ is a CONS in $H(\nu_2^t)$, and the expansion (12) is valid with $\xi_i = \eta(i)$.

Thus Theorems 1 and 2 can be proved by showing that: (a) the orthogonal series expansion of $\hat{y}(t|t)$ has only a fixed finite number of terms for all t; and (b) such a finite orthogonal series can be realized as the output of a finite dimensional recursive system (i.e., a system in state-space form). The states of this finite dimensional system are the additional filter states referred to in Section 2. The following theorem proves (a) for a more general formulation; the proof of (b) must be done separately for continuous and discrete time, and involves the calculation and separability of the Volterra kernels.

Theorem 3: Let $x(t)$, $z(t)$, $t\varepsilon S$ be zero-mean jointly Gaussian second order processes, and assume that $y \,\varepsilon\, L_2(x)$ has an orthogonal series expansion of order q. Let the orthogonal expansion of $\hat{y} \triangleq E\left[y|F^z\right]$ be given by

$$\hat{y} = \sum_{r\geq 0} \sum_{\substack{r_1+\ldots+r_j=r \\ \beta_1<\ldots<\beta_j}} b^{r_1\ldots r_j}_{\beta_1\ldots\beta_j} H_{r_1}(\eta_{\beta_1})\ldots H_{r_j}(\eta_{\beta_j}) \qquad (14)$$

where $\{\eta_\delta, \delta\varepsilon\Delta_1\}$ is a CONS in $H(z)$. Then

$$b^{r_1\ldots r_j}_{\beta_1\ldots\beta_j} = 0, \qquad r > q;$$

that is, $\hat{y}$ also has an orthogonal expansion of order q.

Proof: Consider $H(x,z)$, the linear space spanned by $\{x(t),z(t);t\varepsilon S\}$. Since $\{\eta_\delta,\delta\varepsilon\Delta_1\}$ is an orthonormal set in $H(x,z)$, it can be completed by adding elements $\{\eta_\delta,\delta\varepsilon\Delta_2\}$ in $H(x,z)$ to form the CONS $\{\eta_\delta,\delta\varepsilon\Delta_1 U\Delta_2\}$ in $H(x,z)$. The orthogonal expansion for y can then be rewritten in terms of this CONS in $H(x,z)$; the new expansion is clearly also of order q:

$$y = \sum_{p=0}^{q} \sum_{\substack{p_1+\ldots+p_k=p \\ \gamma_1<\ldots<\gamma_k}} c^{p_1\ldots p_k}_{\gamma_1\ldots\gamma_k} H_{p_1}(\eta_{\gamma_1})\ldots H_{p_k}(\eta_{\gamma_k}) \tag{15}$$

where $\{\gamma_i\}\ \varepsilon\ \Delta_1 U\Delta_2$. Now $\hat{y}$ is the orthogonal projection of y onto $L_2(z)$; that is, by Lemma 1 $\hat{y}$ is just the projection of y onto the space spanned by the products $H_{p_1}(\eta_{\gamma_1})\ldots H_{p_k}(\eta_{\gamma_k})$ with $\gamma_1,\ldots,\gamma_k\varepsilon\Delta_1$. The orthogonality of such products in $L_2(x,z)$ (see Lemma 1) then yields

$$\hat{y} = \sum_{p=0}^{q} \sum_{\substack{p_1+\ldots+p_k=p \\ \gamma_1<\ldots<\gamma_k}} c^{p_1\ldots p_k}_{\gamma_1\ldots\gamma_k} H_{p_1}(\eta_{\gamma_1})\ldots H_{p_k}(\eta_{\gamma_k}) \tag{16}$$

where $\{\gamma_i\}\varepsilon\Delta_1$, thus proving the theorem.

Theorem 3 also holds for nonzero-mean and vector-valued processes, with obvious modifications in the proof. This theorem then applies to $\hat{y}(t|t)$ of Theorems 1 and 2. It remains only to prove that the finite orthogonal series expansion for $\hat{y}(t|t)$ is realizable with a nonlinear recursive system of fixed finite dimension. Consider first the continuous time problem (1)-(3).

Proof of Theorem 1: Assume that y(t) has a finite Volterra series expansion of order q. Then Theorem 3 implies that $\hat{y}(t|t)$ has the orthogonal expansion

$$\hat{y}(t|t) = \sum_{p=0}^{q} \int_0^t \int_0^{s_p} \cdots \int_0^{s_2} f_p(t,s_1,\ldots,s_p)\,d\nu(s_1)\ldots d\nu(s_p) \quad (17)$$

where $\nu(t) \triangleq R^{-\frac{1}{2}} \nu_1(t)$. The projection theorem and the orthogonality of the iterated stochastic integrals [6] imply that, $s_1 < \ldots < s_p < t$,

$$f_p(t,s_1,\ldots,s_p) = \frac{1}{p!} \frac{\partial^p}{\partial s_1 \ldots \partial s_p} E\left[y(t)\nu(s_1)\ldots\nu(s_p)\right] \quad (18)$$

(the proof of (18) is analogous to that of Davis [14, p. 95] for the best linear estimate).[1] A proof identical to that of Brockett [11] for the deterministic case shows that, if the kernels (18) are separable (see (9)), then $\hat{y}(t|t)$ in (17) can be generated as the output of a finite dimensional bilinear system driven by the innovations $\nu(t)$. Hence, Theorem 1 is proved if the kernels in (18) are separable.

Lemma 2: The triangular kernels $f_p(t,s_1,\ldots,s_p)$ given by (18) are separable for $s_1 < \ldots < s_p < t$ under the hypotheses of Theorem 1.

Proof: Let $y(t)$ be given by one k^{th} order term in the finite Volterra series (7); the proof generalizes in the obvious way. Since the kernels of (7) are separable due to the analyticity assumption in (2), we can assume that

$$y(t) = \int_0^t \int_0^{\tau_k} \cdots \int_0^{\tau_2} \gamma_1(\tau_1)\ldots\gamma_k(\tau_k)x(\tau_1)\ldots x(\tau_k)\,d\tau_1\ldots d\tau_k \quad (19)$$

---

[1] In general, whenever the linear innovations $\nu(t)$ in a nonlinear estimation problem form a Wiener process, then an (infinite) orthogonal expansion of the form (17) will hold for the estimate of each L_2-state $y(t)$, and the kernels are calculated via (18). The sum of the first two terms (p=0,1) in (17) is the best linear estimate, the sum of the terms for p=0,1,2 yields the best quadratic estimate, etc. These are not necessarily realizable with finite dimensional recursive filters.

Thus, by the Fubini theorem (see [1],[3])

$$f_p(t,s_1,\dots s_p) = \frac{1}{p!} \int_0^t \int_0^{\tau_k} \cdots \int_0^{\tau_2} \gamma_1(\tau_1)\dots\gamma_k(\tau_k)$$
$$\cdot \frac{\partial^p}{\partial s_1 \dots \partial s_p} E\left[x(\tau_1)\dots x(\tau_k)\nu(s_1)\dots\nu(s_p)\right] d\tau_1 \dots d\tau_k \tag{20}$$

Since $x(\tau_1),\dots,x(\tau_k),\nu(s_1),\dots,\nu(s_p)$ are jointly Gaussian, the expectation in (20) can be expanded via Lemma B.1 of [1], resulting in a sum of products of terms of the form: $E[x(\tau_i)]$, $E[\nu(s_i)]$, $\mathrm{cov}[x(\tau_i),x(\tau_j)]$, $\mathrm{cov}[x(\tau_i),\nu(s_j)]$, and $\mathrm{cov}[\nu(s_i),\nu(s_j)]$. Notice that $E[\nu(s_i)] = 0$, so all products involving such terms are zero. If $\mathrm{cov}[\nu(s_i),\nu(s_j)]$ arises, it results in a term of the form $\frac{\partial^2}{\partial s_i \partial s_j} \mathrm{cov}[\nu(s_i),\nu(s_j)]$, which can be shown to be zero for $s_i \neq s_j$. Also, $\mathrm{cov}[x(\tau_i),x(\tau_j)]$ is the covariance function of the state of the linear system (1); hence, for $\tau_i > \tau_j$,

$$\mathrm{cov}[x(\tau_i),x(\tau_j)] = \exp\left[\int_{\tau_j}^{\tau_i} A(\sigma)d\sigma\right] \mathrm{cov}[x(\tau_j),x(\tau_j)] \tag{21}$$

Finally, consider

$$\begin{aligned} R^{\frac{1}{2}}\mathrm{cov}[x(\tau_i),\nu(s_j)] &= \mathrm{cov}\left[x(\tau_i),z(s_j) - \int_0^{s_j} C(\sigma)\hat{x}(\sigma|\sigma)d\sigma\right] \\ &= \mathrm{cov}\left[x(\tau_i), \int_0^{s_j} C(\sigma)(x(\sigma)-\hat{x}(\sigma|\sigma))d\sigma + v(s_j)\right] \\ &= \mathrm{cov}\left[x(\tau_i), \int_0^{s_j} C(\sigma)(x(\sigma)-\hat{x}(\sigma|\sigma))d\sigma\right] \end{aligned} \tag{22}$$

since $x(\tau_i)$ and $v(s_j)$ are independent. This gives rise in

(20) to

$$\frac{\partial}{\partial s_j} \operatorname{cov}\left[x(\tau_i), \nu(s_j)\right] = R^{-\frac{1}{2}} \operatorname{cov}\left[x(\tau_i), C(s_j)(x(s_j)-\hat{x}(s_j|s_j))\right] \tag{23}$$

$$= C(s_j)R^{-\frac{1}{2}} \operatorname{cov}\left[x(\tau_i), x(s_j)-\hat{x}(s_j|s_j)\right],$$

which is the covariance function of a finite (two-) dimensional linear system with states $x(t)$ and $x(t)-\hat{x}(t|t)$, and is thus also separable.

Lemma B.1 of [1] and the separability of the relevant covariance functions imply that there exist functions $\{\alpha_{\ell i}, \beta_{\ell i}\}$ such that (20) can be written as

$$f_p(t,s_1,\ldots,s_p) = \frac{1}{p!} \int_0^t \int_0^{\tau_k} \cdots \int_0^{\tau_2} \gamma_1(\tau_1)\ldots\gamma_k(\tau_k)$$

$$\cdot \left(\sum_{\ell=1}^{m} \alpha_{\ell 1}(\tau_1)\ldots\alpha_{\ell k}(\tau_k)\beta_{\ell 1}(s_1)\ldots\beta_{\ell p}(s_p) \right) d\tau_1\ldots d\tau_k \tag{24}$$

$$= \frac{1}{p!} \sum_{\ell=1}^{m} \alpha_\ell(t)\beta_{\ell 1}(s_1)\ldots\beta_{\ell p}(s_p).$$

and f_p is separable as claimed; this also completes the proof of Theorem 1.

An example in which the kernels and the recursive estimator are computed explicitly is presented in the next section. The discrete time result which is analogous to Lemma 2 can be used to prove Theorem 2, but for the sake of brevity we will only present an example of the procedure (Section 5).

4 A CONTINUOUS TIME EXAMPLE

Before discussing the example, we present an extension of Theorem 1 to a class of systems in which $y(t)$ contains process noise; the analogous extension of Theorem 2 is proved in the same manner.

Theorem 4: Consider the system (1)-(3), and assume that (2) has a finite Volterra series expansion. Assume that there is an additional state $y_1(t)$ satisfying

$$dy_1(t) = (F(t)y_1(t) + G(t)y(t))dt + H(t)d\tilde{w}(t) \tag{25}$$

where $\tilde{w}$ is a Wiener process and $\{x(0),y(0),y_1(0),w(t_1),v(t_2),\tilde{w}(t_3)\}$ are independent for all t_1,t_2,t_3. Then $\hat{y}_1(t|t)$ can also be computed with a finite dimensional recursive estimator.

Proof: The solution of (25) is

$$\begin{aligned} y_1(t) &= \Phi(t,0)y_1(0) + \int_0^t \Phi(t,s)G(s)y(s)ds + \int_0^t \Phi(t,s)H(s)d\tilde{w}(s) \\ &\triangleq \tilde{y}_1(t) + \int_0^t \Phi(t,s)H(s)d\tilde{w}(s) \end{aligned} \tag{26}$$

where Φ is the state transition matrix for F. Since $\tilde{w}(\cdot)$ and $z(\cdot)$ are independent, $\hat{y}_1(t|t) \triangleq E[y_1(t)|F_t^z] = E[\tilde{y}_1(t)|F_t^z]$, and $\tilde{y}_1(t)$ is just described by a finite Volterra series expansion in x. The theorem then follows from Theorem 1.

Example 1: Consider the scalar system

$$dx(t) = -\alpha x(t)dt + dw_1(t) \tag{27}$$

$$dy(t) = (-\gamma y(t) + x^2(t))dt + dw_2(t) \tag{28}$$

$$dz(t) = x(t)dt + dv(t) \tag{29}$$

with the same assumptions as in Theorem 4. The solution of (28) is

$$y(t) = e^{-\gamma t}\, y(0) + \int_0^t e^{-\gamma(t-\sigma)}\, x^2(\sigma)d\sigma + \int_0^t e^{-\gamma(t-\sigma)} dw_2(\sigma) \tag{30}$$

By Theorems 3 and 4, it follows that

$$\hat{y}(t|t) = f_0(t) + \int_0^t f_1(t,s)d\nu(s) + \int_0^t \int_0^{s_2} f_2(t,s_1,s_2)d\nu(s_1)d\nu(s_2), \tag{31}$$

where $\nu(t) = z(t) - \int_0^t \hat{x}(s|s)ds$. Using (18) to compute the kernels as in Lemma 2, we have (since $y(0)$ and $\tilde{w}$ are independent of ν)

$$f_0(t) = E[y(t)] = e^{-\gamma t} E[y(0)] + \int_0^t e^{-\gamma(t-\sigma)} E[x^2(\sigma)]d\sigma \tag{32}$$

$$\begin{aligned} f_1(t,s) &= \frac{\partial}{\partial s} E[y(t)\nu(s)] = \frac{\partial}{\partial s}\int_0^t e^{-\gamma(t-\sigma)} E[x^2(\sigma)\nu(s)]d\sigma \\ &= 2\int_0^t e^{-\gamma(t-\sigma)} m(\sigma)\frac{\partial}{\partial s} E\left[x(\sigma)\int_0^s (x(\tau)-\hat{x}(\tau|\tau))d\tau\right]d\sigma \\ &= 2\int_0^t e^{-\gamma(t-\sigma)} m(\sigma)\, E[x(\sigma)(x(s)-\hat{x}(s|s))]d\sigma \end{aligned} \tag{33}$$

where $m(\sigma) = E[x(\sigma)] = e^{-\alpha\sigma} E[x(0)]$. It can be shown, using Lemma 2.2 of [1], that

$$E[x(\sigma)(x(s)-\hat{x}(s|s))] = \begin{cases} e^{-\alpha(\sigma-s)}\, P(s), & \sigma \geq s \\ K(s,\sigma)\, .\, P(s), & \sigma < s \end{cases} \tag{34}$$

where $K(s,\sigma) = \exp\left[\alpha(s-\sigma) - \int_\sigma^s P^{-1}(\tau)d\tau\right]$ and $P(t)$ is the Kalman filter error covariance for $x(t)$. Thus

$$f_1(t,s) = 2\left[\int_0^s e^{-\gamma(t-\sigma)} m(\sigma)K(s,\sigma)d\sigma + \int_s^t e^{-\gamma(t-\sigma)} m(\sigma)e^{-\alpha(\sigma-s)}d\sigma\right]P(s) \tag{35}$$

Similarly,

$$f_2(t,s_1,s_2) = \left[\int_0^{s_1} e^{-\gamma(t-\sigma)} K(s_1,\sigma)K(s_2,\sigma)d\sigma \right.$$

$$+ \int_{s_1}^{s_2} e^{-\gamma(t-\sigma)} e^{-\alpha(\sigma-s_1)} K(s_2,\sigma)d\sigma \tag{36}$$

$$\left. + \int_{s_2}^{t} e^{-\gamma(t-\sigma)} e^{-\alpha(\sigma-s_1)} e^{-\alpha(\sigma-s_2)} d\sigma \right] P(s_1)P(s_2)$$

(recall that $0 < s_1 < s_2 < t$).

These kernels are obviously separable, so $\hat{y}(t|t)$ can be realized as the output of a finite dimensional bilinear system driven by the innovations. However, it may not be efficient to realize each term in (31) individually. In fact, one efficient recursive realization of $\hat{y}(t|t)$ is readily derived via the procedure of [1, Example 2.1]; a recursive 3-state filter which computes both $\hat{x}(t|t)$ and $\hat{y}(t|t)$ is constructed as follows. First, augment the state $x(t)$ of (27) with the additional state $\xi(t)$ given by

$$\dot{\xi}(t) = (\alpha - \gamma - P^{-1}(t))\xi(t) + x(t); \quad \xi(0) = 0 \tag{37}$$

Then the Kalman-Bucy 2-state filter for the linear system (27), (37) with observations (29) recursively computes $\hat{x}(t|t)$ and $\hat{\xi}(t|t)$. Finally, $\hat{y}(t|t)$ is computed by

$$d\hat{y}(t|t) = (-\gamma\hat{y}(t|t)+[\hat{x}(t|t)]^2+P(t))dt+2P(t)\hat{\xi}(t|t)d\nu(t)$$

$$\hat{y}(0|0) = E[y(0)] \tag{38}$$

To check that this filter has the series expansion (31), (32), (35), (36) is straightforward.

It should also be noted that if $x(t)$ has zero mean, then the best linear estimate of $y(t)$ given z^t (the first two terms in (31)) is equal to the a priori mean of $y(t)$. This is due to the fact that, in this case, y and z are uncorrelated. However, since y and z are not independent, the best quadratic estimator (which is equal to the conditional mean in this example) can in fact offer significant improvement in estimator performance (see [18] for some case studies and further analysis along these lines).

5 A DISCRETE TIME EXAMPLE

Example 2: Consider the scalar discrete time system

$$x(t+1) = \alpha x(t) + w_1(t) \tag{39}$$

$$y(t+1) = \gamma y(t) + x^2(t) + w_2(t) \tag{40}$$

$$z(t) = x(t) + v(t) \tag{41}$$

with the same assumptions as in Theorem 2; also, w_2 is a discrete time white noise process independent of $x(0)$, $y(0)$, w_1 and v. The solution of (40) is

$$y(t) = \gamma^t y(0) + \sum_{i=0}^{t-1} \gamma^{t-i-1} x^2(i) + \sum_{i=0}^{t-1} \gamma^{t-i-1} w_2(i) \tag{42}$$

By Theorem 3, it follows from (16) that

$$\begin{aligned} \hat{y}(t|t) &= c_0(t) + \sum_{j=0}^{t} c_1(t,j) H_1(\eta(j)) \\ &+ \sum_{\substack{j,k=0 \\ j<k}}^{t} c_{11}(t,j,k) H_1(\eta(j)) H_1(\eta(k)) \\ &+ \sum_{j=0}^{t} c_2(t,j) H_2(\eta(j)) \end{aligned} \tag{43}$$

where $\eta(t) = [P(t|t-1)+1]^{-\frac{1}{2}} [z(t) - \alpha\hat{x}(t-1|t-1)]$ are the normalized innovations. By orthogonality (Lemma 1) and the projection theorem,

$$c_{p_1 \ldots p_k}(t, j_1, \ldots, j_k) = \frac{1}{p_1! \ldots p_k!} E\left[y(t) H_{p_1}(\eta(j_1)) \ldots H_{p_k}(\eta(j_k))\right] \tag{44}$$

These kernels can, as in Example 1, be explicitly evaluated. They are indeed separable, and a 3-state filter can be constructed as follows using the methods of [1],[2]. First, augment the state $x(t)$ of (39) with

$$\xi(t+1) = \frac{\alpha\gamma P(t|t)}{P(t+1|t)} \xi(t) + \frac{\gamma P(t|t)}{P(t+1|t)} x(t); \quad \xi(0) = 0 \tag{45}$$

Then $\hat{x}(t|t)$ and $\hat{\xi}(t|t)$ can be calculated by a 2-state Kalman filter. Finally,

$$\hat{y}(t+1|t+1) = \alpha\hat{y}(t|t) + \hat{x}(t|t)^2 + \delta(t)$$

$$+ 2M(t,t+1)\left[\hat{x}(t|t) + \gamma\hat{\xi}(t|t)\right]\nu(t+1) \tag{46}$$

$$+ \left[\sum_{i=0}^{t} \gamma^{t-i} M(i,t+1)^2\right] \nu(t+1)^2$$

$$\hat{y}(0|0) = E\left[y(0)\right]. \tag{47}$$

$$M(i,t+1) = \frac{\alpha^{t-i+1} P(i|i) \ldots P(t+1|t+1)}{P(i+1|i) \ldots P(t+1|t)}$$

and $\delta(t)$ are deterministic functions, and $\nu(t) = z(t) - \alpha\hat{x}(t-1|t-1)$ is the unnormalized innovations process.

Notice that the recursive optimal estimator (46) contains a final term which is quadratic in the innovations. In general, if $y(t)$ contains a Volterra series of order q in $x(t)$, then the recursive estimator for $\hat{y}(t|t)$ will contain polynomials of degree q in $\nu(t)$. This result was proved in [2], but it also follows naturally from the orthogonal series decomposition (16) of $\hat{y}(t|t)$ -- if $y(t)$ has a Volterra series of order q, (16) will contain terms such as $H_q(\eta(t))$, or polynomials of order q in $\eta(t)$. This phenomenon does not occur in continuous time estimation problems with observations corrupted by "Gaussian white noise", in which the optimal recursive estimator is always linear in the innovations.

In [2], this contrast is explained by means of the different martingale representation theorems in continuous and discrete time [19],[20]. However, a simple explanation is provided by the representation (16). In continuous time, the elements η_γ of the CONS in $L_2(z^t)$ are of the form $\int_0^t \phi_\gamma(s)d\nu_1(s)$, and the series (16) can be expressed in terms of iterated stochastic integrals as in (17). Given separability, the series can then be realized with a finite dimensional bilinear system -- that is, the stochastic differential equations in the realization are linear in $d\nu_1(t)$. In the discrete time case, the elements η_γ of the CONS in $L_2(z^t)$ are given by the normalized discrete time innovations $\eta(t)$; the series (16) then gives rise to a finite Volterra series in the innovations $\nu_2(t)$ which contains polynomials in $\nu_2(t)$. Given the appropriate realizability conditions, this series can be realized by a finite dimensional state-affine system [2],[21] -- that is, the recursive equations in the realization contain polynomials in $\nu_2(t)$. Hence state-affine equations containing polynomials in $\nu_2(t)$ arise in a very natural way as realizations of the finite series expansion of $\hat{y}(t|t)$.

6 ACKNOWLEDGEMENT

This research was supported in part by the DoD Joint Services Electronics Program through the Air Force Office of Scientific Research (AFSC) Contract F49620-77-C-0101, in part by the National Science Foundation under Grant ENG 76-11106, and in part by the Air Force Office of Scientific Research (AFSC) under Grant AFOSR 77-3281.

7 REFERENCES

1. Marcus, S. I. and Willsky, A. S. "Algebraic structure and finite dimensional nonlinear estimation", *SIAM J. Math. Analy.*, **9**, 312-327, (1978).

2. Marcus, S. I. "Optimal nonlinear estimation for a class of discrete-time stochastic systems", *IEEE Trans Aut. Control*, **AC-24**, 297-302, (1979).

3. Marcus, S. I. "Estimation and analysis of nonlinear stochastic systems", Ph.D. Thesis, Dept. of Electrical Engineering, M.I.T., Cambridge, MA., June, (1975).

4. Wiener, N. "The homogeneous chaos", *Amer. J. Math.*, **60**, 897-936, (1938).

5. Cameron, R. H. and Martin, W. T. "The orthogonal development of nonlinear functionals in series of Fourier-Hermite functionals", *Ann. Math.*, **48**, 385-392, (1947).

6. Itô, K. "Multiple Wiener integrals", *J. Math. Soc. Japan*, **13**, 157-169, (1951).

7. Kallianpur, G. "The role of reproducing kernel Hilbert spaces in the study of Gaussian processes", in Advances in Probability and Related Topics, P. Ney, ed., New York: Marcel Dekker, 49-83, (1970).

8. Huang, S. T. and Cambanis, S. "Estimation and prediction of nonlinear functionals of Gaussian processes", Proc. 1977 Johns Hopkins Conf. on Inform. Sciences and Systems, Baltimore, Md., 140-144.

9. Gustafson, D. E. "On optimal estimation and control of linear systems with state-dependent and control-dependent noise", Ph.D. Thesis, Dept. of Aeronautics and Astronautics, M.I.T., Cambridge, MA., June, (1972).

10. Sorenson H. W. and Stubberud, A. R. "Nonlinear filtering by approximation of the a posteriori density", *Int. J. Control*, **8**, 33-51, (1968).

11. Brockett, R. W. "Volterra series and geometric control theory", *Automatica*, **12**, 167-176, (1976).

12. Gilbert, E. G. "Bilinear and 2-power input-output maps: Finite dimensional realizations and the role of functional series", *IEEE Trans. Aut. Control*, **AC-23**, 418-425, (1978).

13. Gilbert, E. G. "Functional expansions for the response of nonlinear differential systems", *IEEE Trans. Aut. Control*, **AC-22**, 909-921, (1977).

14. Davis, M. H. A. "Linear Estimation and Stochastic Control", London: Chapman and Hall, (1977).

15. Davis, M. H. A. "A direct proof of innovations/observations equivalence for Gaussian processes", *IEEE Trans. Inform. Theory,* **IT-24**, 252-254, (1978).

16. Kailath, T. "An innovations approach to least-squares estimation, part I: Linear filtering in additive white noise", *IEEE Trans. Aut. Control,* **AC-13**, 646-655, (1968).

17. Meditch, J. S. "Stochastic Optimal Linear Estimation and Control", New York: McGraw-Hill, (1969).

18. Liu, C.-H. "A comparison of optimal and suboptimal estimators and estimation lower bounds", M.S. Thesis, Dept. of Electrical Engineering, University of Texas at Austin, August, (1978).

19. Segall, A. "Stochastic processes in estimation theory", *IEEE Trans. Inform. Theory,* **IT-22**, 275-286, (1976).

20. Brémaud, P. M. and Van Schuppen, J. H. "Discrete time processes: Parts I and II", preprint, June, (1976).

21. Sontag, E. D. "On the internal realization of polynomial response maps", Ph.D. Dissertation, University of Florida, Gainesville, (1976).

22. Miller, K. S. "Multidimensional Gaussian Distributions", New York: Wiley, (1965).

23. Kroy, W. H. "A higher measurement space filter for passive tracking", Proc. 3rd Sym. on Nonlinear Estimation and Its Applications, San Diego, Calif., 138-141, (1972).

24. Balakrishnan, A. V. "A general theory of nonlinear estimation problems in control systems", *J. Math. Anal. Appl.,* **8**, 4-30, (1964).

CHAPTER 7

PARAMETER ESTIMATION

7.1 SOME BASIC IDEAS IN RECURSIVE IDENTIFICATION

L. Ljung

(Department of Electrical Engineering, Linköping University, S-581 83 Linköping, Sweden)

1 INTRODUCTION

Identification, i.e. to provide a mathematical model of a dynamical system from measured input-output data, is now a standard tool in systems engineering. Many problems require the availability of a model in real time, as the process develops. Typical such applications are adaptive control, failure detection and process monitoring. Models in real time can be obtained by recursive identification algorithms. By "recursive" we here mean that, at time t, the algorithm produces a model by processing the data measured at time t in a fixed (t-independent) and finite number of arithmetic operations making use of prior information condensed into a fixed (t-independent) and finite memory space. Recursive identification algorithms are also of interest in non-real-time situations, when, for example, limited memory space is available in the data acquisition equipment or as a competitive way of minimizing identification criteria.

In this paper we shall discuss approaches to recursive identification for a single-input single-output model. Most of what will be said is valid for more general models, and for a more complete treatment we refer to [1]. For some recent surveys of the field, see [2], [3] and [4].

Here the aim is to display some basic ideas in the construction of recursive identification algorithms. These will cover many commonly used methods. It will be shown how guiding principles are provided by stochastic approximation schemes (stochastic gradient and stochastic Newton methods). For simple models that are linear in the parameters (equation-error models) these ideas lead to the familiar recursive least squares method. For more complex models an "enforced" linearity-in-the-parameter approximation leads to one class of algorithms, while a more careful approximation of the gradient of an associated criterion gives another class of algorithms.

2 A PROTOTYPE ALGORITHM: The least squares method

The least squares method is archetypal to most recursive identification methods. Let us therefore outline this method here as a basis for the discussion.

The system is supposed to be given as

$$y(t)+a_1y(t-1) + \dots + a_ny(t-n)=b_1u(t-1) + \dots + b_nu(t-n)+v(t) \quad (1)$$

where y(t) and u(t) are the output and input, respectively, at time t. Introduce

$$\phi^T(t) = \{-y(t-1),\dots,-y(t-n),u(t-1),\dots,u(t-n)\} \quad (2)$$

$$\theta_0^T = (a_1\dots a_n\ b_1\dots b_n) \quad (3)$$

Then (1) can be written

$$y(t) = \theta_0^T\phi(t) + v(t). \quad (4)$$

Now θ_0 is not known. The input and output of (1) {or (4)} are measured for t=1,2,...N. It is natural to estimate θ so that somehow the difference $|y(t) - \theta^T\phi(t)|$ is minimized. Two approaches can be taken. One is to minimize the sum

$$\frac{1}{N}\sum_{t=1}^{N} |y(t) - \theta^T\phi(t)|^2 \quad (5)$$

with respect to θ at time N. Denote the minimizing value by $\hat{\theta}(N)$. It is well known that this estimate can be written recursively as

$$\hat{\theta}(t) = \hat{\theta}(t-1) + R^{-1}(t)\phi(t)\varepsilon_0(t) \quad (6a)$$

$$R(t) = R(t-1) + \phi(t)\phi^T(t) \quad (6b)$$

$$\varepsilon_0(t) = y(t) - \hat{\theta}^T(t-1)\phi(t) \quad (6c)$$

This is the least squares method.

The other approach is to consider (assuming stationarity of the processes $y(\cdot)$ and $\phi(\cdot)$) the minimum with respect to θ of

$$V(\theta) = E|y(t) - \theta^T\phi(t)|^2, \tag{7}$$

which, since it is quadratic in θ, can be minimized by solving

$$\frac{d}{d\theta}V(\theta) = E\phi(t)\{y(t) - \theta^T\phi(t)\} = 0. \tag{8}$$

Apply the Robbins-Monro scheme [5] to (8):

$$\hat{\theta}(t) = \hat{\theta}(t-1) + \gamma(t)\phi(t)\varepsilon_0(t) \tag{9a}$$

$$\varepsilon_0(t) = y(t) - \hat{\theta}^T(t-1)\phi(t) \tag{9b}$$

A good and common choice of $\gamma(t)$ is

$$\gamma(t) = \frac{1}{r(t)}; \; r(t) = r(t-1) + |\phi(t)|^2 \tag{10}$$

Notice the similarity with (6). Equations (9) can be understood as a stochastic gradient method to minimize $V(\theta)$. Now, it is known that the gradient method of minimizing functions is quite inefficient, especially close to the minimum. Instead, a suitable search direction is the Newton direction

$$f = -[V''(\theta)]^{-1}V'(\theta)$$

where $V''(\theta)$ is the second derivative matrix (the Hessian) of $V(\theta)$. Since $V''(\theta) = E\phi(t)\phi^T(t)$, an obvious estimate of $V''(\theta)$ at time t is

$$\frac{1}{t}\sum_{k=1}^{t}\phi(k)\phi^T(k) = \frac{1}{t}R(t).$$

A stochastic approximation algorithm using the descent direction f instead of the gradient direction could be called a "Stochastic Newton algorithm". Clearly the recursive least squares method (6) can be interpreted as a stochastic Newton algorithm for the criterion (7). All that has been said about the model (1) {or (4)} and estimation of its parameters is relevant for most recursive identification schemes. In this paper we shall give a general model structure and reduce it to (4) (Section 3). A straight-forward application of (6) or (9) will lead to one class of identification algorithms (Section 4). Convergence of those is tied to positive realness properties. A more careful application of the guiding idea from (6), (9) leads to a second class of algorithms (Section 5). They have good global convergence properties.

3 A GENERAL MODEL

Consider a system described by

$$\{A(q^{-1})\}y(t) = \frac{B(q^{-1})}{F(q^{-1})}\,u(t) + \frac{C(q^{-1})}{D(q^{-1})}\,e(t) \tag{11}$$

where q^{-1} is the backward shift operator and F, B, A, C, D are polynomials in q^{-1} of orders n_f, n_b etc., such that A(O)=F(O)=C(O)=D(O)=1 and B(O)=O. We shall suppress the index of n for convenience. The sequence $\{e(t)\}$ is supposed to be white noise.

The structure (11) is of course overly general. In applications one or several of the polynomials will be unity. We may note that model (1) is obtained with F=C=D=1. F=D=1 gives the so called ARMAX model. Model reference (or output error) models are obtained with A=C=D=1. The generalized least squares model according to Clarke corresponds to F=C=1.

We shall let (11) represent a description of the true system as well as of the model structure. Let the true parameters be denoted by subscript O. Introduce

$$z_O(t) = \frac{B_O(q^{-1})}{F_O(q^{-1})}\,u(t) \tag{12}$$

$$v_O(t) = \frac{C_O(q^{-1})}{D_O(q^{-1})}\,e(t) \tag{13}$$

and

$$\phi_u^o(t) = \{-z_o(t-1),\ldots,-z_o(t-n),u(t-1),\ldots,u(t-n)\}^T \tag{14}$$

$$\theta_u^o = (a_1^o\ldots a_n^o,\ b_1^o\ldots b_n^o)^T \tag{15}$$

$$\phi_e^o(t) = \{-v_o(t-1),\ldots,-v_o(t-n),e(t-1),\ldots,e(t-n)\}^T \tag{16}$$

$$\theta_e^o = (d_1^o\ldots d_n^o\ c_1^o\ldots c_n^o)^T \tag{17}$$

$$\phi_y(t) = \{-y(t-1),\ldots,-y(t-n)\}^T \tag{18}$$

$$\theta_y^o = (a_1^o,\ldots,a_n^o)^T. \tag{19}$$

Introduce also

$$\phi_o(t) = \begin{pmatrix} \phi_y(t) \\ \phi_u^o(t) \\ \phi_e^o(t) \end{pmatrix} \qquad \theta_o = \begin{pmatrix} \theta_y^o \\ \theta_u^o \\ \theta_e^o \end{pmatrix} \tag{20}$$

Then the true system (11) can be written

$$y(t) = \theta_o^T\phi_o(t) + e(t). \tag{21}$$

Compare with (4)!

For identification purposes a problem with expression (21) is that the definition of $\phi_o(t)$ relies upon knowledge of θ_o and the sequence $\{e(t)\}$, which are unknown. In recursive identification we replace θ_o in the definition of $\phi_o(t)$ with the current estimate $\hat{\theta}(k)$, k=1...t. Procedures to determine such estimates are discussed in the following two sections.

Denote the respective estimates of the polynomials by

$\hat{A}(q^{-1},k)$ etc. Define $\phi(t+1)$ from $\phi(t)$ recursively as

$$\hat{F}(q^{-1},t)z(t) = \hat{B}(q^{-1},t)u(t) \tag{22}$$

$$\varepsilon(t) = y(t) - \hat{\theta}^T(t)\phi(t) \tag{23}$$

$$\hat{D}(q^{-1},t)v(t) = \hat{C}(q^{-1},t)\varepsilon(t) \tag{24}$$

$$\phi_u(t+1) = \{-z(t)\ldots-z(t-n+1),u(t),\ldots u(t-n+1)\}^T \tag{25}$$

$$\phi_\varepsilon(t+1) = \{-v(t),\ldots-v(t-n+1),\varepsilon(t),\ldots\varepsilon(t-n+1)\}^T \tag{26}$$

and

$$\phi(t+1) = \begin{pmatrix} \phi_y(t+1) \\ \phi_u(t+1) \\ \phi_\varepsilon(t+1) \end{pmatrix} \tag{27}$$

We may then write the model

$$y(t+1) = \theta^T\phi(t+1) + \varepsilon(t+1) \tag{28}$$

4 A LINEARITY-IN-THE-PARAMETER APPROXIMATION

The straightforward way to estimating $\hat{\theta}(t)$ is to use algorithm (6) or (9) as it stands, using the definition (22) - (27) of $\phi(t)$.

We shall call this approach LIP, linear-in-the Parameters, since it corresponds to a straightforward application of model (28) which appears to be linear in the parameters. This is not quite the case, however, since in the definition of $\phi(t)$ there are hidden dependencies on θ. More about this will be said in Section 5.

For F=D=1 this method is known as Extended least squares (or Panuskas method, approximate maximum likelihood, the extended matrix method, RML1, cf [4]). For A=C=D=1 it is the parallel model-reference identification method, see [6].

The convergence of this method will now be analysed with the methods of [7], [8], to which we refer for details. Let

$\bar{\phi}(t,\theta)$ be the vector that would be obtained from (22-27) with a constant model θ. Assume that it has reached stationarity. Let $\bar{\varepsilon}(t,\theta) = y(t) - \theta^T\bar{\phi}(t,\theta)$.

Let

$$f(\theta) = E\bar{\phi}(t,\theta)\bar{\varepsilon}(t,\theta) \tag{29}$$

$$G(\theta) = E\bar{\phi}(t,\theta)\bar{\phi}^T(t,\theta) \tag{30}$$

with expectation over $\{e(t)\}$. The differential equation

$$\dot{\theta} = R^{-1}f(\theta) \tag{31}$$

$$\dot{R} = G(\theta) - R$$

is now associated with the asymptotic properties of (6), as explained in [7], [8].

Straightforward calculations, cf [8], show that

$$\bar{\varepsilon}(t,\theta) = \left[\frac{D_O(q^{-1})}{C_O(q^{-1})}\,\phi_y(t)\right]^T \Delta\theta_y + \left[\frac{D_O(q^{-1})}{C_O(q^{-1})F_O(q^{-1})}\,\bar{\phi}_u(t,\theta)\right]^T \Delta\theta_u + \left[\frac{1}{C_O(q^{-1})}\,\bar{\phi}_\varepsilon(t,\theta)\right]^T \Delta\theta_e + e(t) \tag{32}$$

where $\Delta\theta = \theta_O - \theta$, with obvious partitioning. Therefore convergence of the algorithm (6) or (9) will be inherently tied to positive realness of the transfer functions indicated in (32). This is explained in detail in [8].

5 THE CRITERION MINIMIZATION APPROACH

It is clear that the problems with convergence in the

previous approach stem from the fact that the implicit dependence of θ upon $\phi(t)$ was ignored. The proper criterion to consider is of course, cf(7),

$$V(\theta) = E|y(t) - \theta^T\bar{\phi}(t,\theta)|^2 = E|\bar{\varepsilon}(t,\theta)|^2 \tag{33}$$

in the notation of the previous section. By direct calculation it is verified that

$$-\frac{d\bar{\varepsilon}(t,\theta)}{d\theta_y} = \frac{D(q^{-1})}{C(q^{-1})}\phi_y(t) \triangleq \bar{\psi}_y(t,\theta) \tag{34}$$

$$-\frac{d\bar{\varepsilon}(t,\theta)}{d\theta_u} = \frac{D(q^{-1})}{C(q^{-1})F(q^{-1})}\bar{\phi}_u(t,\theta) \triangleq \bar{\psi}_u(t,\theta) \tag{35}$$

$$-\frac{d\bar{\varepsilon}(t,\theta)}{d\theta_e} = \frac{1}{C(q^{-1})}\bar{\phi}_\varepsilon(t,\theta) \triangleq \bar{\psi}_\varepsilon(t,\theta). \tag{36}$$

Hence C, D, F denote the polynomials associated with the parameter vector θ. The equation to be solved, analogously to (8), therefore is

$$-\frac{d}{d\theta}V(\theta) = E\bar{\psi}(t,\theta)\bar{\varepsilon}(t,\theta) = 0 \tag{37}$$

where

$$\bar{\psi}(t,\theta) = \begin{pmatrix} \bar{\psi}_y(t,\theta) \\ \bar{\psi}_u(t,\theta) \\ \bar{\psi}_\varepsilon(t,\theta) \end{pmatrix} \tag{38}$$

In order to apply the ideas of section 3, define, recursively in t, the vector $\psi(t)$ as follows:

$$[1+\hat{C}(q^{-1},t)]\psi_y(t) = [1+\hat{D}(q^{-1},t)]\phi_y(t) \tag{39}$$

$$[1+\hat{C}(q^{-1},t)][1+\hat{A}(q^{-1},t)]\psi_u(t) = [1+\hat{D}(q^{-1},t)]\phi_u(t) \tag{40}$$

$$[1+\hat{C}(q^{-1},t)]\psi_\varepsilon(t) = \phi_\varepsilon(t) \tag{41}$$

$$\psi(t) = \begin{bmatrix} \psi_y(t) \\ \psi_u(t) \\ \psi_\varepsilon(t) \end{bmatrix} \tag{42}$$

Now the stochastic gradient and stochastic Newton approaches lead to algorithms (9) and (6) respectively, in which $\phi(t)$ is replaced by $\psi(t)$ in (9a), (10), (6ab) (but of course not in (9b), (6c)).

The method is known as RML, recursive maximum likelihood, for the case F=D=1 and recursive generalized least squares cf [1], [4] in the case F=C=1. For A=C=D=1 it corresponds to the extended Kalman filter with no process noise, see [1].

The associated differential equation with this algorithm is

$$\begin{aligned} \dot{\theta} &= R^{-1}E\bar{\psi}(t,\theta)\bar{\varepsilon}(t,\theta) = -R^{-1}\frac{d}{d\theta}V(\theta) \\ \dot{R} &= E\bar{\psi}(t,\theta)\bar{\psi}^T(t,\theta) - R \end{aligned} \tag{43}$$

Here R is a positive definite matrix and $V(\theta)$ is positive. Therefore $V(\theta)$ is a Lyapunov function for the differential equation (43), assuring stability of the stationary points that correspond to local minima of $V(\theta)$. It then follows from [7] that $\hat{\theta}(t)$ as defined by the described algorithm converges with probability one to a point that gives a local minimum of $V(\theta)$. We may note that this holds even if the true system cannot be described within the chosen model structure. This is a good global convergence result, in a way the best one can hope for. We may note that if $\{e(t)\}$ is a normal process, then $V(\theta)$ is the expected value of the negative log-likelihood function for the estimation problem.

Questions of uniqueness of the local minima of $V(\theta)$ are of course of interest in this context. The answer will depend on the actual structure. It is known, [9], that for B=0 and F=D=1 all local minima give the correct description for the system. It is also known [10], that for F=C=1 there exist "false" local minima.

6 CONCLUSIONS

With the common recursive least squares identification algorithm as a starting point, it has been shown how there are two basic approaches to recursive identification of systems given in input-output form. For a general model structure, many common methods can be classified as special cases in either of the approaches.

7 REFERENCES

1. Ljung, L. "On recursive prediction error identification algorithms", Report LiTH-ISY-I-0226, Linköping University, Sweden, Aug. 1978. Submitted to *Automatica*.

2. Isermann, R., Baur, U., Bamberger, W., Kneppo, P., and Siebert, H. "Comparison of six on-line identification and parameter estimation methods", *Automatica* **10**, 81-103 (1974).

3. Saridis, G.N. "Comparison of six on-line identification algorithms", *Automatica* **10**, 69-79 (1974).

4. Söderström, T., Ljung, L., and Gustavsson, I. "A theoretical analysis of recursive identification methods", *Automatica* **14**, 231-244 (1978).

5. Robbins, H. and Monro, S. "A stochastic approximation method", *Ann. Math. Stat.* **22**, 400-407 (1951).

6. Landau, I.D. "Unbiased recursive identification using model reference adaptive techniques", *IEEE Trans. Autom. Contr.* **AC-21**, 194-202 (1976).

7. Ljung, L. "Analysis of recursive, stochastic algorithms", *IEEE Trans. Automatic Control* **AC-22**, 551-575 (1977).

8. Ljung, L. "On positive real transfer functions and the convergence of some recursive schemes", *IEEE Trans. Automatic Control* **AC-22**, 539-551 (1977).

9. Åström, K.J. and Söderström, T. "Uniqueness of the maximum likelihood estimates of the parameters of an ARMA model", *IEEE Trans. Automatic Control,* **AC-19**, 769-773 (1974).

10. Söderström, T. "Convergence properties of the generalized least squares identification method", *Automatica* **10**, 617-626 (1974).

7.2 A COMPARATIVE STUDY BETWEEN EXTENDED LEAST SQUARES AND EXTENDED KALMAN FILTER IN LINEAR SYSTEM IDENTIFICATION

Ph. de Larminat and C. Doncarli

(Laboratoire d'Automatique de l'Ouest, Université de Nantes, E.N.S.M., 1 rue de la Noë, 44072 Nantes, Cedex, France)

In this paper, the authors present a particular formulation of the non-linear extended state equations describing the states and parameters of a linear process. Different linear approximations of these non-linear equations are given, which lead, either to the Ordinary Least Squares (O.L.S.), or to the Extended Least Squares (E.L.S.), or to the Extended Kalman Filter (E.K.F.). This provides a fruitful comparison between these methods. Furthermore, extensions to the multivariate case are investigated, according to the choice of the possible canonical forms of the model.

1 INTRODUCTION

The paper is organized as follows: the general non-linear extended state equations of the system are presented in section 2. A first linear approximation provides the Ordinary Least Squares method (Sect. 3). The second provides the Extended Least Squares (Sect. 4) method. The Extended Kalman Filter is obtained (Sect. 5) by the last linearisation. The multivariable canonical forms are presented in section 6 and the results are extended to multi-output systems. Some experimental results (on simulation and real processes) are provided in section 7.

2 THE PROCESS AND THE EXTENDED STATE EQUATION

Consider a mono-input u(k), mono-output y(k) discrete-time linear stochastic n^{th} order system:

$$\begin{aligned} y(k) &= a_1 y(k-1) + \dots + a_n y(k-n) \\ &+ b_1 u(k-1) + \dots + b_n u(k-n) \qquad (1) \\ &+ c_1 \nu(k-1) + \dots + c_n \nu(k-n) + \nu(k) \end{aligned}$$

To define an extended state equation, the usual way is to convert (1) into the state form, and to define the extended state as the state itself, plus the parameters, a_i, b_i, c_i.

Here, it is proposed to define directly the extended state by the following vector:

$$X(k) \triangleq [a_1 \ldots a_n \; b_1 \ldots b_n \; c_1 \ldots c_n \; \nu(k-1) \ldots \nu(k-n)]^T \quad (2)$$

This definition, together with (1), gives the extended state equations:

$$X(k+1) = \Phi\, X(k) + \Gamma\, \nu(k) \quad (3)$$

$$y(k) = \psi_k(X(k)) + \nu(k) \quad (4)$$

where ϕ and Γ are appropriate matrices of zeros and ones, and

$$\psi_k(X(k)) = \sum_{i=1}^{n} (a_i\, y(k-i) + b_i\, u(k-i) + c_i\, \nu(k-i)) \quad (5)$$

One can see that the state equation (3) is linear, and stationary. In the measurement equation (4), ψ_k is defined from the known values $y(k-i)$, $u(k-i)$, viewed as given coefficients; the non-linearities of ψ_k come from the state products $c_i.\nu(k-i)$.

The well-known Extended Kalman Filter [1,2] consists here of a linearization of (4) by a first order Taylor's development. But any other linear approximation of (4) may be used. Consider an approximation of (4), of the general form:

$$y(k) = z(k) + H_k\, X(k) + \nu(k) + r_k \quad (6)$$

where $z(k)$ and H_k are to be defined, and r_k is a residue, considered as zero for filtering. Kalman Filter Equations can be applied to the system (3), (6), to generate estimates $\hat{X}$ of the extended state X.

We investigate three different linear approximations of (4).

3 APPROXIMATION OF THE ORDINARY LEAST SQUARES (O.L.S.)

If terms in equation (6) are chosen as follows:

$$z(k) = 0$$

$$H_k = \left[y(k-1) \ldots y(k-n)\ u(k-1) \ldots u(k-n),\ 0 \ldots 0,\ 0 \ldots 0\right],$$

the residue r_k is

$$r_k = \sum_{i=1}^{n} c_i\ \nu(k-i)$$

Consider the algorithm obtained by using the above values of $z(k)$ and H_k in the filtering equations. The following points can be observed:

- The components c_i in the linearized state equation are not observable, and always keep their initial value $\hat{c}_{i/0}$

- The equations of the evolution of the estimated parameters $\hat{a}_i$, $\hat{b}_i$ are the same as those of the real-time version of the ordinary least squares method. So, the well known properties of stability, and convergence to biased values will be observed (except if $\hat{c}_{i/0} = c_i$, or if $\nu(k) = 0$).

4 APPROXIMATION OF THE EXTENDED LEAST SQUARES (E.L.S.)

A better choice in (6) is:

$$z(k) = 0$$

$$H_k = \left[y(k-1) \ldots y(k-n)\ u(k-1) \ldots u(k-n)\ \hat{\nu}(k-1) \ldots \hat{\nu}(k-n)\ 0 \ldots 0\right]$$

Then:

$$r_k = \sum_{i=1}^{n} c_i \left[\nu(k-i) - \hat{\nu}(k-i)\right]$$

Here, $\hat{\nu}(k-i)$ denotes any estimate of $\nu(k-i)$.

Different forms may be proposed. For instance:

$$[\hat{\nu}(k-1) \ldots \hat{\nu}(k-n)] = [\hat{\nu}(k-1)_{/k-1} \ldots \hat{\nu}(k-n)_{/k-1}]$$

$$[\hat{\nu}(k-1) \ldots \hat{\nu}(k-n)] = [\hat{\nu}(k-1)_{/k-2} \ldots \hat{\nu}(k-n)_{/k-n-1}]$$

$$[\hat{\nu}(k-1) \ldots \hat{\nu}(k-n)] = [\hat{\nu}(k-1)_{/k-1} \ldots \hat{\nu}(k-n)_{/k-n}]$$

The first form would be the most natural one in the present context. On the other hand it can be shown by developing the filtering equations that the second form leads to the Extended Least Squares equations of Panuska [4] and that the third one leads to its modified version, studied by P. C. Young [5], and C. Doncarli [6].

In all the cases, the classical properties of E.L.S. are then observed, particularly the convergence to the true values (if convergence takes place), and this point is connected with the fact that $r_k \to 0$ if $\hat{X}(k) \to X(k)$. Moreover, it has been proved [7] that the third E.L.S. version is globally stable, in the sense that for any bounded data u(k) y(k) (random or deterministic) and for any c_i, the estimates $\hat{\nu}(k)_{/k}$ are mean square bounded. In fact, this result is a preliminary to the analysis of convergence via the Ordinary Differential Equation (O.D.E.) [8,9] method, which assumes that all the signals are bounded. It is demonstrated [10,11] via O.D.E., that a sufficient condition for convergence is: $(\frac{1}{C(z)} - \frac{1}{2})$ real positive. If this condition does not hold, cases will exist [10] where E.L.S. do not converge. However, the same algorithm remains always stable in the above sense.

5 APPROXIMATION OF EXTENDED KALMAN FILTER (E.K.F.)

Equation (6) is now chosen to be the first order Taylor's development of (5):

$$z(k) = - \sum_{i=1}^{n} \hat{c}_{i/k-1} \cdot \nu(k-i)_{/k-1}$$

$$H_k = [y(k-1) \ldots y(k-n) \;\; u(k-1) \ldots u(k-n) \;\; \hat{\nu}(k-1)_{/k-1} \ldots \hat{\nu}(k-n)_{/k-1}$$
$$\hat{c}_{1/k-1} \ldots \hat{c}_{n/k-1}]$$

Then:

$$r_k = \sum_{i=1}^{n} \left[c_i - \hat{c}_{i/k-1}\right] \cdot \left[\nu(k-i) - \hat{\nu}(k-i)_{/k-1}\right]$$

The algorithm that emerges by this choice of z(k) and H_k is now an Extended Kalman Filter (E.K.F.) which has been shown [12] to be asymptotically equivalent to the Recursive Maximum Likelihood (R.M.L.) algorithm [13]. One can also see [14] that the invariant imbedding equations for recursive minimisation of the likelihood function are exactly the same as the E.K.F. ones. Due to the nice properties of the maximum likelihood estimates (asymptotic consistency and efficiency under normality hypothesis), one can presume that this third algorithm will be better, and that is corroborated by the fact that r_k is here a second order residue when $\hat{X}(k) \to X(k)$. As for stability, it is not proved. But contrary to other E.K.F. versions, which are known to diverge easily, the present one appears to have very good stability properties: Many experiments were made [12] with intentional aberrant initial values, and it was impossible to observe any divergence.

6 MULTIVARIABLE CASE

6.1 Presentation of the model

Consider the system under the innovation form:

$$\begin{cases} x(k+1) = F\,x(k) + G\,u(k) + L\,\nu(k) \\ y(k) = H\,x(k) + \nu(k) \end{cases} \tag{7}$$

where:

y(k) is the output vector (dimension m)
ν(k) is the innovation vector (dimension m)
x(k) is the state vector (dimension N)
u(k) is the input vector (dimension p)

An infinite number of representations of (7) type can be obtained by a change of base and some canonical forms lead to a minimum number of parameters and/or this set of parameters is unique. In any case, we can always cancel out the state, in the system(7), and a recurrent matrix equation is obtained (operator form). The most general form of this equation is:

$$\begin{aligned} A(0)\; y(k) &= A(1)\; y(k-1) + \ldots\ldots + A(n)\; y(k-n) \\ &+ B(1)\; u(k-1) + \ldots\ldots + B(n)\; u(k-n) \\ &+ C(1)\; \nu(k-1) + \ldots\ldots + C(n)\; \nu(k-n) \\ &+ C(0)\; \nu(k) \end{aligned} \tag{8}$$

where n is the maximum observability index [16].

This lead to certain remarks:

1) $A(0) = C(0)$: this is a direct consequence of the measurement equation (7).

2) When (7) is written under a special canonical form [17], it is shown [6] that:

$A(0)$ and $C(0)$ are lower triangular matrices, with a unity diagonal.

$A(i)$, $B(i)$, $C(i)$ are sparse matrices having null parameters whose position is determined by the values of the indices of observability of each component of the output. This structure is not destroyed by a post multiplication by a lower triangular matrix.

3) It will be, in some cases, necessary to factorise the matrix R (covariance of ν): $R = T\, D\, T^T$, where T is a lower triangular matrix and D a diagonal one. So, we can define a sequence $\tilde{\nu}(k) = T^{-1}\, \nu(k)$, which has a diagonal covariance matrix. The products $C(i).\nu(k-i)$ can be written $C(i).T.\tilde{\nu}(k-i)$, now called $\tilde{C}(i).\tilde{\nu}(k-i)$, where $\tilde{C}(i)$ has the same structure as $C(i)$, but $\tilde{C}(0) \neq A(0)$.

4) It will be possible to consider another sequence $\tilde{\nu}(k) = C(0).\nu(k)$. The products $C(i).\nu(k-i)$ can be written $C(i).C^{-1}(0).\tilde{\nu}(k-i)$, called also $\tilde{C}(i)\; \tilde{\nu}(k-i)$, where $\tilde{C}(i)$ has the same strucutre as $C(i)$, but $\tilde{C}(0) = 1$. In this case, $\tilde{\nu}(k)$ does not have a diagonal covariance matrix.

5) It is possible to pre-multiply all the factors of equation (8) by $A^{-1}(0)$, or $C^{-1}(0)$, but the matrices $\tilde{A}(i) = A^{-1}(0).A(i)$, $\tilde{B}(i)$, $\tilde{C}(i)$ are not sparse here.

In conclusion, the equation (8) can be written under several forms, having the properties summarised in Table I.

6.2 *Identification method*

With the same point of view as in sect. 2 the matrix equation (8) is seen as a non-linear measurement equation, and several linear approximations are possible. The extended state is defined with the unknown parameters, followed by the innovations.

For example, let us consider a 2 inputs/2 outputs system, with a maximum index of observability equal to 2. The most general form of the equation (8) is:

$$\begin{bmatrix} 1 & 0 \\ a(0) & 1 \end{bmatrix} y(k) = \begin{bmatrix} a_{11}(1) & a_{12}(1) \\ a_{21}(1) & a_{22}(1) \end{bmatrix} y(k-1) + \begin{bmatrix} a_{11}(2) & a_{12}(2) \\ a_{21}(2) & a_{22}(2) \end{bmatrix} y(k-2)$$

$$+ \begin{bmatrix} b_{11}(1) & b_{12}(1) \\ b_{21}(1) & b_{22}(1) \end{bmatrix} u(k-1) + \begin{bmatrix} b_{11}(2) & b_{12}(2) \\ b_{21}(2) & b_{22}(2) \end{bmatrix} u(k-2)$$

$$+ \begin{bmatrix} c_{11}(1) & c_{12}(1) \\ c_{21}(1) & c_{22}(1) \end{bmatrix} \nu(k-1) + \begin{bmatrix} c_{11}(2) & c_{12}(2) \\ c_{21}(2) & c_{22}(2) \end{bmatrix} \nu(k-2)$$

$$+ \begin{bmatrix} 1 & 0 \\ c(0) & 1 \end{bmatrix} \nu(k)$$

the extended state is defined as follows:

Table I

form 8.1	R Symmetric	A(O) triangular	C(O) triangular	A(i), B(i), C(i) sparse
form 8.2	R Symmetric	A(O) = I	C(O) = I	A(i), B(i), C(i) characterless
form 8.3	R Symmetric	A(O) triangular	C(O) = I	A(i), B(i), C(i) sparse
form 8.4	R diagonal	A(O) triangular	C(O) triangular	A(i), B(i), C(i) sparse
form 8.5	R diagonal	A(O) = I	C(O) triangular	A(i), B(i), C(i) characterless
form 8.6	R diagonal	A(O) triangular	C(O) = I	A(i), B(i), C(i) characterless

$$X^T(k) = \Big[a_{11}(1)\ a_{12}(1)\ a_{11}(2)\ a_{12}(2)\ b_{11}(1)\ b_{12}(1)\ b_{11}(2)\ b_{12}(2)\ c_{11}(1)\ c_{12}(1)\ c_{11}(2)\ c_{12}(2)\Big] \text{ Block 1}$$

$$\Big[a_{21}(1)\ a_{22}(1)\ a_{21}(2)\ a_{22}(2)\ b_{21}(1)\ b_{22}(1)\ b_{21}(2)\ b_{22}(2)\ c_{21}(1)\ c_{22}(1)\ c_{21}(2)\ c_{22}(2)\Big] \text{ Block 2}$$

$$\Big[a(0)\quad c(0)\Big] \text{ Block 3}$$

$$\Big[\nu_1(k-1)\ \nu_2(k-1)\ \nu_1(k-2)\ \nu_2(k-2)\Big] \text{ Block 4}$$

The dimension of X(k) directly depends on the form of the equation (8). Block 3 can be absent (form 8.2), or reduced to one parameter (forms 8.3, 8.5, 8.6); Block 2 can be reduced if A(i), B(i), C(i) are sparse matrices (forms 8.1, 8.3, 8.4). In every case, the evolution of the extended state is described by a generalised linear equation (3). Different linear approximations (6) of (8) will be investigated now.

6.3 *Approximation of ordinary least squares*

For simplicity, the method is presented on the previous example. The choice of O.L.S. approximation is:

$z(k) = 0$

$$H_k = \begin{bmatrix} y^T(k-1) & y^T(k-2) & u^T(k-1) & u^T(k-2) & 0\ 0\ 0\ 0 \\ 0\ 0 & 0\ 0 & 0\ 0 & 0\ 0 & 0\ 0\ 0\ 0 \end{bmatrix} \text{ Block 1,}$$

written $\begin{bmatrix} z^T(k) \\ 0 \end{bmatrix}$

$$\begin{bmatrix} 0\ 0 & 0\ 0 & 0\ 0 & 0\ 0 & 0\ 0\ 0\ 0 \\ y^T(k-1) & y^T(k-2) & u^T(k-1) & u^T(k-2) & 0\ 0\ 0\ 0 \end{bmatrix} \text{ Block 2,}$$

written $\begin{bmatrix} 0 \\ z^T(k) \end{bmatrix}$

$$\begin{bmatrix} 0 & 0 \\ y_1(k) & 0 \end{bmatrix} \text{ Block 3}$$

$$\begin{bmatrix} 0\ 0 & 0\ 0 \\ 0\ 0 & 0\ 0 \end{bmatrix} \text{ Block 4}$$

Some comments can be made for all the forms of equation (8):

1) As in the monovariable case, the residue is not a negligible vector, thus the estimations will be biased.

2) The filtering equations are decoupled, because the extended state is not completely observable, and the linearized system can be reduced. As a consequence, R and C(i) cannot both be estimated, and, without a priori knowledge, R is chosen equal to 1.

Define:

$$\theta_1^T = [a_{11}(1)\ a_{12}(1)\ a_{11}(2)\ a_{12}(2)\ b_{11}(1)\ b_{12}(1)\ b_{11}(2)\ b_{12}(2)]$$

$$\theta_2^T = [a_{21}(1)\ a_{22}(1)\ a_{21}(2)\ a_{22}(2)\ b_{21}(1)\ b_{22}(1)\ b_{21}(2)\ b_{22}(2)\ a(0)]$$

The linearized system is reduced to the decoupled equations:

$$\begin{cases} \theta_1(k+1) = \theta_1(k) \\ y_1(k) = Z^T(k)\ \theta_1(k) + \nu_1(k) \end{cases}$$

$$\begin{cases} \theta_2(k+1) = \theta_2(k) \\ y_2(k) = [Z^T(k)\ y_1(k)]\ \theta_2(k) + \nu_2(k) \end{cases}$$

The important properties of the several forms of equation (8) are, in this case, the structure of A(O), and the sparse matrices.

Forms 8-2 and 8-5

θ_1 and θ_2 have the same dimension, and the observation matrices are the same. So, the computation of covariance of the filter is the same for each component of the output, which can be treated together. There is no sparse matrix.

Forms 8-3 and 8-1

A(i), B(i) and C(i) are sparse matrices. θ_1 and θ_2 do not have the same dimension, and the observation matrices are not the same. However the observation matrices are very close, and the algorithm can be organized [18] in order to reduce the computations.

The other forms are uninteresting for this O.L.S. approximation.

6.4 Approximation of extended least squares

The method is presented on the same example. The choice of E.L.S. approximation is:

$z(k) = 0$

$$H_k = \begin{bmatrix} y^T(k-1) & y^T(k-2) & u^T(k-1) & u^T(k-2) & \hat{\nu}^T(k-1)_{/k-1} & \hat{\nu}(k-2)_{/k-2} \\ 0\;0 & 0\;0 & 0\;0 & 0\;0 & 0\;0 & \end{bmatrix} \begin{matrix} \text{Block} \\ 1 \end{matrix}$$

$$\text{written} \begin{bmatrix} Z_1^T(k) \\ 0 \end{bmatrix}$$

$$\begin{bmatrix} 0\;0 & 0\;0 & 0\;0 & 0\;0 & 0\;0 & 0\;0 \\ y^T(k-1) & y^T(k-2) & u^T(k-1) & u^T(k-2) & \hat{\nu}^T(k-1)_{/k-1} & \hat{\nu}^T(k-2)_{/k-2} \end{bmatrix} \begin{matrix} \text{Block} \\ 2 \end{matrix}$$

$$\text{written} \begin{bmatrix} 0 \\ Z_1^T(k) \end{bmatrix}$$

$$\begin{bmatrix} 0 & 0 \\ y_1(k) & \hat{\nu}_1(k) \end{bmatrix} \quad \begin{matrix} \text{Block} \\ 3 \end{matrix}$$

$$\begin{bmatrix} 0\;0 & 0\;0 \\ 0\;0 & 0\;0 \end{bmatrix} \quad \begin{matrix} \text{Block} \\ 4 \end{matrix}$$

The extended state is not completely observable. The extended system can be reduced:

$$\theta_1^T = [a_{11}(1)\ a_{12}(1)\ a_{11}(2)\ a_{12}(2)\ b_{11}(1)\ b_{12}(1)\ b_{11}(2)\ b_{12}(2)$$
$$c_{11}(1)\ c_{12}(1)\ c_{11}(2)\ c_{12}(2)]$$

$$\theta_2^T = [a_{21}(1)\ a_{22}(1)\ a_{21}(2)\ a_{22}(2)\ b_{21}(1)\ b_{22}(1)\ b_{21}(2)\ b_{22}(2)$$
$$c_{21}(1)\ c_{22}(1)\ c_{21}(2)\ c_{22}(2)\ a(0)\ c(0)]$$

the linearized system can be written:

$$\begin{cases} \theta_1(k+1) = \theta_1(k) \\ y_1(k) = Z_1^T(k)\ \theta_1(k) + \nu_1(k) \end{cases}$$

$$\begin{cases} \theta_2(k+1) = \theta_2(k) \\ y_2(k) = \left[Z_1^T(k) \,\vdots\, y_1(k),\ \hat{\nu}_1(k) \right] \theta_2(k) + \nu_2(k) \end{cases}$$

The equations of the filter are decoupled with the estimation of C(i) possible. The discrimination among the forms of the equation (8) depends on Block 3, and on the structure of C(O) (with the joint problem of the estimation $\hat{\nu}_1(k)$). When C(O) $\neq$ 1, it is necessary to filter the first component before the second.

form 8.1:

$\nu_2(k)$ is not independent of $y_1(k)$. So the estimation will be biased (R undiagonal).

form 8.2:

The Block 3 is cancelled, and, although R is undiagonal, $\nu_2(k)$ is independent of $Z_1(k)$. So, the estimation is unbiased, and the computation of covariance of the filter is the same (with the choice R = 1) for all the components of the output. All those components can be treated together.

form 8.3:

The estimation is biased (as for the form 8.1).

form 8.4:

R is diagonal, $\nu_2(k)$ is independent of $y_1(k)$ and $\hat{\nu}_1(k)$, which must be estimated first, by filtering the first component before the second. So the estimates are unbiased; the components are treated sequentially, and the matrices of observation are very close to each other, so the computation can be organized. This form presents sparse matrices.

form 8.5 and 8.6:

The estimations are unbiased but there is no facility of computation.

In conclusion, only the forms 8.2 and 8.4 will be recommended for the E.L.S. approximation.

6.5 Approximation of Extended Kalman Filter

In the same example, the choice of E.K.F. approximation is:

$$z(k) = - \sum_{i=1}^{n} \hat{C}(i)_{/k-1} \, \hat{\nu}(k-i)_{/k-1}$$

$$H_k = \begin{bmatrix} y^T(k-1) & y^T(k-2) & u^T(k-1) & u^T(k-2) & \hat{\nu}^T(k-1)_{/k-1} & \hat{\nu}^T(k-2)_{/k-1} \\ 0\;0 & 0\;0 & 0\;0 & 0\;0 & 0\;0 & 0\;0 \end{bmatrix} \text{Block 1}$$

$$\begin{bmatrix} 0\;0 & 0\;0 & 0\;0 & 0\;0 & 0\;0 & 0\;0 \\ y^T(k-1) & y^T(k-2) & u^T(k-1) & u^T(k-2) & \hat{\nu}^T(k-1)_{/k-1} & \hat{\nu}^T(k-2)_{/k-1} \end{bmatrix} \text{Block 2}$$

$$\begin{bmatrix} 0 & 0 \\ y_1(k) & \hat{\nu}_1(k) \end{bmatrix} \text{Block 3}$$

$$\begin{bmatrix} c_{11}{}^{(1)}{}_{/k-1} & c_{12}{}^{(1)}{}_{/k-1} & c_{11}{}^{(2)}{}_{/k-1} & c_{12}{}^{(2)}{}_{/k-1} \\ c_{21}{}^{(1)}{}_{/k-1} & c_{22}{}^{(1)}{}_{/k-1} & c_{21}{}^{(2)}{}_{/k-1} & c_{22}{}^{(2)}{}_{/k-1} \end{bmatrix} \text{Block 4}$$

All the states are observable and the linearized system cannot be reduced. The equations of the filter are not decoupled, and R must be estimated in parallel. So, the possible forms are those where C(0) = 1, or R is diagonal. The most interesting form is the form (8.4), with sparse matrices, and R diagonal.

The components can be treated sequentially (no matrix inversion), but no other organisation of the computation is proposed.

In conclusion, we can draw Table II.

Table II

	O.L.S	E.L.S.	E.K.F.
FORM. 8.1	recommended	uninteresting	uninteresting
FORM. 8.2	recommended	recommended	possible
FORM. 8.3	recommended	uninteresting	possible
FORM. 8.4	possible	recommended	recommended
FORM. 8.5	recommended	possible	possible
FORM. 8.6	possible	possible	possible

The forms which produce sparse matrices will be used when the observability indices are known, and quite different. When they are unknown, or very close to each other, the forms giving the least computation will be chosen.

7 EXPERIMENTAL RESULTS

7.1 Simulation: Monovariable case

The simulated system is described by the following equation:

$$y(k) = 1.4\, y(k-1) - 0.49\, y(k-2) + 0.9\, u(k-1) - 0.7\, \nu(k-1) + \nu(k)$$

The input $u(k)$ is a filtered white sequence and the standard deviation of $\nu(k)$ is equal to 0.1. There is no a priori knowledge, and the coefficients equal to 0 are also identified.

One can see that $C(z) = 1 - 0.7\ z^{-1}$ is real positive.

The comparison between E.L.S. and E.K.F. is done with fifteen samples of input/output records, and the results are presented in Table III.

Table III

	True value	average	Mean square deviation	
$\hat{a}_1$	1.4	1.4012	4.7×10^{-3}	ELS
		1.3999	3.45×10^{-3}	EKF
$\hat{a}_2$	-0.49	-0.4909	4.01×10^{-3}	ELS
		-0.4899	2.93×10^{-3}	EKF
$\hat{b}_1$	0.9	0.9026	1.60×10^{-2}	ELS
		0.9023	0.47×10^{-2}	EKF
$\hat{b}_2$	0	-0.0055	1.2×10^{-2}	ELS
		-0.0026	0.91×10^{-2}	EKF
$\hat{c}_1$	-0.7	-0.675	5.35×10^{-2}	ELS
		-0.670	6.3×10^{-2}	EKF
$\hat{c}_2$	0	-0.031	6.5×10^{-2}	ELS
		-0.041	5.9×10^{-2}	EKF

As expected, the mean square deviation is better with E.K.F. than with E.L.S.

7.2 *Real process: multivariable case*

The process is the rolling-pitching motion of a ship. The models, found with E.L.S. and E.K.F. are very close together, and the validation of the results is the comparison between the correlation function of the output of the process and of the models (Fig. 1).

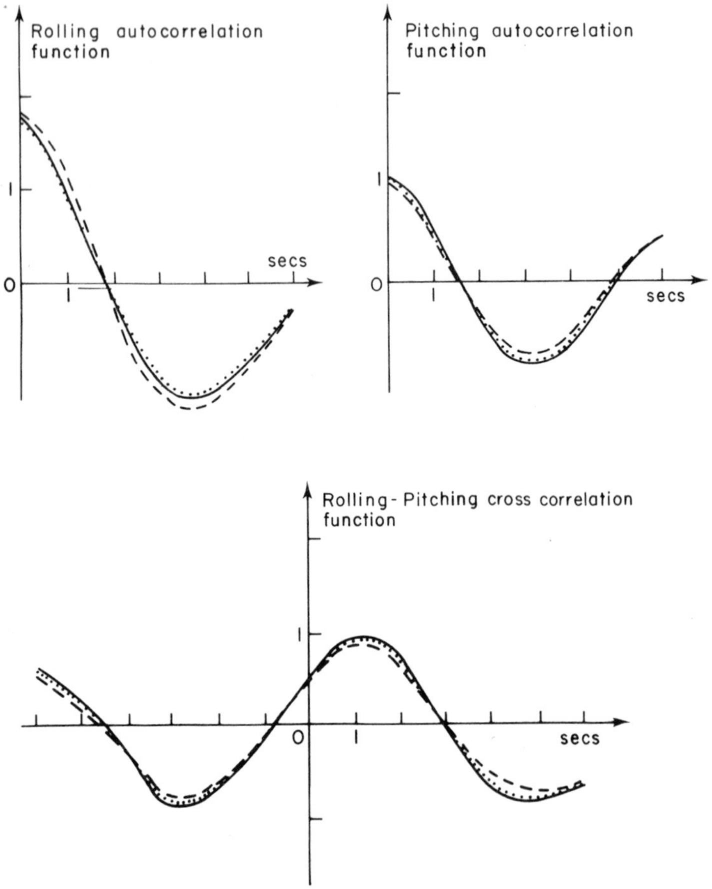

Fig. 1 Correlation functions

(Output of real process ———
Output of E.L.S. model ------
Output of E.K.F. model .....)

8 CONCLUSION

We have presented a new version of the Extended State Equations for modelling linear systems. These yield an Extended Kalman Filter which seems pårticularly stable, and which can be compared with the classical E.L.S. and E.K.F. methods. Extensions to the multivariate case are provided, with a discussion on the canonical forms to be retained, according to the choice of the method.

9 REFERENCES

1. Jazwinski, A. H. "Stochastic process and filtering theory", Academic Press, (1970).

2. de Larminat, Ph. and Thomas, Y. "Automatique des systèmes linéaires Tome 2: Identification", Flammarion Science, 173, (1977).

3. de Larminat, Ph. "Sur l'identification par filtrage non linéaire", Thèse D.E., Nantes (1971).

4. Panuska, V. "An adaptive recursive least squares identification algorithm", 8th I.E.E.E. Symp. Adaptive Process, (1969).

5. Young, P. C. Comments on "On-line identification of linear dynamic systems, with applications to Kalman filtering", *I.E.E.E. Trans.*, **AC-17**, 269-270, April (1972).

6. Doncarli, C. "Sur l'identification des processus stochastiques multivariables", Thèse D.E., Nantes (1977).

7. Doncarli, C. and de Larminat, Ph. "Analyse de la stabilité globale d'un algorithme d'identification récursive des systèmes linéaires stochastiques discrets", Revue Rairo, Serie Jaune: Septembre (1978).

8. Ljung, L. "Convergence of recursive stochastic algorithm", Proceeding of IFAC stochastic control symposium, Budapest (1974).

9. Ljung, L. "Analysis of recursive, stochastic algorithms", *I.E.E.E. Trans.*, **AC-22**, 4, 551-575, Aug. (1977).

10. Ljung, L., Söderström, T. and Gustavsson, I. "Counter examples to general convergence of commonly used recursive identification method", *I.E.E.E. Trans.*, **AC-20**, 643-652, Oct. (1975).

11. Ljung, L. "On positive real transfer functions and the convergence of some recursive schemes", *I.E.E.E. Trans.*, **AC-22**, 4, 539-551, Aug. (1977).

12. de Larminat, Ph. and Doncarli, C. "Sur le comportement asymptotique d'un algorithme d'identification par filtrage de Kalman étendu", Note interne, Laboratoire d'Automatique de l'E.N.S.M., (1977).

13. Söderström, T., Ljung, L. and Gustavsson, I. "A theoretical analysis of recursive identification methods", *Automatica,* **14**, 3, 231-244, May (1978).

14. Sage, A. P. and Melsa, J. L. "System identification", Academic Press, **80**, (1971).

15. Diep, D. "Contribution à l'étude d'algorithmes d'identification par filtrage non linéaire", Thèse 3eme cycle, Université de Nantes, Oct. (1978).

16. Brunovsky, P. "A classification of linear controllable system", *Cybernetica,* **3**, 6, (1970).

17. Guidorzi, R. "Canonical structures in the identification of multivariable systems", 3th IFAC Symp. The Hague (1973).

18. Linard, A., de Larminat, Ph. and Doncarli, C. "A real time identification method for linear multivariable stochastic process", System Science V. International Conference Wroclaw, Poland (1978).

7.3 CONSISTENT ESTIMATES OF PARAMETERS IN CONTINUOUS-TIME SYSTEMS

A. Bagchi

(Department of Applied Mathematics, Twente University of Technology, 7500 AE Enschede, P. O. Box 217, The Netherlands)

1 INTRODUCTION

Consistency of maximum likelihood type estimates of parameters for discrete-time stochastic dynamical systems has been studied in the literature [1,2]. The present paper is concerned with the corresponding question in continuous-time dynamical systems. Estimation of unknown parameters in diffusion-type equations has been extensively studied in the thesis of Le Breton [3]. For linear dynamical systems with noisy observation, likelihood functional for estimating unknown parameters has been derived in Balakrishnan [4], when the observation noise covariance is known. A proof has also been given for the weak consistency of the maximum likelihood estimates. The proof needs uniform boundedness in the parameters of the third gradient of the likelihood functional. This result has, however, not been established in [4]. Our purpose is to remove this inadequacy, and furthermore, to strengthen the conclusion by proving that the estimates are indeed strongly consistent. Our proof continues to hold when the observation noise covariance is unknown. As suggested in [5], a modified likelihood functional is to be used in that situation.

2 PROBLEM STATEMENT

Let (Ω,β,P) be the basic probability space. Consider the following continuous-time stochastic linear dynamical system

$$x(t;\omega) = \int_0^t Ax(s;\omega)ds + \int_0^t Bu(s)ds + \int_0^t FdW_1(s;\omega) \tag{2.1}$$

$$Y(t;\omega) = \int_0^t Cx(s;\omega)ds + \int_0^t GdW_2(s;\omega) \tag{2.2}$$

where $u(t)$ is a p-dimensional "input" function, $x(t;\omega)$ and $Y(t;\omega)$ are n- and m-dimensional "state" and "output" functions respectively; A, B, C are $n \times n$, $n \times p$, $m \times n$ constant but partially unknown matrices; $W_1(t;\omega)$ and $W_2(t;\omega)$ are $n \times 1$ and $m \times 1$ independent Wiener processes and F, G are $n \times n$, $m \times m$ constant but partially unknown matrices with GG^* having an inverse, where "*" denotes the transpose. Let Θ denote the vector of all the unknown system parameters with Θ_0 denoting their true values. We want to determine an estimate $\hat{\Theta}_T(\omega)$ of Θ_0 based on $Y(t;\omega)$, $0 \le t \le T$, that is consistent in the sense that $\hat{\Theta}_T(\omega) \to \Theta_0$ a.s. at $T \to \infty$.

3 MAXIMUM LIKELIHOOD ESTIMATION

We assume that the system is stable and completely observable; that is all the eigenvalues of A lie strictly in the left half of the complex plane, and rank $[C^* \mid A^* C^* \mid \dots \mid (A^{n-1})^* C^*] = n$. Suppose that the system has reached the steady state. Assume that GG^* is completely known; there then is no loss of generality in assuming that GG^* is the identity matrix I.

A detailed discussion on what is meant by the likelihood functional for this problem can be found in [4]. We give here two alternative expressions for this likelihood functional. The first such expression has been derived in [4] and we only explain the different terms involved. The dependence on Θ is explicitly mentioned from now on. All the functions appearing below will be vector-valued, square integrable on $[0,T]$ and therefore elements of appropriate Hilbert spaces with $\|\cdot\|$ and $[\cdot,\cdot]$ denoting norm and inner product there. Let $L(\Theta)$ denote the Volterra operator:

$$L(\Theta)f = g;\ g(t) = C\int_0^t \exp.[(A-PC^*C)(t-s)]\ PC^*f(s)ds \tag{3.1}$$

where P satisfies the algebraic Riccati equation

$$AP + PA^* + FF^* - PCC^*P = 0. \tag{3.2}$$

Let

$$m(\Theta;t) = \int_0^t C\exp.[A(t-s)]Bu(s)ds. \qquad (3.3)$$

Define $\hat{x}(\Theta;t;\omega)$ by

$$\hat{x}(\Theta;t;\omega) = \int_0^t A\hat{x}(\Theta;s;\omega)ds + \int_0^t Bu(s)ds + \int_0^t PC^* dY(s;\omega) - \int_0^t PC^* C\hat{x}(\Theta;s;\omega)ds \qquad (3.4)$$

and

$$Z(\Theta;t;\omega) = Y(t;\omega) - \int_0^t C\hat{x}(\Theta;s;\omega)ds. \qquad (3.5)$$

Then $Z(\Theta_0;t;\omega)$, denoted $Z_0(t;\omega)$ from now on, is the so called "Innovation process". It is a Wiener process with identity covariance.

The likelihood functional for our problem is then

$$H(\Theta;Y(\cdot;\omega);T) = \exp. - \tfrac{1}{2}\{\|m(\Theta;\cdot) + L(\Theta)(dY(\cdot;\omega) - m(\Theta;\cdot))\|^2 - 2[m(\Theta;\cdot) + L(\Theta)(dY(\cdot;\omega) - m(\Theta;\cdot)), dY(\cdot;\omega)]\}. \qquad (3.6)$$

Defining the Volterra operator $K(\Theta)$ by

$$K(\Theta)f = g; \; g(t) = \int_0^t C\exp.[A(t-s)]PC^* f(s)ds. \qquad (3.7)$$

we see easily that $(I+K(\Theta))^{-1} = I - L(\Theta)$ and

$$dY(\cdot;\omega) - m(\Theta_0;\cdot)dt = (I+K(\Theta_0))dZ_0(\cdot;\omega). \qquad (3.8)$$

Define $q(\Theta;Y(\cdot;\omega);T) = -\frac{2}{T}\log H(\Theta;Y(\cdot;\omega);T)$, so that

$$\text{(L.F.1)} \quad q(\Theta;Y(\cdot;\omega);T) = \frac{1}{T}\{\|m(\Theta;\cdot)+L(\Theta)(dY(\cdot;\omega)-m(\Theta;\cdot))\|^2 - 2[m(\Theta;\cdot)+L(\Theta)(dY(\cdot;\omega)-m(\Theta;\cdot)),dY(\cdot;\omega)]\}.$$

We now give an alternative expression for $q(\Theta;Y(\cdot;\omega),T)$ derived originally in [6]. We outline the main steps. Define

$$M(\Theta;t-s) = -C\exp.[A(t-s)]PC^{*}\chi(t-s), \tag{3.9}$$

where $\chi(t)$ is the indicator function: $\chi[(t)](\sigma) = 1$ for $\sigma < t$ and 0 for $\sigma \geq t$. Using the Kalman filter representation, we easily see that the process $Y(t;\omega)$ is Gaussian with mean

$$\overline{m}(t) = \int_0^t m(\theta_0;\sigma)\,d\sigma$$

and covariance matrix

$$R(s,t) = \min.(s,t)I - \int_0^s\int_0^t K(\Theta_0;u,v)\,du\,dv$$

where $K(\Theta_0;u,v)$ is given by

$$K(\Theta_0,u,v) = M(\Theta_0;u-v) + M^*(\Theta_0;v-u) - \int_0^T M(u-\sigma)M^*(v-\sigma)\,d\sigma.$$

One can show that $K(\Theta_0;u;v)$ is square integrable on $[0,T]\times[0,T]$ and that 1 does not belong to the spectrum of the Fredholm operator K with kernel $K(\Theta_0;u;v)$. Furthermore, K is nuclear. Let μ_Y be the measure induced on the space $\mathbb{C}^m[0,T]$ of $\mathbb{R}^m$-valued continuous functions on $[0,T]$ by the process $Y(t;\omega)$, $0 \leq t \leq T$, and μ be the Wiener measure there on. Following a well-known result of Shepp [7], μ_Y is absolutely continuous with respect to μ. The corresponding R-N derivative obtained in [7] gives us an alternative form for the likelihood functional for our problem:

$$H(\Theta;Y(\cdot;\omega);T) = \left[d(\Theta;1)\right]^{-\frac{1}{2}} \exp.\{\tfrac{1}{2}\int_0^T \|m(\Theta;t)\|^2 dt$$

$$-\tfrac{1}{2}\int_0^T\int_0^T \left[dY(s;\omega) - m(\Theta;s)ds\right]^* H(\Theta;s,t)\left[dY(t;\omega) - m(\Theta;t)dt\right] \tag{3.10}$$

$$+ \int_0^T m(\Theta;t)^*\left[dY(t;\omega) - m(\Theta;t)dt\right]\},$$

where $d(\Theta;1) = \prod_j (1-\lambda_j)$, $\{\lambda_j\}$ being the eigenvalues of the operator $K(\Theta)$, the double integral is the double Wiener integral as defined in [7], and $H(\Theta;s,t)$ is the Fredholm resolvent of $K(\Theta)$ at 1. This yields an alternative expression for $q(\Theta;Y(\cdot;\omega);T)$:

(L.F.2) $$q(\Theta;Y(\cdot;\omega);T) = -\frac{2}{T}\text{Log}H(\Theta;Y(\cdot;\omega);T)$$

with $H(\Theta;Y(\cdot;\omega);T)$ given by (3.10).

4 CONDITIONS ON SYSTEM AND INPUT

We consider Θ to be in a suitable compact neighbourhood $\overline{N}$ of Θ_0 to be specified later.

Condition I As assumed already, suppose that for Θ in $\overline{N}$, C-A is completely observable.

Condition II Assume further that the eigenvalues of A are in the left half of the complex plane for all Θ in $\overline{N}$; that is, the system is stable. These two conditions imply that $(A-PC^*C)$ is also a stable matrix for all Θ in $\overline{N}$.

Condition III Assume that the input $u(.)$ is such that

$$\lim_{T\to\infty} \frac{1}{T}\int_0^T u(s)u(s+t)^* ds$$

exists and is a continuous function of t in every finite interval. Assume further that $\sup_t \|u(t)\| < \infty$. It follows then that the averages

$$\lim_{T\to\infty} \frac{1}{T}\int_0^T \left[m(\Theta;t),m(\Theta;t)\right]dt$$

and similar limits occurring subsequently, exist for Θ in $\overline{N}$. Denote the above limit by $\|m(\Theta;\cdot)\|_{av}^{2}$.

5 CONSISTENCY CONSIDERATION OF THE ESTIMATES

We now come to the main part of our paper. Our estimate $\hat{\Theta}_T(\omega)$ is a minimum of $q(\Theta;Y(\cdot;\omega);T)$ in a suitable neighbourhood of Θ_0. We prove that minimizing this functional $q(\Theta;Y(\cdot;\omega);T)$ in an appropriate neighbourhood of Θ_0 will yield consistent estimates of all the unknown parameters of the system under certain sufficient conditions.
Let Θ_i denote the i-th component of Θ and let $\nabla_\Theta q(\Theta;Y(\cdot;\omega);T)$ be the gradient vector with the i-th component $q_i(\Theta;Y(\cdot;\omega);T)$.

Let $Q(\Theta;Y(\cdot;\omega);T)$ be the matrix with the ij-th component $q_{ij}(\Theta;Y(\cdot;\omega);T) = \frac{\partial^2}{\partial\Theta_i\partial\Theta_j} q(\Theta;Y(\cdot;\omega);T)$. We write $q(\Theta;T) = Eq(\Theta;Y(\cdot;\omega);T)$ and $q(\Theta) = \lim_{T\to\infty} q(\Theta;T)$, and use similar notations for the partial derivatives of q.

<u>Lemma 1</u>. $\nabla_\Theta q(\Theta_0) = 0$.

<u>Proof</u>. We use (L.F.1) for $q(\Theta;Y(\cdot;\omega);T)$. Then

$$q_i(\Theta_0;Y(\cdot;\omega);T) = -\frac{2}{T}\int_0^T \left[f_i(t;\omega), dZ_0(t;\omega)\right] \qquad (5.1)$$

where $f_i(t;\omega) = \left[(I-L(\Theta_0))m_i(\Theta_0;\cdot) + L_i(\Theta_0)(I+K(\Theta_0))d\Theta_0(\cdot;\omega)\right](t)$.

Taking expectation, $q_i(\Theta_0;T) = 0$, and the result follows.

<u>Lemma 2</u>. $\nabla_\Theta q(\Theta;Y(\cdot;\omega);T) \to 0$ a.s. as $T \to \infty$.

<u>Proof</u>. In the expression for $q_i(\Theta_0;Y(\cdot;\omega);T)$ in (5.1), the condition $\sup_t.\|u(t)\| < \infty$ and the stability requirements on the system imply that $\sup_t.E\|f_i(t;\omega)\|^2 = M_i < \infty$. The argument in [9, pp. 125-126] can now be directly used to prove our assertion.

<u>Lemma 3</u>. $Q(\Theta_0;Y(\cdot;\omega);T) \to Q(\Theta_0)$ a.s. as $T \to \infty$.

Proof. After little manipulation, we can write

$$q_{ij}(\Theta_0;Y(\cdot;\omega);T) = \frac{2}{T}\{ [(I-L(\Theta_0))m_i(\Theta_0;\cdot)+L_i(\Theta_0)(I+K(\Theta_0))dZ_0(\cdot;\omega),$$

$$[(I-L(\Theta_0))m_i(\Theta_0;\cdot)+L_j(\Theta_0)(I+K(\Theta_0))dZ_0(\cdot;\omega)]$$

$$-[(I-L(\Theta_0))m_j(\Theta_0;\cdot) - L_i(\Theta_0)m_j(\Theta_0;\cdot) - L_j(\Theta_0)m_i(\Theta_0;\cdot)$$

$$+ L_{ij}(\Theta_0)(I+K(\Theta_0))dZ_0(\cdot;\omega),dZ_0(\cdot;\omega)]\}.$$

The term involving the second inner product on the right hand side goes a.s. to 0 as $T \to \infty$, using the same argument as in the preceding lemma. Let υ_i denote the operator $L_i(\Theta_0)(I+K(\Theta_0))$.

We show that

$$\frac{1}{T}[\upsilon_i dZ_0(\cdot;\omega),\upsilon_j dZ_0(\cdot;\omega)] \xrightarrow[T\to\infty]{a.s.} \lim_{T\to\infty} \frac{1}{T} E[\upsilon_i dZ_0(\cdot;\omega),\upsilon_j dZ_0(\cdot;\omega)].$$

It is enough to prove this for $i = j$. Omitting the suffix i, we note that υ is a volterra operator with kernel $\upsilon(t-s)$, a function only of the difference of the arguments. Define

$$\xi(t;\omega) = \int_0^t \upsilon(t-s)dZ_0(s;\omega), \quad 0 \le t \le T.$$

Stability requirements on the system imply that $\{E\|\xi(t;\omega)\|^2\}$ is bounded, uniform in t for $t \ge 0$. Extend $Z_0(t;\omega)$ to the whole real line $-\infty < t < \infty$, by possibly enlarging the probability space. Define

$$\eta(t;\omega) = \int_{\infty}^t \upsilon(t-s)dZ_0(s;\omega), \quad \zeta(t;\omega) = \int_{-\infty}^0 \upsilon(t-s)dZ_0(s;\omega)$$

so that $\xi(t;\omega) = \eta(t;\omega) - \zeta(t;\omega)$. A and $(A-PC^*C)$ being stable matrices, η and ζ are both well-defined. The process $\eta(\cdot;\omega)$ is strictly stationary and ergodic, while $E\|\zeta(t;\omega\|^2 \to 0$ as $t \to \infty$. We have even stronger result. Each component of $\zeta(t;\omega)$ consists of finite sums of the form:

$$e^{\lambda t}t^{\alpha} \int_{-\infty}^0 e^{-\mu s}s^{\beta}dW(s;\omega), \ \lambda < 0, \ \mu < 0; \ \alpha,\beta \ \text{constants and } W \text{ is}$$

some one-dimensional Brownian motion. $\int_{-\infty}^{0} (e^{-\mu s} s^{\beta})^2 ds < \infty$ implies that $\int_{-\infty}^{0} e^{-\mu s} s^{\beta} dW(s;\omega)$ is defined a.s. It follows that $\zeta(t;\omega) \to 0$ a.s. as $t \to \infty$. Now

$$E\xi(t;\omega)\xi(t;\omega)^* = \int_0^t \upsilon(t-s)\upsilon(t-s)^* ds \xrightarrow[t\to\infty]{} \int_0^{\infty} \upsilon(s)\upsilon(s)^* ds = \Sigma, \text{ say.}$$

Using properties of $\zeta(\cdot;\omega)$ and ergodicity of $\eta(\cdot;\omega)$,

$$E\|\eta(t;\omega)\|^2 = \lim_{t\to\infty} E\eta(t;\omega)^*\eta(t;\omega) = \lim_{t\to\infty} E\xi(t;\omega)^*\xi(t;\omega) = \mathrm{Tr}.\Sigma.$$

We now prove the key result

$$\frac{1}{T}\int_0^T \|\xi(t;\omega)\|^2 dt \to \mathrm{Tr}.\Sigma \qquad \text{a.s. at } T \to \infty.$$

Since A and $(A-PC^*C)$ are stable, $\int_1^{\infty} \|\zeta(t;\omega)^*\| dt < \infty$ a.s. This implies $\sum_1^{\infty} \|\zeta(n;\omega)^*\| < \infty$ a.s. Furthermore,

$$\sup_{n\geq 1} \frac{1}{n}\|\eta(n;\omega)\| \leq \sup_{N\geq 1} \frac{1}{N}\sum_{k=1}^{N} \|\eta(k;\omega)\| = M < \infty \quad \text{a.s.}$$

because $\lim_{N\to\infty} \frac{1}{N}\sum_1^N \|\eta(k;\omega)\| < \infty$ a.s. due to ergodicity of the sequence $\{\eta(k;\omega)\}$. Therefore,

$$\sum_{n=1}^{\infty} \frac{1}{n}\|\eta(n;\omega)\| \|\zeta(n;\omega)^*\| \leq M \sum_1^{\infty} \|\zeta(n;\omega)^*\| < \infty \text{ a.s.}$$

so that $\int_1^{\infty} \frac{1}{t}\|\eta(t;\omega)\| \|\zeta(t;\omega)^*\| dt < \infty \Longrightarrow \int_1^{\infty} \frac{1}{t}\eta(t;\omega)\zeta(t;\omega)^* dt < \infty.$

Invoking Kronecker's lemma for integrals,

$$\frac{1}{T}\int_0^T \eta(t;\omega)\zeta(t;\omega)^* dt \to 0 \text{ a.s. as } T \to \infty.$$

Splitting $\xi(t;\omega)$ as $\eta(t;\omega) - \zeta(t;\omega)$ and using properties of $\zeta(\cdot;\omega)$,

$$\frac{1}{T}\int_0^T \|\xi(t;\omega)\|^2 dt = \mathrm{Tr}.\frac{1}{T}\int_0^T \xi(t;\omega)\xi(t;\omega)^* dt \to \mathrm{Tr}.\lim_{T\to\infty}\int_0^T \eta(t;\omega)\eta(t;\omega)^* dt$$

$$= \mathrm{Tr}.E\eta(t;\omega)\eta(t;\omega)^* = \mathrm{Tr}.\Sigma = \lim_{T\to\infty}\frac{1}{T}\int_0^T E\|\xi(t;\omega)\|^2 dt.$$

To complete the proof, we have to show that

$$\lim_{T\to\infty}\frac{1}{T}\left[(I-L(\Theta_0))m_i(\theta_0)),L_j(\Theta_0)(I+K(\Theta_0))dZ_0(\cdot;\omega)\right] = 0 \text{ a.s.}$$

With $\alpha(t) = \left[(I-L(\Theta_0))m_i(\Theta_0;\cdot)\right](t)$, the expression on the left hand side can be written $\frac{1}{T}\int_0^T \left[\alpha(t),\xi(t,\omega)\right]dt$. Assumptions on the input and the system imply that $\int_1^\infty \|\alpha(t)\|dt < \infty$ and $\|\alpha(t)\| \le \alpha$ for all $t \ge 0$. Splitting $\xi(t;\omega)$ into $\eta(t;\omega) - \zeta(t;\omega)$ and using the technique employed just above, we get the desired result.

Lemma 4. Let $Q(\Theta_0)$ be positive definite. There is a neighbourhood of Θ_0 such that no other value of Θ yields a response identical to the one observed for all t for any ω omitting a set of zero probability.

Proof. See [4], pp. 210-212.

Remark. We take a compact neighbourhood in which the above lemma holds. Denot this by $\overline{N}$ and consider Θ in $\overline{N}$.

Lemma 5. Assuming that conditions on system and input are satisfied,

$$\sup_T .\left|q(\Theta;Y(\cdot;\omega);T)\right| < M(\omega)$$

where $M(\omega)$ is independent of $\Theta \in \overline{N}$ and finite for a.e. ω. Similar results hold for partial derivatives of q.

Proof. We use the expression (L.F.2) for $q(\Theta;Y(\cdot;\omega);T)$. Thus

$$q(\Theta;Y(\cdot;\omega);T) = \frac{1}{T}\log d(\Theta;1) - \frac{1}{T}\int_0^T \|m(\Theta;t)\|^2 dt$$

$$+ \frac{1}{T}\int_0^T\int_0^T \left[dY(s;\omega) - m(\Theta;s)ds\right]^* H(\Theta;s,t)\left[dY(t;\omega) - m(\Theta;t)dt\right]$$

$$- \frac{2}{T}\int_0^T m(\Theta;t)\left[dY(t;\omega) - m(\Theta;t)dt\right].$$

Let $\hat{\hat{x}}(t;\omega) = \int_0^t \exp.\left[A(t-s)\right] PC^* dZ_0(s;\omega)$. Then

$$\frac{1}{T}\int_0^T m(\Theta;t)^*\left[dY(t;\omega) - m(\Theta;t)dt\right] = \frac{1}{T}\int_0^T m(\Theta;t)^*\left[m(\Theta_0;t) - m(\Theta;t)\right]dt +$$

$$+ \frac{1}{T}\int_0^T m(\Theta;t)^* C_0\hat{\hat{x}}(\Theta_0;t;\omega)dt + \frac{1}{T}\int_0^T m(\Theta;t)^* dZ_0(t;\omega),$$

and

$$\frac{1}{T}\int_0^T\int_0^T \left[dY(s;\omega) - m(\Theta;s)ds\right]^* H(\Theta;s,t)\left[dY(t;\omega) - m(\Theta;t)dt\right]$$

$$= \frac{1}{T}\int_0^T\int_0^T \left[m(\Theta_0;s) - m(\Theta;s) + C_0\hat{\hat{x}}(\Theta_0;s;\omega)\right]^* H(\Theta;s,t)\left[m(\Theta_0;t) - m(\Theta;t)\right.$$

$$\left. + C_0\hat{\hat{x}}(\Theta_0;t;\omega)\right] ds dt +$$

$$+ \frac{1}{T}\int_0^T\int_0^T \left[m(\Theta_0;s) - m(\Theta;s) + C_0\hat{\tilde{x}}(\Theta_0;s;\omega)\right]^* H(\Theta;s,t)\,ds\,dZ_0(t;\omega) +$$

$$+ \frac{1}{T}\int_0^T\int_0^T dZ_0(s;\omega)^* H(\Theta;s,t)\left[m(\Theta_0;t) - m(\Theta;t) + C_0\hat{\hat{x}}(\Theta_0;t;\omega)\right]dt +$$

$$+ \frac{1}{T}\int_0^T\int_0^T dZ_0(s;\omega)^* H(\Theta;s,t)\,dZ_0(t;\omega).$$

We show the uniform boundedness in Θ of the last term, which is a double Wiener integral. The method of proof can be used to show uniform boundedness of the terms involving ordinary Wiener integral. The remaining terms are deterministic and can be handled easily using continuity in Θ and compactness of $\bar{N}$

The double Wiener integral appearing above remains unchanged if we replace H by $\tilde{H} = \frac{1}{2}(H+H^*)$, the symmetric part. $\tilde{H}(\Theta;s,t)$ is, for fixed Θ, continuous on $0 \leq s, t \leq T$ except possibly on the diagonal $s = t$. At each diagonal point (t,t), $\tilde{H}(\Theta;t+0, t+0)$, $\tilde{H}(\Theta;t+0, t-0)$, $\tilde{H}(\Theta,t-0,t+0)$ and $\tilde{H}(\Theta;t-0,t-0)$ all exist and are equal. Define $\tilde{H}(\Theta;t,t)$ to be this common limiting value, $\tilde{H}(\Theta;t,s)$ is continuous throughout. Furthermore, the first and second partial derivative of $\tilde{H}$ w.r.t s and t exist a.e. Using the integration by parts formula given in [7].

$$\frac{1}{T}\int_0^T\int_0^T dZ_0(s;\omega)^*\tilde{H}(\Theta;s,t)dZ_0(t;\omega)=\frac{1}{T}Z_0(T;\omega)^*\tilde{H}(\Theta;T,T)Z_0(T;\omega)+$$

$$-\frac{1}{T}Z_0(T;\omega)^*\int_0^T\frac{\partial\tilde{H}(\Theta;T,t)}{\partial t}Z_0(t;\omega)dt-\frac{1}{T}\int_0^T Z_0(s;\omega)^*\frac{\partial\tilde{H}(\Theta;s,T)}{\partial s}dsZ_0(T;\omega)+$$

$$+\frac{1}{T}\int_0^T\int_0^T Z_0(s;\omega)^*\frac{\partial^2\tilde{H}(\Theta;s,t)}{\partial s\partial t}Z_0(t;\omega)dsdt.$$

$\tilde{H}$ and its partial derivatives are, for fixed s, t, continuous in Θ. Let

$$\sup_{\Theta\varepsilon\bar{N}}.\|\tilde{H}(\Theta;T,T)\| \leq \alpha(T), \quad \sup_{\Theta\varepsilon\bar{N}}.\left\|\frac{\partial\tilde{H}(\Theta;T,t)}{t}\right\| \leq \beta_1(T,t),$$

$$\sup_{\Theta\varepsilon\bar{N}}.\left\|\frac{\partial\tilde{H}(\Theta;s,T)}{\partial s}\right\| \leq \beta_2(s,T), \quad \sup_{\Theta\varepsilon\bar{N}}.\left\|\frac{\partial^2\tilde{H}(\Theta;s,t)}{\partial s\partial t}\right\| \leq \gamma(s,t)$$

Using the stability property of the system, one can show that $\alpha(T)$,

$$\int_0^T \beta_1(T,t)^2 t\,dt,\quad \int_0^T \beta_2(s,T)^2 s\,ds \text{ and } \int_0^T\int_0^T \gamma(s,t)^2 st\, dsdt \text{ are all}$$

bounded by numbers independent of T for all T > 0. Then

$$E\left|\frac{1}{T}\int_0^T\int_0^T dZ_0(s;\omega)^* H(\Theta;s,t)\,dZ_0(t;\omega)\right| < L, \text{ uniformly in } \Theta,$$

where L is a constant independent of T. It follows that there exists a random variable $L(\omega)$, finite for a.e. ω, and independent of T such that

$$\sup_{\Theta\varepsilon\bar{N}}.\left|\frac{1}{T}\int_0^T\int_0^T dZ_0(s;\omega)^* H(\Theta;s,t)\,dZ_0(t;\omega)\right| < L(\omega).$$

Similar arguments used for other terms will establish the lemma. In proving consistency, we have to use this lemma for $J(\Theta;Y(\cdot;\omega);T)$, the gradient (Fréchet derivative) of $Q(\Theta;Y(\cdot;\omega);T)$ w.r.t. Θ.

We use the above lemmas to prove our main result, namely, that the root of the gradient of $q(\Theta;Y(\cdot;\omega);T)$ converges to Θ_0 a.s. The proof is standard and can be established with slight variation of the argument in [4], pp. 213-217.

Theorem. Suppose that $Q(\Theta_0)$ is positive definite and the stability requirements are satisfied. Then given $\delta > 0$ and for almost all ω, the gradient $\nabla_\Theta q(\Theta;Y(\cdot;\omega);T)$ has a root in a sphere of radius δ and about Θ_0 for all $T > T_0(\delta;\omega)$.

6 CONCLUSION

We have established strong consistency of the maximum likelihood estimates of unknown parameters in a continuous-time stochastic dynamical system when the observation noise covariance is known. When the observation noise covariance is unknown, a modified likelihood functional suggested in [5], is to be used to estimate the unknown parameters. The consistency proof given here will go through in that situation without any alteration. Positive definiteness of $Q(\Theta_0)$ is the sufficient condition for the estimates to be consistent; and it is usually referred to as the "identifiability condition".

Our consistency proof is similar to the classical argument of Cramér [9]. In statistics, objections are raised about Cramér's method for two reasons. Firstly, the existence of partial derivatives of q up to third order forces the method to have limited applicability. In our present problem, however, this condition can be easily verified. Secondly, Cramér's method only establishes the consistency of some root of the equation $\nabla_\Theta q(\Theta;Y(\cdot;\omega);T) = 0$, but not necessarily that of the absolute minimum of $q(\Theta;Y(\cdot;\omega);T)$. Since $Q(\Theta_0)$ is positive definite, the argument of Huzurbazar [10] can be used to show that $\nabla_\Theta q(\Theta;Y(\cdot;\omega);T) = 0$ has at most one consistent solution and $q(\Theta;Y(\cdot;\omega);T)$ has a relative minimum for such a solution. These two results taken together still do not imply that a solution of $\nabla_\Theta q(\Theta;Y(\cdot;\omega);T) = 0$ that makes $q(\Theta;Y(\cdot;\omega);T)$ an absolute minimum is necessarily consistent. This objection is, however, only of theoretical interest. From the computational point of view, we can only find local minima of $q(\Theta;Y(\cdot;\omega);T)$. Therefore, if we know a neighbourhood of Θ_0 in which $q(\Theta;Y(\cdot;\omega);T)$ has only one minimum and we use a minimization algorithm that converges to that point, we arrive at a constructive method for finding the estimate.

7 REFERENCES

1. Ljung, L. "On consistency for prediction error identification methods", Report 7405, Division of Automatic Control, Lund Institute of Technology, Sweden, March 1974.

2. Bagchi, A. "Consistent estimates of parameters in noisy dynamical systems", *Int. J. Control,* **26**, (1977).

3. Le Breton, A. "Sur L'estimation de parameters dans les modeles differentiels stochastiques multidimensionnels", Doctoral Dissertation, University of Grenoble, France, (1976).

4. Balakrishnan, A. V. "Stochastic differential systems", Lecture Notes in Economics and Mathematical Systems, **84**, (1973).

5. Bagchi, A. "Continuous time systems identification with unknown noise covariance, *Automatica,* **11**, (1975).

6. Woo, K. T. "Identification of noisy stems", Ph.D. Dissertation, University of California, Los Angeles, U.S.A., (1970).

7. Shepp, L. A. "Radon-Nikodyn derivative of Gaussian measures", *Annals of Mathematical Statistics*, **37**, (1966).

8. Gihman, I. I. and Skorohod, A. V. "Stochastic differential equations", Springer-Verlag, Berlin, (1972).

9. Cramér, H. "Mathematical methods of statistics", Princeton University Press, Princeton, (1946).

10. Huzurbazar, V. S. "The likelihood equation, consistency and the maxima of the likelihood functions", *Annals of Eugenics*, **14**, (1948)

7.4 CONSISTENT ORDER-ESTIMATES OF AUTOREGRESSIVE PROCESSES BY SHORTEST DESCRIPTION OF DATA

J. Rissanen

(IBM Research Laboratory, 5600 Cottle Road, San Jose, California 95193, USA)

1 INTRODUCTION

Because the maximum likelihood criterion is useless in determining structural parameters, such as the order in autoregressive moving average processes, attempts have been made to introduce an additional structure-dependent term into the criterion. Of these the best-known criterion is AIC [1], introduced and studied by Akaike. In this criterion the structure-dependent term is twice the number of the parameters in the model. Although AIC is an interesting and useful criterion, R. Shibata's [2] analysis of order-determination of autoregressive processes by AIC shows that the order cannot be estimated consistently with this criterion.

In [3] we introduced another principle for parameter estimation; it calls for minimization of the estimated average number of bits needed to write down the observed sequence, each observed real number being truncated to a desired accuracy. The broad reasoning behind this principle, which can be made precise by Gibbs' theorem, is that the most economical way to describe observations - of any kind whatsoever - is possible if and only if the true machinery generating the observed values is used. Hence, in cases where the description length can be explicitly expressed relative to an assumed model, a minimization of this length with respect to the model parameters, be they real-valued or not, should lead to good estimates. It is clear that the description length of the parameters themselves must be included in the total length, and this obviously gives the term penalizing an over-parameterization.

Asymptotically, the criterion derived for fairly general classes of nonlinear dynamic models is as follows:

$$U(x,\theta) = (N-k)\log \sigma + (k+1)\log (N+2),$$

where $N\log\sigma$ is the asymptotic loglikelihood term with opposite sign; and k and N denote the numbers of real-valued parameters in the vector θ and in the observed data vector x, respectively. This is done in Section 2. In Section 3 we specialize in autoregressive processes and show by arguments suggested by E. Hannan (private correspondence) that indeed the minimum length criterion leads to consistent order estimates. In Section 4 we report estimation results of computer simulated processes which confirm the findings of the analysis in Section 3.

2 MINIMUM LENGTH CRITERION

We consider models for the observed sequence $x = \{x_i\}_1^N$ of the type

$$x_t = f_\theta(x_{t-1},\ldots,x_{t-n}, e_t,\ldots,e_{t-m})$$
$$x_t = e_t = 0 \text{ for } t\leq 0, \tag{2.1}$$

where f_θ is such that e_t can be solved in terms of the other variables, and θ is a vector of the parameters to be estimated. In general, θ consists of integer-valued structural parameters such as n and m, and real-valued system parameters ξ with as a rule k=n+m+1 components:

$$\theta = (n,m,\xi) \;,\; \xi = (\xi_0,\ldots,\xi_k), \tag{2.2}$$

where ξ_0, in particular, denotes the variance parameter r of the zero-mean normal distribution modelled for e_t. The process $\{e_t\}$ is modelled as an independent process for reasons discussed in [3].

As we intend to describe the sequence x in terms of the $e=\{e_t\}$-sequence and the parameters, we must write the numbers e_t with some precision $\pm\varepsilon/2$, and the i'th parameter ξ_i with precision $\pm\delta_i/2$.

By assigning a codeword of length $-\log h_r(e_t)$ for e_t, where $h_r(e_t)$ is the modelled probability of the truncated outcome e_t, the sequence $e=(e_1,\ldots,e_N)$ can be described with about

$$L(e|\theta) = \frac{N}{2}\log 2\pi \frac{r}{\varepsilon^2} + \frac{1}{2}\sum_{i=1}^{N} e_i^2/r \qquad (2.3)$$

units, the unit depending on the base of the logarithm. If e_t in fact has such a distribution this coding gives the shortest mean length.

We next must describe the parameters θ, and once an expression $L(\theta)$ for the length is arrived at the total length for describing x is of the form

$$L(x,\theta) = L(e|\theta) + L(\theta).$$

It is illuminating to compare this total length, where e may be exchanged for x in view of the invertibility of f_θ, with the loglikelihood of the posterior probability in Bayes' approach:

$$\ell(\theta|x) = (\ell(x|\theta) + \ell(\theta)) - \ell(x).$$

In Bayes' approach it is necessary to visualize the "true" value for the parameter vector to be a sample of a random variable with an a priori distribution whose existence is a highly controversial issue of long standing, see e.g. [8]. In the current context such an assumption is particularly difficult to justify. After all, (2.1) is just our model of who-knows-what-machinery generating the data, and whatever parameter there is characterizing that machinery it seems more likely than not to be an individual object without any stochastic origin. In our view, in contrast, it is always meaningful to speak about the description length of any object, stochastic or individual alike, which can be described in the first place. And if we are trying to estimate something that cannot be described then our estimation problem needs rethinking!

Returning to the description of θ we clearly can describe n and m with about log nm units, which is normally negligible. With the agreed truncation levels we can describe ξ_i with about $\log(|\xi_i|/\delta_i)$ units, and, hence, the length of describing all the parameters is about

$$L(\theta) = \sum_{i=0}^{k} \log \frac{|\xi_i|}{\delta_i} = \frac{1}{2} \sum_{i=0}^{k} \log \left(\frac{\xi_i}{\delta_i}\right)^2 . \tag{2.4}$$

The total description length of the sequence x is now

$$L(x,\theta) = \frac{N}{2} \log 2\pi \frac{\xi_0}{\varepsilon^2} + \frac{1}{2} \sum_{i=0}^{N} \frac{e_i^2}{\xi_0} + \frac{1}{2} \sum_{i=0}^{k} \log \left(\frac{\xi_i}{\delta_i}\right)^2 . \tag{2.5}$$

The objective is to find the model parameters $\theta=(n,m,\xi)$ such that $L(x,\theta)$ is minimized. To illustrate how added information about the parameters could be taken advantage of without a priori distributions suppose that we knew the parameters to be positive and ordered thus: $\xi_0<\xi_1<\ldots<\xi_k$. Then by encoding the first as above and the others by the differences we get a shorter length than (2.4). This is, of course, not a realistic example; it was chosen just to illustrate our point that we do not need to be a Bayesian in order to get a similar and in fact an even stronger effect, stronger because we have a richer assortment of ways at our disposal to take advantage of a priori information.

Returning to (2.5) we obtain at the minimum:

$$\begin{aligned} \frac{\partial L}{\partial \xi_0} &= \frac{N+2}{2\xi_0}(1 - \sigma/\xi_0) = 0 \\ \frac{\partial L}{\partial \xi_i} &= \frac{N+2}{2\xi_0} \frac{\partial \sigma}{\partial \xi_i} + \xi_i^{-1} = 0, \ i=1,\ldots,k, \end{aligned} \tag{2.6}$$

where

$$\sigma = \frac{1}{N+2} \sum_{j=1}^{N} e_j^2 . \tag{2.7}$$

Observe that for all values of the parameters ξ_i, $i=1,\ldots,k$, the first equation in (2.6) holds at the minimizing point $\hat{\xi}_0=\sigma$. We also need the second partials evaluated at the minimum $\hat{\xi}_i$, which are obtained by some calculations as follows:

$$\frac{\partial^2 L}{\partial \hat{\xi}_0^2} = \frac{N+2}{2}\hat{\sigma}^{-2}, \quad \frac{\partial^2 L}{\partial \hat{\xi}_0 \partial \hat{\xi}_i} = 0, \; i > 0,$$

$$\frac{\partial^2 L}{\partial \hat{\xi}_i \partial \hat{\xi}_j} = \frac{N+2}{2}\frac{\partial^2 \log \hat{\sigma}}{\partial \hat{\xi}_i \partial \hat{\xi}_j} - \frac{\delta(i,j)}{\hat{\xi}_i^2} + \frac{2}{N+2}\frac{1}{\hat{\xi}_i \hat{\xi}_j}, \; i,j > 0, \tag{2.8}$$

where $\delta(i,j)$ is 1 for $i=j$ and 0 for $i \neq j$.

We next determine optimum truncation levels δ_i, which we do by adapting an idea used by Boulton and Wallace in the context of classification problems, [4]. When each $\hat{\xi}_i$ is truncated to the level $\pm \delta_i/2$ a certain error, say Δ_i, results, and the estimate lies in the interval

$$\left| \hat{\xi}_i - \frac{\delta_i}{2}, \hat{\xi}_i + \frac{\delta_i}{2} \right| .$$

These errors are assumed to be uniformly distributed in their intervals. Next, $L(x,\theta)$ is expanded in Taylor's series about $\hat{\xi}$, which with (2.8) results within second order terms in:

$$L(x,\theta) \cong L(x,\hat{\theta}) + \frac{N+2}{4}\left[\frac{\Delta_0^2}{\sigma^2} + \sum_{i,j=1}^{k} \frac{\partial^2 \log \sigma}{\partial \hat{\xi}_i \partial \hat{\xi}_j}\Delta_i \Delta_j\right]$$

$$+ \frac{1}{N+2}\sum_{i,j=1}^{k}\frac{\Delta_i \Delta_j}{\hat{\xi}_i \hat{\xi}_j} - \frac{1}{2}\sum_{i=1}^{k}\frac{\Delta_i^2}{\hat{\xi}_i^2},$$

where $\hat{\theta}=(n,m,\hat{\xi})$. The expected value of $L(x,\theta)$ with respect to the uniform distribution for Δ_i is within second order terms:

$$EL(x,\theta) \cong L(x,\hat{\theta}) + \frac{N+2}{48}\left[\frac{\delta_0^2}{\sigma^2} + \sum_{i=1}^{k}\frac{\partial^2 \log \sigma}{\partial \hat{\xi}_i^2}\delta_i^2\right] + \frac{N}{24(N+2)}\sum_{i=1}^{k}\frac{\delta_i^2}{\hat{\xi}_i^2} .$$

This is minimized for

$$
\left.
\begin{aligned}
\delta_0 &= \sigma \sqrt{\frac{24}{N+2}} \\
\delta_i &= \left\{ \frac{N+2}{2} \frac{\partial^2 \log \sigma}{\partial \hat{\xi}_i^2} - \frac{N}{12(N+2)\hat{\xi}_i^2} \right\}^{-1/2} \\
&= \left\{ \frac{24}{(N+2)\partial^2 \log \sigma / \partial \hat{\xi}_i^2} \right\}^{1/2} \quad + O(N^{3/2}/\hat{\xi}_i^2),
\end{aligned}
\right\} \quad (2.9)
$$

for $i=1,\ldots k$. The corresponding minimum is given asymptotically as

$$U(x,\theta) = (N-k)\log \sigma + (k+1)\log(N+2). \quad (2.10)$$

Strictly speaking (2.10) is valid only at the optimized parameter values $\hat{\xi}_i$, and hence it should be used to optimize the numbers n and m, or k, only. However, because for each n and m the first term for large N is dominant in both (2.5) and (2.10) we clearly may use (2.10) as a criterion to be minimized over all parameters θ. A closely related criterion was derived by Bayesian arguments in [9].

From (2.9) we see that for any non-zero estimates $\hat{\xi}_i$ positive values for δ_i exist when N is taken large enough, provided that the double partial derivatives do not vanish. A non-real solution for δ_i is an indication that the parameter ξ_i is too small to have a noticeable effect on the dominant first term in (2.10) for the given number of observations N, and we should then set that parameter to zero, and perhaps reduce k by one.

3 AR-PROCESSES

Let $\{x_i\}_1^N$ be a sequence of observations from a stationary random process, also denoted by $\{x_i\}$, which is generated by an autoregressive scheme:

$$x_t + a_1 x_{t-1} + \dots + a_n x_{t-n} = w_t, \quad a_n \neq 0. \tag{3.1}$$

The process $w=\{w_t\}$ is an independent, zero mean stationary gaussian process of some variance. We normalize the observed sequence such that $Ex_t^2=1$. Stationarity of $\{x_i\}$ implies that the roots of the characteristic polynomial of (3.1) are inside the unit circle. We intend to show that the minimum length criterion permits a consistent estimation of the order n in the sense that the probability of the order estimate, say k, to be n goes to 1 as $N\to\infty$.

We fit models of type

$$\begin{aligned} &x_t + a_1(k)x_{t-1} + \dots + a_k(k)x_{t-k} = e_t, \quad t>0, \\ &x_t = e_t = 0 \text{ for } t\leq 0, \end{aligned} \tag{3.2}$$

where k and the numbers $a_i(k)$ are to be estimated by minimization of the criterion (2.10):

$$U(x,k,a) = (N-k)\log\sigma + (k+1)\log(N+2).$$

First, for each k the first term asymptotically dominates and the best estimates $\hat{a}_i(k)$ are the ones that minimize

$$\sigma = \sigma_k = \frac{1}{N}\sum_{t=1}^{N} e_t^2. \tag{3.3}$$

And these can be seen to be the solutions to the Yule-Walker equations, [5], [6]:

$$\frac{1}{N}\sum_t x_t x_{t-j} = -\sum_{i=1}^{k} a_i(k)\,\frac{1}{N}\sum_t x_{t-i}x_{t-j}, \quad j=1,2,\dots,k, \tag{3.4}$$

where the undefined values x_t are 0. These can be written as the vector equation with a Toeplitz-matrix R_k

$$\begin{pmatrix} r_0 & r_1 & \cdots & r_{k-1} \\ r_1 & r_0 & \cdots & r_{k-2} \\ \vdots & & & \vdots \\ r_{k-1} & \cdots & \cdots & r_0 \end{pmatrix} \begin{pmatrix} a_1(k) \\ \vdots \\ a_k(k) \end{pmatrix} = - \begin{pmatrix} r_1 \\ \vdots \\ r_k \end{pmatrix} + O\left(\frac{1}{N}\right), \tag{3.5}$$

where

$$r_i = \frac{1}{N} \sum_t x_t x_{t-i}. \tag{3.6}$$

For each N and k the matrix in (3.5) is positive definite with probability one, and Durbin [7] has given recurrence equations for the solution $\hat{a}_i(k)$ for $k=1,2,\ldots$ within the error term $O(1/N)$, which we drop from here on. These equations not only have a computational value but also give highly useful properties of the other quantities involved. Of these we need the minimized value for σ_k, which first can be written as

$$\hat{\sigma}_k = r_0 + \sum_{i=1}^{k} \hat{a}_i(k) r_i,$$

and then, recursively, [7], as

$$\hat{\sigma}_{k+1} = \hat{\sigma}_k(1-\hat{a}_{k+1}^2(k+1)), \quad \hat{\sigma}_0 \triangleq r_0. \tag{3.7}$$

Moreover, $\hat{\sigma}_{k+1}>0$ for all k, because R_k in (3.5) is positive definite with probability 1.

The estimates themselves satisfy the recursions:

$$\hat{a}_{k+1}(k+1) = -\left(r_{k+1} + \sum_{i=1}^{k} r_{k+1-i}\hat{a}_i(k)\right) \Big/ \hat{\sigma}_k$$

$$\hat{a}_i(k+1) = \hat{a}_i(k) + \hat{a}_{k+1}(k+1)\hat{a}_{k+1-i}(k), \quad i=1,\ldots,k. \tag{3.8}$$

Next, we write the criterion at the minimizing parameter vector $\hat{a}(k)=(\hat{a}_i(k),\ldots,\hat{a}_k(k))$ as follows

$$U(x,k,\hat{a}(k)) = \log \hat{\sigma}_k + \frac{k}{N}\log N. \tag{3.9}$$

Suppose, first, that $k<n$. Consider the difference

$$U(x,n,\hat{a}(n)) - U(x,k,\hat{a}(k)) = \sum_{i=k+1}^{n} \log(1-\hat{a}_i^{\,2}(i)) + \frac{n-k}{N}\log N. \tag{3.10}$$

Because $\hat{a}_n(n)\to a_n$ a.s. and $a_n\neq 0$, the first term converges to a negative number. Because the second term goes to zero, the limit of the difference is negative a.s.

Suppose, next, that $k>n$. As above

$$U(x,k,\hat{a}(k)) - U(x,n,\hat{a}(n)) = \sum_{i=n+1}^{k} \log(1-\hat{a}_i^{\,2}(i)) + \frac{k-n}{N}\log N. \tag{3.11}$$

By expanding the logarithms (take them as natural logarithms)

$$\log(1-\hat{a}_i^{\,2}(i)) = -\hat{a}_i^{\,2}(i) + O(\hat{a}_i^{\,4}(i))$$

Because $\hat{a}_i(i)$ is asymptotically normal, its square has a χ^2-distribution, and

$$-N\sum_{i=n+1}^{k} \log(1-\hat{a}_i^{\,2}(i))$$

has asymptotically a χ^2-distribution with k-n degrees of freedom. Accordingly, the probability that the absolute value of the first term in (3.11), which is negative a.s., exceeds the second term; i.e., the probability that the difference (3.11) is negative, is the tail from $(k-n)\log N$ of the mentioned χ^2-distribution. And this clearly goes to zero as $N\to\infty$, proving that with probability 1 the difference (3.11) is positive. Thus $U(x,k,\hat{a}(k))$, as a function of k, has a global minimum at $k=n$ a.s.

4 COMPUTER SIMULATIONS

In order to obtain experience of order estimation in autoregressive processes by the shortest length criterion we generated sequences x by computer according to the following equation

$$x_t = 0.55x_t + 0.35x_{t-2} + w_t,$$

where $\{w_t\}$ was a simulated independent gaussian process of zero mean and unit variance.

As the criterion we used (2.10). We wanted, in particular, to find out how large would N have to be taken in order to get the order right with reasonable certainty. With N=100, the order came out incorrectly as 1 fairly often, but with N=500 the 60 runs that we repeatedly and independently ran gave the optimum order correctly as 2 every time.

In comparison we also determined the order with N=500 by minimizing Akaike's AIC-criterion

$$N \log \sigma_k + 2k.$$

The results were 56 times the correct order k=2, 3 times k=3, and once k=4.

5 REFERENCES

1. Akaike, H. "Information theory and an extension of the maximum likelihood principle", 2nd International Symposium and Information Theory, B. N. Petrov and F. Caski, eds., Akademiai Kiado, Budapest, 267-281,(1973).

2. Shibata, R. "Selection of the order of an autoregressive model by Akaike's information criterion", *Biometrica,* **63**, 1, 117-26, (1976).

3. Rissanen, J. "Modelling by shortest data description", *Automatica,* **14**, (1978).

4. Wallace, C. S. and Boulton, D. M. "An information measure for classification", *The Computer Journal,* **11**, 2, 185, (1968).

5. Anderson, T. W. "The statistical analysis of time series", New York: Wiley and Sons, (1963).

6. Hannan, E. J. "Multiple time series", John Wiley and Sons, Inc., (1970).

7. Durbin, J. "The fitting of time series models", *Rev. Inst. Internat. Statist.*, **28**, 233-244, (1960).

8. Cox, D. R. and Hinkley, D. V. "Theoretical statistics", Chapman and Hall, London, (1974).

9. Schwarz, G. "Estimating the dimension of a model", *Ann. Stat.*, **6**, 2, (1978).

7.5 SEQUENTIAL GENERATION OF OPTIMAL TEST-SIGNALS FOR PARAMETER ESTIMATION IN DYNAMIC SYSTEMS

M. B. Zarrop

(Department of Computing and Control, Imperial College, 180 Queens Gate, London, SW7 2BZ, UK)

1 INTRODUCTION

The problem of designing experiments for estimating parameters in dynamic systems has been treated by several authors [1] - [6]. This paper takes Mehra's approach [6] and considers the problem of sequentially generating test signals for parameter estimation in linear, single input, single output, discrete time systems using a frequency domain approach based on the theory of optimal experiments in regression analysis developed by Kiefer and Wolfowitz [7] and others. The input signals are power constrained and are optimal in the sense that system information is maximised where the criterion employed is the determinant of the Fisher information matrix (D-optimality).

The organisation of the paper is as follows. In Section 2, the input design problem is formulated. The Kiefer-Wolfowitz Equivalence Theorem is stated in Section 3 and the basic sequential design algorithm is set out in Section 4. An extension to the Equivalence Theorem (Section 5) leads to a class of algorithms (S-algorithms) each member of which generates a sequence of designs converging to a D-optimum (Section 6). Some of these sequential design procedures allow the removal of "bad" design frequencies. In Section 7 a number of sequential design algorithms are compared computationally.

2 THE INPUT DESIGN PROBLEM

Consider the linear, time-invariant, single-input, single-output discrete-time dynamic system

$$y_k = \frac{B(z^{-1})}{A(z^{-1})}u_{k-s} + \frac{D(z^{-1})}{C(z^{-1})}e_k \quad k = 1,\ldots,N \qquad (1)$$

where $\{e_k\}$ is a zero-mean, unit-variance Gaussian white noise sequence, z is the unit forward shift operator and

$$A(z) = 1 + \sum_{j=1}^{n} a_j z^j; \quad B(z) = \sum_{j=0}^{m} b_j z^j$$
$$C(z) = 1 + \sum_{j=1}^{r} c_j z^j; \quad D(z) = \sum_{j=0}^{q} d_j z^j. \tag{2}$$

It is assumed that there are no pole-zero cancellations in (1) and that A(z), C(z), D(z) have no zeros on the closed unit disc.

The integers, m, n, q, r, s are assumed known and the (m+n+q+r+2) - vector β of unknown parameters is to be estimated from input-output data. Estimation accuracy can be assessed by forming the covariance matrix of the parameter estimator. Subject to weak regularity conditions ([8], pp. 35-37), the covariance matrix of any efficient and unbiased estimator is equal to the inverse of M_β, the Fisher information matrix.

For the system (1) - (2)

$$M_\beta = \left[\begin{array}{c|c} M(u) & O \\ \hline O & R \end{array}\right] \tag{3}$$

where the matrix R corresponds to the noise parameters and is independent of the input u [9]. The matrix M(u) corresponding to the system parameters is of order p(=m+n+1) and is given by

$$M(u) = \sum_{k=1}^{N} g_k g_k^T \tag{4}$$

where

$$g_k = h(z^{-1}) u_k$$
$$h_i(z) = B(z)C(z)z^{(s+i)}/[D(z)A^2(z)], \; i = 1, \ldots, n \tag{5}$$
$$\text{or} \quad -C(z)z^{(s+i-n-1)}/[D(z)A(z)], \; i = n+1, \ldots, p$$

The average information matrix per sample is defined via the elementwise limit

$$\bar{M}(\xi) \underset{=}{\Delta} \lim_{N\to\infty} N^{-1}M(u) = \text{Re} \int_{\Omega} h(e^{j\omega})h^{*}(e^{j\omega})d\xi(\omega) \tag{6}$$

where it is assumed that the input sequence satisfies suitable regularity conditions [6] [10]. In (6), Ω denotes the set $\{\omega | 0 \leq \omega \leq \pi\}$, ξ is the (one-sided) cumulative input power distribution function, Re denotes "real part" and the superscript * denotes complex conjugate transpose.

In the context of experiment design, ξ is called a design measure (or design). The problem of optimal design is to choose ξ from some allowable set Ξ_1 to minimize a suitable scalar function of $\bar{M}(\xi)$.

Optimal designs which are independent of the system and noise parameters exist only for special classes of systems [11] [12]. In general, however, M_β depends on the true value of β which is unknown. To resolve this problem it is appropriate to adopt a Bayesian viewpoint and regard β as a random variable with a prior probability distribution. Payne and Goodwin [13] have shown that the use of the prior mean $\bar{\beta}$ rather than the true value β gives a good approximation to M_β provided the prior distribution is sufficiently sharp and $\bar{\beta}$ is sufficiently close to β. This is the approach adopted here.

<u>Definition 1</u>

Let Ξ_1 be the set of non-decreasing, right-continuous design measures of bounded variation, corresponding to normalized input power, i.e.

$$\int_{\Omega} d\xi(\omega) = 1 \tag{7}$$

Denote by ξ_ω a single-frequency design in Ξ_1 and let $\mathcal{D}_1$ be the subset of Ξ_1 containing only designs with purely discrete spectra. A design in $\mathcal{D}_1$ with ℓ spectral frequencies can be characterized by the array

$$\begin{Bmatrix} \lambda_1, \ldots, \lambda_\ell \\ \omega_1, \ldots, \omega_\ell \end{Bmatrix}$$ where the positive weights $\lambda_1, \ldots, \lambda_\ell$ sum to unity.

Design measures $\mathcal{D}_1$ having a discrete spectrum are of particular importance [6] because they can be realised using sums of sinusoidal test signals. It is not necessary, however, to use sinusoids to achieve a given input spectrum. There exist design measures that are exactly realisable using binary signals [14] and any spectrum can be approximated by a periodic binary signal [15].

Definition 2

Let $\lambda(\xi,\omega) \triangleq \xi(\omega) - \xi(\omega-)$ for all $\xi \in \Xi_1$, $\omega \in \Omega$ where $\xi(\pi) = 1$, $\xi(0-) = 0$. If ξ has a component with a discrete spectrum as above then

$$\lambda(\xi,\omega) = \lambda_k \text{ if } \omega = \omega_k,\ k = 1, \ldots, \ell$$
$$= 0 \text{ otherwise}$$

Definition 3

Let $\Xi_{(p)}$ denote the subset of Ξ_1 corresponding to "persistently exciting" input signals [6] [10], i.e. the set containing those designs ξ for which the p×p matrix $\bar{M}(\xi)$ is nonsingular [16].

Definition 4

The design ξ^* is D-optimal in Ξ_1 if

$$\xi^* = \arg \inf_{\Xi_1} \det \bar{M}^{-1}(\xi) \tag{8}$$

Clearly, $\xi^* \in \Xi_{(p)}$.

D-optimality is achieved asymptotically by several sequential design procedures [12] [17] [18] introduced into the static experiment design field since 1970 and based on the Kiefer-Wolfowitz theory [19].

3 THE KIEFER-WOLFOWITZ EQUIVALENCE THEOREM

The point of departure for the Kiefer-Wolfowitz theory is the equivalence between D-optimality and a criterion based on the variance of $y_k(\hat{\theta})$, the output derived using the estimated system transfer function.

Definition 5

The generalized variance $d(\omega,\xi)$ is defined as

$$d(\omega,\xi) = \lim_{N\to\infty} \frac{1}{N} \operatorname{var} \left[C(z^{-1})D^{-1}(z^{-1})y_k(\hat{\theta})\right]$$
$$= \operatorname{tr}\left[\bar{M}^{-1}(\xi)\bar{M}(\xi_\omega)\right], \text{ using (6)} \tag{9}$$

Definition 6

The design $\xi^* \in \Xi_1$ is said to be G-optimal if ξ^* minimizes $\max_{\omega\in\Omega} d(\omega,\xi)$ in Ξ_1.

The equivalence theorem [6] [7] can now be stated.

Theorem 1 (Equivalence Theorem)

The following statements are equivalent:

(i) The normalized design ξ^* is D-optimal;

(ii) ξ^* is G-optimal;

(iii) $\max_{\omega\in\Omega} d(\omega,\xi^*) = p$. □

Note that, if ξ is not D-optimal, then $d(\omega,\xi) > p$ for some $\omega \in \Omega$.

4 SEQUENTIAL DESIGN PROCEDURES

The basic sequential design procedure involves the addition of a single frequency to the design spectrum at each step, i.e. choose $\omega_k \in \Omega$, $\alpha_k \in [0,1]$ so that, at step k+1,

$$\xi_{k+1} = (1-\alpha_k)\xi_k + \alpha_k\xi_{\omega_k}$$

Note that if $\xi_0 \in \Xi_{(p)}$ then $\xi_k \in \Xi_{(p)}$ for $k \geq 1$.

At step k+1 the determinant update is given by

$$|\bar{M}(\xi_{k+1})| = |\bar{M}(\xi_k)|\phi(\omega_k,\beta_k,\xi_k) \tag{10}$$

where

$$\beta \triangleq \alpha/(1-\alpha) \in [0,\infty] \tag{11a}$$

$$\phi(\omega,\beta,\xi) \triangleq (1+\beta)^{-p}\{1+\beta d(\omega,\xi)+\beta^2 g(\omega,\xi)\} \tag{11b}$$

$$g(\omega,\xi) \triangleq \frac{1}{4}\{d^2(\omega,\xi) - \| h^T(e^{j\omega})\bar{M}^{-1}(\xi)h(e^{j\omega})\|^2\} \tag{11c}$$

5 AN EXTENSION OF THE EQUIVALENCE THEOREM

Using (10) and (11)

$$\left(\frac{\partial}{\partial\beta}|\bar{M}(\xi_{k+1})|\right)_{\beta_k=0} = |\bar{M}(\xi_k)|\{d(\omega_k,\xi_k) - p\} \tag{12}$$

Theorem 1 ensures that the righthand side of (12) can be made positive if ξ_k is not D-optimal. This motivates the following definition.

Define the set

$$S(\omega_0,\beta_0,\xi) = \{\omega_0\}_\xi \times \mathcal{B}(\beta_0) \text{ for any } \xi \in \Xi_1$$

where

$$\mathcal{B}(\beta_0) = \{\beta | 0 \leq \beta \leq \beta_0\} \text{ for any } \beta_0 > 0$$

and $\{\omega_0\}_\xi$ is a single-element set containing any ω_0 satisfying

$$d(\omega_0,\xi) \begin{cases} = p \text{ if } \xi \text{ is D-optimal} \\ > p \text{ otherwise} \end{cases}$$

Definition 7

A compact set P of pairs (ω,β) is called an S-set for ξ if there exists a pair (ω_0,β_0) such that

$$P \supseteq S(\omega_0,\beta_0,\xi) \tag{13}$$

Not all S-sets are dependent on a specific design measure, e.g. $\mathcal{B}(\beta_0)\times\Omega$ is an S-set for any member of Ξ_1.

It follows from (13) that if $S_2 \supseteq S_1$ and S_1 is an S-set for ξ, then so is the set S_2. This inclusion property of S-sets is crucial for establishing a class of globally convergent sequential design algorithms.

The following result extends the Equivalence Theorem 1:

Theorem 2 [20]

If $P(\xi)$ is an S-set for each ξ in $\Xi_{(p)}$, then the following statements are equivalent

(i) The normalized design ξ^* is D-optimal

(ii) ξ^* minimizes $\max\limits_{P(\xi)} \phi(\omega,\beta,\xi)$ in Ξ_1

(iii) $\max\limits_{P(\xi^*)} \phi(\omega,\beta,\xi^*) = 1$ □

The proof is based on the sequential design procedure for which, using (11),

$$\phi(\omega_k,\beta_k,\xi_k) = \max_{P(\xi_k)} \phi(\omega,\beta,\xi_k)$$

at the (k+1)th stage.

6 S-ALGORITHMS AND GLOBAL CONVERGENCE

The following algorithms have feasible sets which are S-sets:

Algorithm 1

(i) Choose ω_k s.t. $d(\omega_k,\xi_k) = \max\limits_{\omega\in\Omega} d(\omega,\xi_k)$

(ii) Choose β_k s.t. $\phi(\omega_k,\beta_k,\xi_k) = \max_{\beta\in\mathcal{B}} \phi(\omega_k,\beta,\xi_k)$ where

$\mathcal{B} = \{\beta \mid 0\le\beta\le\infty\}$.

Denoting the feasible set for Algorithm 1 by $S_i(\cdot)$, then

$$S_1(\xi_k) = \{(\omega,\beta) \mid \omega=\omega_f,\ d(\omega_f,\xi_k) = \max_{\omega'\in\Omega} d(\omega',\xi_k),\beta\in\mathcal{B}\} \tag{14}$$

and

$$S_1(\xi_k) \supseteq S(\omega_f,\beta_0,\xi_k),\ \beta_0 \text{ arbitrary} \tag{15}$$

This algorithm is due to Fedorov [12] who proves global convergence in the case of static systems. Mehra proposes this sequential procedure for dynamic systems [6].

Algorithm 2

Choose ω_k, β_k s.t. $\phi(\omega_k,\beta_k,\xi_k) = \max_{\Omega\times\mathcal{B}} \phi(\omega,\beta,\xi_k)$, i.e.

$$S_2(\xi_k) = \Omega \times \mathcal{B} \supset S(\omega_0,\beta_0,\xi_k),\ (\omega_0,\beta_0) \text{ arbitrary} \tag{16}$$

Algorithm 3

Atwood [18] has shown that the removal of power from a frequency in the current design may lead to an improved cost value over any possible frequency addition.

Allowing for this possibility in Algorithm 1 leads to

$$S_3(\xi_k) = S_1(\xi_k) \cup S_r(\xi_k) \supseteq S_1(\xi_k) \tag{17}$$

where

$$S_r(\xi_k) = \{(\omega,\beta) \mid \omega\in\Omega,\ \beta\in[-\lambda(\xi_k,\omega),0)\} \tag{18}$$

Algorithm 4

Allowing for such power removal in Algorithm 2 yields the feasible set

$$S_4(\xi_k) = \{(\omega,\beta)\,|\,\omega\in\Omega, \beta\in[-\lambda(\xi_k,\omega),\infty]\}$$

$$= S_2(\xi_k) \cup S_r(\xi_k) \supseteq S_2(\xi_k) \quad (19)$$

The relations (15), (16), (17) and (19) imply that Algorithms 1 - 4 have feasible sets that are S-sets for each ξ_k. This suggests the following definition:

Definition 8

An algorithm is an S-algorithm if the feasible set $S(\xi)$ is an S-set for each ξ in $\Xi_{(p)}$.

The main convergence result can now be stated briefly.

Theorem 3 [20]

Every S-algorithm generates a sequence of design measures in Ξ_1 converging to a D-optimum. □

The result follows in a straightforward way from Theorem 2.

Algorithms 1 - 4 are S-algorithms and global convergence is therefore guaranteed.

7 NUMERICAL RESULTS

Consider the system

$$y_k = (b_0+b_1z^{-1})u_k + (1-0.06z^{-4})^{-1}e_k \quad (20)$$

where

$b_0 = 1$, $b_1 = 0.3$, i.e. $p = 2$

A pattern search algorithm (PAT) was used assuming an initial design with one frequency and then two frequencies with equal weights. The runs terminated when the search step size fell below 10^{-10}. This method was compared with the sequential design algorithms 1 - 4 using the same initial designs and incorporating at each iteration the frequency round-off

$$\left\{\begin{matrix} \cdots & \lambda_i & \cdots & \lambda_k & \cdots \\ \cdots & \omega_i & \cdots & \omega_k & \cdots \end{matrix}\right\} \longrightarrow \left\{\begin{matrix} \cdots & \lambda_i+\lambda_k & \cdots \\ \cdots & \omega' & \cdots \end{matrix}\right\}$$

whenever $|\omega_i-\omega_k|\leq 0.1$ where ω' is given by

$$\omega' = \frac{\lambda_i\omega_i+\lambda_k\omega_k}{\lambda_i+\lambda_k} \tag{21}$$

It can be shown [20] that, to the first order in $(\omega_i-\omega_k)$, the information matrix is unchanged by this procedure at each stage. In each case and for each interation the single frequency update was calculated using a grid search with grid size 0.1 ($\simeq \pi/32$). The algorithm terminated when the relative cost decrease was less than 10^{-4}. The machine used was the CDC 6400 at Imperial College, London. The results are given in Tables I and II.

The pattern search algorithm may achieve a D-optimum provided that the number of frequencies is assumed sufficiently large. For (20), a single frequency design ensures identifiability but cannot yield D-optimality ([21], pp. 74-76). Thus, apart from possible problems of local minima, it is necessary, in general, to initiate the pattern search with p frequencies as it may not be possible to achieve optimality with input spectra that are "merely" persistently exciting, i.e. that contain ℓ frequencies where $p/2 \leq \ell < p$ [6]. The tables show that the choice of initial design does not significantly affect the performance of the sequential design procedures which are assured of achieving a D-optimum.

The possibility of removing spectral frequencies in Algorithms 3 and 4 clearly leads to faster convergence and to more sparse spectra. In general, Algorithm 3 takes less CP time, most of which is consumed in repeated grid searches for the frequency updates.

Table I

Initial design: $\omega = 1$; $\lambda = 1$

Algorithm	Cost	No. of Frequencies	No. of Iterations	CP Secs.
PAT	1.160	1	-	0.342
1	0.798	3	60	2.035
2	0.798	3	61	2.382
3	0.796	2	14	0.515
4	0.792	2	13	0.545

Table II

Initial design: $\omega_1 = 1$, $\omega_2 = 2$; $\lambda_1 = 0.5$, $\lambda_2 = 0.5$

Algorithm	Cost	No. of Frequencies	No. of Iterations	CP Secs.
PAT	0.792	2	-	1.372
1	0.798	4	68	2.366
2	0.798	4	68	2.708
3	0.792	2	15	0.577
4	0.792	2	15	0.633

Optimal Design (N is length of test signal)

Frequency	Weight	N var $\hat{b}_0$	N var $\hat{b}_1$	Cost
$\pi/4$ ($\simeq$ 0.785)	½	0.890	0.890	0.792
$3\pi/4$ ($\simeq$ 2.356)	½			

7 REFERENCES

1. Aoki, M. and Staley, R. M. "On input signal synthesis in parameter identification", *Automatica*, **6**, 431-440, (1970).

2. Goodwin, G. C., Murdoch, J. C. and Payne, R. L. "Optimal test signal design for linear SISO system identification", *Int. J. Control*, **17**, 1, 45-55, (1973).

3. Viort B. "D-optimal designs for dynamic models: Part I Theory", Dept. of Statistics, Univ. of Wisconsin, Tech. Rpt. 314, (1972).

4. Kalaba, R. E. and Spingarn, K. "Optimal inputs for nonlinear process parameter estimation", *IEEE Trans. Aerospace and Electronic Systems*, **AES-10**, 3, 339-345, (1974).

5. Mehra, R. K. "Optimal input signals for parameter estimation in dynamic systems - A survey and new results", *IEEE Trans. Auto. Control*, **AC-19**, 6, 753-768, (1974).

6. Mehra, R. K. "Frequency-domain synthesis of optimal inputs for linear system parameter estimation", Trans. ASME, Journal of Dynamic Systems, Measurement and Control, June, (1976).

7. Kiefer, J. and Wolfowitz, J. "Optimum designs in regression problems", *Ann. Math. Stat.*, **30**, 271-294, (1959).

8. Silvey, S. D. "Statistical inference", Penguin, (1970).

9. Payne, R. L. "An A-priori estimate for the information matrix of a SISO linear system", Publication 72/4, Dept. of Computing and Control, Imperial College, London, (1972).

10. Ljung, L. "Characterization of the concept of "Persistently Exciting" in the frequency domain", Div. of Aut. Control, Lund Inst. of Technology, Report 7119, (1971).

11. Levin, M. J. "Optimal estimation of impulse response in the presence of noise", *IRE Trans. Circuit Theory*, **CT-7**, 50-56, (1960).

12. Fedorov, V. V. "Theory of optimal experiments", Academic Press, (1972).

13. Payne, R. L. and Goodwin, G. C. "A Bayesian approach to experiment design with applications to linear multivariable

dynamic systems", IMA Conference on Computational Problems in Statistics, University of Essex, (1973).

14. Martins de Carvalho, J. L. "Binary sequences with prescribed autocorrelations", PhD Thesis, University of London, Dept. of Computing and Control, Imperial College, London, (1977).

15. Van den Bos, A. "Construction of binary multifrequency test signals", 1st IFAC Symposium on Identification, Prague, (1967).

16. Rothenberg, T. J. "Identification in parametric models", *Econometrica,* **39**, 577-591, (1971).

17. Wynn, H. P. "The sequential generation of D-optimum experimental designs", *Ann. Math. Stat.,* **41**, 1655-1664, (1970).

18. Atwood, C. L. "Sequences converging to D-optimal designs of experiments", *Annals of Statistics,* **1**, 342-352, (1973).

19. St. John, R. C. and Draper, N. R. "D-optimality for regression designs", *Technometrics,* **17**, 1, 15-23, (1975).

20. Zarrop, M. B. "Sequential generation of optimal test signals for parameter estimation in dynamic systems", publication 78/29, Dept. of Computing and control, Imperial College, London, (1978).

21. Zarrop, M. B. "Optimal experiment design for dynamic system identification", PhD Thesis, University of London, Dept. of Computing and Control, Imperial College, London, (1977).

7.6 PERIODIC MEAN-VALUE PROCESSES: ESTIMATION-ORIENTED INFORMATION ANALYSIS*

S. Bittanti

(Politecnico di Milano, Istituto di Elettrotecnica ed Elettronica, 20133 Milano, Piazza Leonardo da Vinci 32, Italy)

1 INTRODUCTION

Data selection is an important problem in Estimation Theory. Basically, some trade-off between the information contained in the data and the resulting complexity for the estimation algorithm has to be reached. If the aim of estimation is parameter estimation, the Fisher Information Matrix [1] may be regarded as a fair evaluation of the (achievable) accuracy. This interpretation of the Fisher Information Matrix is well known; it relies upon the Cramér-Rao inequality [2] and has been discussed in a large number of papers and books. The interested reader is referred to [3]-[5]. As far as complexity is concerned, the requirement that there be a limited number of measurement points is of primary relevance. The set of all those time points at which the values of the observable time series are actually taken into consideration as data in the estimation procedure will be referred to as the measurement schedule set S. Due to limitations in the data storage capacity, S may not coincide with the entire set of points where measuring the process realization is allowed. In this connection, observe that S is usually dealt with in the literature as an interval. However, the opportunity of scattering the measurement schedule set along the time axis should be taken into consideration as an effective tool to increase information without increasing complexity [6]-[10]. This of course is relevant when the signal-to-noise ratio is a non-constant function of time, as typically happens for non-stationary processes.

In the present paper, discrete-time normal processes with periodic mean-value are considered. Since the celebrated Schuster's work on hidden periodicities [11], this family of

*The research has been supported by the Centro di Teoria dei Sistemi (C.N.R.), Via Ponzio, 34/5, 20133 Milano (Italy).

processes has been widely considered in different areas, see e.g. [12]-[16]. The purpose of our analysis is to find the Fisher Information Matrices corresponding to all the admissible measurement schedule sets. For technical reasons, the problem turns out to be properly stated by making reference to the so-called Average Information Matrix. This matrix, denoted by M(S), is defined on the basis of the Fisher Information Matrix associated with a measurement schedule set S, $\mathcal{M}(S)$, in the following way

$$M(S) = \begin{cases} \frac{1}{|S|}\mathcal{M}(S) & \text{, if S is non-empty and finite} \\ \lim_{\nu\to+\infty} \frac{1}{\nu}\, \mathcal{M}(S_\nu) & \text{, if S is infinite and the limit exists.} \end{cases} \tag{1}$$

Here, $|\cdot|$ denotes cardinality and for any integer ν, $0<\nu \leq |S|$, S_ν is the set of the first ν elements of S. The time-averaging implicit in the above definition makes possible a meaningful comparison among information matrices associated with different cardinality measurement schedule sets. Notice that the case $|S|=+\infty$, included in the definition, is of fundamental interest for the identifiability problem. Indeed, the existence of a (locally) consistent parameter estimator for a given measurement schedule set S, $|S|=+\infty$, essentially depends upon the non-singularity of the associated Average Information matrix [17].

In this paper, two problems are discussed and solved in some detail. First of all, the Average Information Matrices Set for the processes of interest is found. Then given a matrix belonging to this set, the "inverse" problem of deriving a measurement schedule set whose Average Information Matrix is equal to the given one is solved.

The plan of the paper is as follows: in Section 2 the assumptions and the problems statement are formally given, together with some simple preliminary results. The stated problems are solved in Section 3, while a number of applications of the derived results to identification are presented in Section 4. All the proofs can be found in [18]. Here, for the sake of conciseness, they will be either briefly sketched or completely omitted.

2 PROBLEMS STATEMENT

Denoting by N the non-negative integers, let $\{z(k) | k \in N\}$ be a p-dimensional stochastic process whose mean value $y(k,\theta)=E[z(k)]$ depends upon a q-dimensional parameter vector θ. The following assumptions are made.

A1 - y is a time-periodic function of a given period n, $n>1$: $y(k+n,\theta)=y(k,\theta)$, $\forall\ k\in N$; furthermore, y is assumed to be differentiable with respect to θ.

A2 - $\{z(k)-y(k,\theta) | k\in N\}$ is a zero-mean normal process, namely, for any finite set $A \subset N$, the random variables $z(k)-y(k,\theta)$, $k\in A$, are zero-mean jointly normally distributed.

A3 - The covariance kernel of $\{z(k)-y(k,\theta) | k\in N\}$ is given by $E[(z(r)-y(r,\theta))\ (z(s)-y(s,\theta))'] = I\delta(r-s)$, where I is the identity matrix, δ is the Kronecker function and $'$ denotes transpose.

Notice that in A3 the hypothesis that the covariance matrix of $z(k)-y(k,\theta)$ is the identity is introduced for the sake of simplicity. All the results given in Section 3 can be straightforwardly generalized to the case when the covariance matrix is any constant positive definite matrix.

By definition, the Fisher Information Matrix associated with a measurement schedule set S is a qxq matrix whose (i,j) element is given by

$$M(S)_{ij}=E\left[\frac{\partial L(S,\theta)}{\partial\theta_i}\ \frac{\partial L(S,\theta)}{\partial\theta_j}\right] \tag{2}$$

where $L(S,\theta)$ is the logarithm of the likelihood function relative to S. For conciseness, the dependence of M upon θ is deleted in the adopted notation. On the basis of Assumptions A1-A3 and definition (2), it is an easy matter to derive the following expression for $M(S)$:

$$M(S) = \Sigma_{k\in S}\ \eta(k,\theta)'\,\eta(k,\theta), \tag{3}$$

where $\eta(k,\theta)$ is the pxq matrix whose i-th column is given by the i-th sensitivity coefficient vector

$$\eta_i(k,\theta) = \frac{\partial}{\partial\theta_i}\, y(k,\theta)\quad,\ i=1,=,\ldots,q. \tag{4}$$

Finally, we introduce the family $\hat{P}(N)$ constituted by all the finite non-empty subsets of N and by all the infinite subsets such that the associated Average Information Matrix does exist:

$$\hat{P}(N) = \{S | S \in N \text{ and } M(S) \text{ exists}\}. \tag{5}$$

Clearly, given any set $S \in \hat{P}(N)$, formulas (1), (3) and (4) enable one to evaluate the corresponding matrix M(S). Letting

$$\hat{W} = \{M(S) \; | S \in \hat{P}(N)\} \tag{6}$$

be the family of all the obtainable Average Information Matrices, the questions dealt with in this paper are precisely stated as follows:

Problem 1: Qualify $\hat{W}$.

Problem 2: Given a matrix $\bar{M} \in \hat{W}$, find a set $S \subset N$ such that $M(S) = \bar{M}$

3 MAIN RESULTS

Problems 1 and 2 are deeply analysed in [18]. Here, only the main results can be presented. With this purpose, we introduce the family of the Average Information Matrices associated with the subsets of the elementary interval $[0,n-1]$, i.e.

$$W_n = \{M(S) | \; S \subset [0,n-1], \; S \neq \phi\},$$

and the family of the Average Information Matrices associated with the singletons of 0,n-1, i.e.

$$W_s = \{M(\{k\}) \; | \; k \in [0,n-1]\}.$$

Then, the solutions to Problems 1 and 2 can be given.

3.1 Solution to Problem 1

The family $\hat{W}$ of all the obtainable Average Information Matrices is given by the convex hull of family W_n; furthermore, this convex hull coincides with the convex hull of W_s; in short

$$\hat{W} = CH\left[W_n\right] = CH\left[W_s\right], \tag{7}$$

where CH $[\cdot]$ denotes convex hull.

This result strongly qualifies the infinite family $\hat{W}$, by showing that $\hat{W}$ simply reduces to a polytope whose vertices are the Average Information Matrices relative to the singletons of $[0,n-1]$.

3.2 Solution to Problem 2

Let $\bar{M}$ be an obtainable Average Information Matrix; then, in view of (7), $\bar{M}$ can be written as follows

$$\bar{M} = \Sigma_{j\varepsilon J}\ \gamma^j\ M(S^j) \quad ; \quad S^j \subset [0,n-1];\ S^j \neq \phi\ ;$$

$$\Sigma_{j\varepsilon J}\ \gamma^j = 1; \gamma^j > 0, \forall j\varepsilon J.$$

A general algorithm leading to a set $S \subset N$ such that $M(S) = \bar{M}$ is the following one.

Step 1

Evaluate

$$\alpha^j = \gamma^j \left(\Sigma_{i\varepsilon J}\ \gamma^i \frac{|S^j|}{|S^i|} \right)^{-1} , \quad \forall j\varepsilon J$$

Note that $\Sigma_{j\varepsilon J}\ \alpha^j = 1$ and that $0 < \alpha^j \leq 1, \forall j\varepsilon J$.

Step 2

Pick up any element, say $\bar{j}$, of J; for simplicity in notation, we identify J with the set of the first s integers, $J = \{1,2,\ldots,s\}$, and we take $\bar{j}=s$. By letting $\alpha^0 = 0$, evaluate

$$\beta^j = (1-\alpha^1 - \alpha^2 \ldots - \alpha^{j-1})^{-1}\ \alpha^j\ , \quad j=1,2,\ldots,s-1.$$

Note that $0 < \beta^j \leq 1$. To any $j = 1,2,\ldots,s-1$, associate a sequence $\{\delta(k) | k \geq 2\}$ of non-negative integers as follows:

$$\delta^j(2) = \begin{cases} 0, \text{ if } |\beta^j| < |\beta^j - 1| \\ 1, \text{ if otherwise;} \end{cases}$$

$$\delta^j(k+1)=\begin{cases}\delta^j(k) & \text{if } \left|\beta^j-\frac{\delta^j(k)}{k}\right|<\left|\beta^j-\frac{\delta^j(k)+1}{k}\right| \\ \delta^j(k)+1, & \text{otherwise;}\end{cases}\quad, k\geq 2.$$

It can be proved that

$$\lim_{k\to+\infty} \delta^j(k)/_k=\beta^j \quad, j=1,2,\ldots,s-1.$$

Hence there is a $\bar{k}$ such that

$$\Sigma_{j\in[1,s-1]}\delta^j(k) \leq k,\ k\geq\bar{k}$$

Step 3

To any $j = 1,2,\ldots,s-1$, associate the infinite subset of the integers $T^j = \{k|k\geq\bar{k}$ and $\delta^j(k)>\delta^j(k-1)\}$.

Given any subset A of the integers, define the mapping $p : A\to N$ as $p(a)=k$, if a is the k-th element of A.

Then, let

$$H^1 = T^1 \quad ; \quad \bar{H}^1 = N - H^1 \; ;$$

$$H^j = \{k|k \in \bar{H}^{j-1} \text{ and } p(k) \in T^j\} \quad , j=2,3,\ldots,s-1 \; ;$$

$$\bar{H}^j = N - \bigcup_{i\in[1,j]} H^i \quad , j=2,3,\ldots,s-1 \; ;$$

$$H^s = \bar{H}^{s-1}.$$

The family $\{H^j|j\in J\}$ is a partition of the integers.

Step 4

For any fixed n, define the following translation operator

$$\Delta_h(A) = \{k \;|k = a+(h-\bar{k})n,\ a\in A\} \;,$$

where A is any subset of the integers. Define

$$S_j = \bigcup_{h\in H^j} \Delta_h(S^j).$$

and finally let

$$S = \bigcup_{j \in J} S_j . \square$$

Then it can be shown that the set S satisfies the condition $M(S)=\bar{M}$. The described procedure is now illustrated with reference to a simple example.

Example 1

Given a period n=2, assume that

$$S^1 = \{0,1\} \ , \ S^2 = \{0\}$$

and

$$\bar{M} = \frac{2}{3} \ M(S^1) + \frac{1}{3} \ M(S^2).$$

Then, Step 1 leads to the coefficients $\alpha^1=\alpha^2=1/2$. By taking e.g., $\bar{j}=2$, and by applying the definition of $\delta^j(k)$ given in Step 2, one finds that

$$\delta^1(k) = \begin{cases} \frac{k}{2} & , \ k \text{ even, } k>2 \\ \\ \frac{k-1}{2} & , \ k \text{ odd, } \ k>2. \end{cases}$$

Sequence δ^1 is increasing with k. The set H^1 defined in Step 3 is the set of points where this sequence is strictly increasing. Therefore,

$$H^1 = \{2k \mid k \varepsilon N, \ k \geq 0\}$$

$$H^2 = \{2k+1 \mid k \varepsilon N, \ k \geq 0\}.$$

Finally, the sets S_1 and S_2 defined in Step 4 turn out to be the disjoint infinite sets illustrated in Fig. 1; their union is a set whose Average Information Matrix is given by (8) (this statement can be a posteriori verified on the basis of (1) and (3)).

O	1	2	3	4	5	6	7	8	9	1O	11	12	13	14	15
⊙	⊙	X	·	⊙	⊙	X	·	⊙	⊙	X	·	⊙	⊙	X	· · · ·

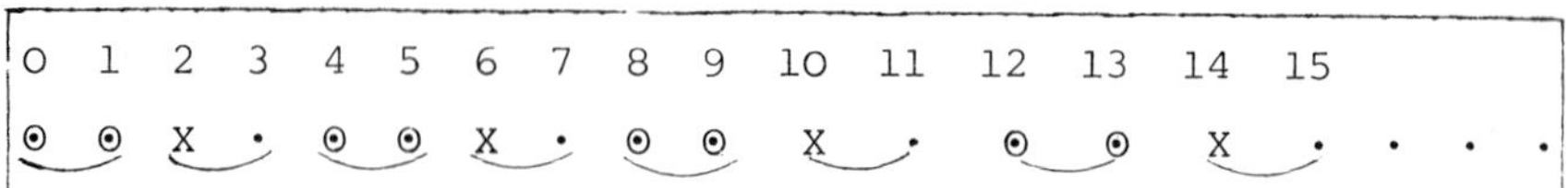

$$S_1 = \left\{ 4k + h \mid k \in N; \ h\epsilon\{0,1\} \right\} , \ S_2 = \left\{ 4k + 2 \mid k \in N \right\}$$

Fig. 1   The sets S_1 and S_2 for example 1

4   APPLICATIONS TO IDENTIFICATION

In this section, the relevance of the results of the preceding section as regards Optimal Experiment Design and the Identifiability problem is discussed.

4.1   Optimal experiment design

A problem in Experiment Design is the selection of the time points where to observe a stochastic process in order to achieve an accurate estimation of the unknown parameters of its distribution. Once a measurement schedule set has been selected, the accuracy in estimation will obviously increase if it is allowed to take a further observation. Hence, a reasonable formulation of the problem is as follows: given the number ν of admissible observations, find their best time allocation. In turn, this problem can be restated as an optimization problem by introducing

$$J(S) = h(M(S))$$

as performance index, where h is a mapping from the qxq symmetric positive semi-definite matrices into the non-negative reals. In order that J(S) be a significant scalar index of the amount of information associated with S to the end of parameter estimation, suitable conditions on the mapping h are to be introduced. Typically, one choses a matrix norm. More generally we assume that

A4 - h is a convex mapping

A5 - $h(\lambda M) = g(\lambda)h(M)$, where λ and $g(\lambda)$ are scalars and $g(\lambda) > O$ if $\lambda > O$.

Many criteria encountered in practice fall in this class.

Under Assumptions A1-A5, the Optimal Measurement Schedule design problem can be explicitly solved. Indeed,

$$J(S) = h(\mathcal{M}(S)) = h(\nu M(S)) = g(\nu)\ h(M(S)).$$

Since ν is given, the problem is equivalent to finding a set S^o, $|S^o| = \nu$, which maximizes

$$J(S) = h(M(S))$$

in the family of all the measurement schedule sets whose cardinality is equal to ν. The scalar $J(S)$ can be seen as the average information per-time-unit relative to S. Now, from (7) and Assumption A4, it follows that the maximum of $J(S)$ for $S\varepsilon W$ is achieved when S coincides with one of the vertices of the polytope $CH[W_s]$, i.e. when the measurement schedule set reduces to a singleton. Let k^o be the optimal singleton, $J(\{k^o\}) \geq J(\{k\})$, $\forall k\varepsilon[0,n-1]$; define

$$S^o = \{k^o,\ k^o+n,\ k^o+2n,\ldots k^o+(\nu-1)n\}. \tag{8}$$

Assumption A1 and expression (3) entail that

$$M(S^o) = M(\{k^o\}).$$

Consequently,

$$J(S^o) \geq J(S),\ \forall S\subset N, |S| = \nu$$

Therefore, the set S^o is the optimal measurement schedule set of cardinality ν. It amounts to picking up a datum once a period, at the point where the signal-to-noise ratio is maximum.

4.2 *Identifiability*

Let S be set such that $M(S)$ does exist, i.e. $S\varepsilon\hat{P}(N)$. In view of (7), $M(S)$ can be given the form

$$M(S) = \Sigma_{k\varepsilon S^*} \lambda_k M(\{k\})\ ;\ S^* \subset [0,n-1];\ S^* \neq \phi$$
$$\Sigma_{k\varepsilon S^*} \lambda_k = 1;\ \lambda_k > 0, \forall k\varepsilon S^*. \tag{9}$$

Introduce the matrix:

$$M(S^*) = \Sigma_{k\varepsilon S^*} \frac{1}{|S^*|} M(\{k\}). \tag{10}$$

All the Average Information Matrices are positive semi-definite. Hence, it can be proved that the positive coefficients linear combination (9) is non-singular if and only if the positive coefficients linear combination (10) is non-singular.

This means that, given any measurement schedule set S (admitting an associated Average Information Matrix), there exists a set S^* contained in the interval $[0,n-1]$ such that M(S) is non-singular iff $M(S^*)$ is non-singular. Consequently, for the non-singularity analysis, the attention may be focused on the matrices of W_n.

Here, only the cases of major interest, $S = [0,n-1]$ and S reduced to a singleton, e.g. $S = \{0\}$, are briefly discussed. With this purpose, consider the Finite Fourier Transform of the mean-value sensitivity coefficients (4):

$$\eta_i(k,\theta) = \Sigma_{h\varepsilon[0,n-1]} F_i(h) \exp(jkh \frac{2\pi}{n}),$$

where the dependence of $F_i(h)$ upon θ is dropped out for simplicity. Define the vectors

$$F_i' = |F_i(0)' \ F_i(1)' \ldots F_i(n-1)'|$$

$$F_i = \Sigma_{h\varepsilon[0,n-1]} F_i(h)$$

and the families

$$\Phi = \{F_i | i = 1,2,\ldots,q\} \tag{11}$$

$$\psi = \{F_i | i = 1,2,\ldots,q\}. \tag{12}$$

Then, it can be proved that the unknown parameter θ is identifiable by the measurement schedule set $S = [0,n-1]$ or by $S = \{0\}$ if and only if Φ or ψ, respectively, is a set of linearly independent vectors.

4.3 System identification

In this paper, we assumed that the observations are generated by the process $z(k) = y(k,\theta)+w(k)$ where w is a white normal process and y is a time periodic deterministic function. In particular, $y(k,\theta)$ may be the output of a dynamic system

$$x(k+1) = f(x(k),u(k),\theta)$$

$$y(x) = y(x(k),u(k),\theta)$$

subject to the periodicity constraint

$$x(0) = x(n).$$

Therefore, the above conclusions can be straightforwardly applied to the parameter estimation of a dynamic system in periodic regime, with output additive noise.

As for Optimal Experiment Design, for any fixed periodic input, the best measurement schedule set of cardinality ν will be of the form (8). Whenever not only the measurement allocation set but also the input function u is matter of choice, then a joint optimal input and measurement schedule design arises. If the system is linear and affected by a scalar parameter ($\theta \varepsilon R^1$), the optimal input can be expressed in a closed form (see [10]).

As far as Identifiability is concerned, the families of vectors (11) and (12) depend now upon the input function. For linear systems, sets of frequencies can be found [18] such that the Average Information Matrix relative to a given measurement schedule set is non-singular provided that the input has a non-vanishing spectrum over one of these sets.

5 ACKNOWLEDGEMENT

The author is grateful to Professor G. Guardabassi for stimulating discussions of the present subject.

6 REFERENCES

1. Fisher, R. A. "Contributions to mathematical statistics", Papers 10, 11 and 38, J. Wiley and Sons, New York, (1950).

2. Cramér, H. "Mathematical methods of statistics", Princeton University Press, (1946).

3. Rao, C. R. "Linear statistical inference and its application", J. Wiley and Sons, New York, (1965).

4. Edwards, A. W. F. "Likelihood", Cambridge University Press, (1972).

5. Fedorov, V. V. "Theory of optimal experiments", Academic Press, New York, (1972).

6. Åström, K. J. "On the choice of sampling rates in parametric identification of time series", *Information Science*, **1**, 273-278, (1969).

7. Goodwin, G. C., Zarrop, M. B. and Payne, R. L. "Coupled design of test signals, sampling intervals, and filters for system identification", *IEEE Trans. on Automatic Control*, **AC19**, No. 6, 748-752, (1974).

8. Mehra, R. K. "Optimization of measurement schedules and sensor designs for linear dynamic systems", *IEEE Trans. on Automatic Control*, **AC21**, No. 1, 55-64, (1976).

9. Bittanti, S. "System cycling for parameters estimation: optimal input and measurement schedule designs", Division of Eng. and Applied Physics, Harvard University, Techn. Report 666, (1977).

10. Bittanti, S. "On optimal experiment design for parameters estimation of dynamic systems under periodic operation", IEEE Conf. on Decision and Control, New Orleans, (1977).

11. Schuster, A. "On the investigation of hidden periodicities with applications to a supposed 26-days period of meteorological phenomena", Terr. Mag. 3.

12. Whittle, P. "The simultaneous estimation of a time series' harmonic components and covariance strucutre", Trab. Estab., 3, 43-57, (1952).

13. Walker, A. M. "On the estimation of a harmonic component in a time series with stationary independent residuals", Biometrika, 58, 21-34, (1971).

14. Walker, A. M. "On the estimation of a harmonic component in a time series with stationary dependent residuals, Adv. Appl. Prob., 8, 767-780, (1973).

15. Gardner, W. A., and Franks, L. E. "Characterization of cyclostationary random signal processes", *IEEE Trans. on Information Theory*, **IT-21**, No. 1, 4-14, (1975).

16. Bloomfield, P. "Fourier analysis of time series: an introduction", J. Wiley and Sons, (1976).

17. Tse, E. "Information matrix and local identifiability of parameters", Proc. Joint Automatic Control Conference, (1973).

18. Bittanti, S. "Fisher information matrices analysis for periodic mean-value normal processes", Centre d'Automatique de l'Ecole des Mines de Paris, Réport R/55, (1977).

7.7 IDENTIFICATION OF TIME-INVARIANT LINEAR SYSTEMS FROM NON-STATIONARY CROSS-SECTIONAL DATA

R. L. Goodrich

(Abt Associates Inc., and Harvard University, USA)

and

P. E. Caines

(Division of Applied Sciences, Harvard University, USA)

1 INTRODUCTION

This paper is concerned with the identification of linear, time-invariant, discrete-time systems where parameter estimates are functions of observations taken during non-stationary behaviour. The data are assumed to consist of an ensemble of independent sample paths of the system under examination. We describe this situation by saying we have a cross-sectional sample of independent, longitudinal, sample-paths. We present results concerning the asymptotic behaviour of maximum likelihood parameter estimates in the case where the number T of observations on the longitudinal behaviour is fixed but the cross-sectional sample size N tends to infinity.

The majority of work on system identification has been concerned with identification from single-sample longitudinal data. However, a large range of identification problems require system parameters to be identified from cross-sectional non-stationary data. Examples may be cited from technical, social, economic and medical fields. At the end of this paper we briefly describe a psychological application.

The organization of this paper is as follows: In Section 2, we describe the dynamic and probabilistic structure of the systems to be identified. In Section 3 we present the main consistency and asymptotic normality theorem for our specified class of parameter estimators. Section 4 contains a new identifiability result which is proved in [2]. The result reveals a sufficient condition (full rank disequilibrium, see

Section 3 below) for the main theorem to hold. The implementation of the identification algorithm is described in Section 5. Finally, in Section 6, an application of the theory and techniques of this paper is briefly described.

2 FORMULATION OF THE CROSS-SECTIONAL IDENTIFICATION PROBLEM

We shall consider a linear time-invariant dynamical system with an input process w and a zero-mean output process y, where w is a wide-sense stationary zero-mean (orthogonal) random process whose value at the k-th instant for the i-th cross-sectional sample is denoted $w_{k,i}$, $1 \le k \le T$, $1 \le i \le N$. The extension to the case of multivariable w and y processes is straightforward as is shown in [1]. For simplicity of presentation, and because of space limitations, we confine our entire exposition to the case of scalar w and y.

We define the polynomials

$$a(z) \underline{\Delta} \sum_{j=0}^{p} a_j z^j, \quad b(z) \underline{\Delta} \sum_{j=0}^{q} b_j z^j,$$

where z is the z-transform variable. Without loss of generality, we set $a_0 = b_0 = 1$ in order to obtain the required form of identifiability which is defined in Section 3 below.

Let $y_i(z)$, $w_i(z)$ denote the formal z-transforms of the processes y and w for the i-th cross-sectional sample. Then on the i-th cross-sectional sample the process y is related to the process w via the equation

$$a(z)y_i(z) = b(z)w_i(z), \tag{2.1a}$$

together with a set of initial conditions whose distribution as random variables must be specified.

Since we make no restrictive assumptions on the distribution of the initial conditions for (2.1a) it is not true that the process y_i is stationary even in the special case where (2.1a) is asymptotically stable. Hence it is possible that the transient behaviour of (2.1a) dominates the observations over the entire interval [1,T].

We assume that there is an observed output process z_i whose i-th cross-sectional sample is the sum of the output process y_i plus a zero-mean disturbance process v_i plus a random variable r_i which is constant along samples (i.e., is constant with respect to k). In symbols

$$z_{k,i} = y_{k,i} + v_{k,i} + r_i, \quad 1 \leq k \leq T, \quad 1 \leq i \leq N. \tag{2.1b}$$

We shall assume that (w,v) and r are independent, that they have a zero-mean joint Gaussian distribution, and that the covariance matrix of the process (w,v) is given by

$$E\begin{bmatrix} w_{k,i} \\ v_{k,i} \end{bmatrix} \begin{bmatrix} w_{\ell,j} & v_{\ell,j} \end{bmatrix} = \begin{bmatrix} \sigma_w^2 & \sigma_{wv} \\ \sigma_{wv} & \sigma_v^2 \end{bmatrix} \delta_{k\ell}\delta_{ij}, \quad \begin{matrix} 1 \leq k,\ell < T, \\ 1 \leq i,j \leq N. \end{matrix} \tag{2.2a}$$

The resulting system (2.1) has the observable state space representation

$$x_{k+1,i} = Fx_{k,i} + Gw_{k,i}, \tag{2.2b}$$

$$1 \leq i \leq N, \quad 1 \leq k \leq T,$$

$$z_{k,i} = Hx_{k,i} + v_{k,i} + r_i, \tag{2.2c}$$

where

$$F = \begin{bmatrix} a_1 & a_2 & \cdots & a_{d-1} & a_d \\ 1 & 0 & \cdots & 0 & 0 \\ 0 & 1 & \cdots & 0 & 0 \\ & & & \vdots & \vdots \\ \cdots & \cdots & \cdots & 1 & 0 \end{bmatrix}, \qquad (d \times d)$$

and d = max (p,q). When $p \leq q$ we have $a_d = 0$.

Also

$$G' = [g_d, g_{d-1}, \ldots, g_1], \quad g_1 = 1, \qquad (1 \times d)$$

where the $\{g_k;\ 1 \le k \le d\}$ are the first d coefficients of the impulse response of the system (2.1a) and

$$H = [0, \ldots, 0, 1]. \qquad (1 \times d)$$

Often we shall drop the cross-sectional index i from the subscripts of random variables, e.g., $x_{k,i}$ will be written as x_k, etc.

Finally we assume that (x_1, r) is independent of (v,w) and that

$$\begin{bmatrix} x_1 \\ r \end{bmatrix} \sim N\left(\begin{bmatrix} 0 \\ 0 \end{bmatrix}, \begin{bmatrix} V_{1|0} & d_r \\ d'_r & \sigma_r^2 \end{bmatrix}\right).$$

Let $(z_1^T)_i$ denote the random vector of i'th cross-sectional sample values $(z_{1,i}, z_{2,i}, \ldots, z_{T,i})$. Then the distribution of $(z_1^T)_i$ is defined by the parameter vector θ composed of the distinct elements of $(F, G, \sigma_w^2, \sigma_{wv}, \sigma_v^2, V_{1|0}, d_r, \sigma_r^2)$.

The cross-sectional identification problem is to estimate θ from the data $(z_1^T)_i$, $1 \le i \le N$.

We remark that it is possible to elaborate this set-up by allowing the process w to have a non-zero time-varying mean value and by allowing x_1 to have a non-zero mean value $\bar{x}_1$. In this case the output process y and the observation process z will possess time-varying non-zero mean values. (See [1].)

3 IDENTIFICATION METHOD AND ASYMPTOTIC PROPERTIES OF THE ESTIMATORS

In this section we construct the likelihood function on N observed output sequences $\{z_{k,i};\ 1 \leq k \leq T,\ 1 \leq i \leq N\}$ of the system (2.3) and obtain the strong consistency and asymptotic normality of the maximum likelihood estimator θ_N of the parameter vector θ (under certain stated identifiability conditions). Relaxing the Gaussian assumptions but retaining (i) all second-order properties of the processes (x_1, w, v, r) and (ii) the ergodicity of the process $\{(z_1^T)_i,\ 1 \leq i \leq \infty\}$ we obtain the asymptotic properties described above for θ_N as a prediction error estimator [3-5].

The likelihood function is constructed via the innovations process $\nu \triangleq \{\nu_{k,i} \triangleq z_{k,i} - H\hat{x}_{k,i|k-1,i};\ 1 \leq i \leq N,\ 1 \leq k \leq T\}$ of the observed process z. This technique has been used previously for system identification from longitudinal data (see e.g., [6]). Further, Rosenberg [7] uses a closely related technique in an application to the identification of a specific (time varying parameter) cross-sectional econometric model.

The likelihood function $f(z_{T,i}, z_{T-1,i}, \ldots, z_{1,i}|\theta)$ is the probability density of the observations $(z_1^T)_i$ when θ parameterizes the system (2.2). Now $f(z_{T,i}, z_{T-1,i}, \ldots, z_{1,i}|\theta)$ can be decomposed, using the chain rule for conditional probability densities, as

$$f(z_{T,i}, z_{T-1,i}, \ldots, z_{1,i}|\theta) =$$

$$f(z_{T,i}|z_{T-1,i}, z_{T-2,i}, \ldots, z_{1,i}, \theta) f(z_{T-1,i}|z_{T-2,i}, z_{T-3,i}, \ldots z_{1,i}, \theta) \ldots$$

$$f(z_{2,i}|z_{1,i}, \theta)\ f(z_{1,i}|\theta).$$

The set of random variables $\{(z_1^T)_i;\ 1 \leq i \leq N\}$ are i.i.d. So the innovation

$$z_{k,i} - E(z_{k,i}|z_{k-1,i}, \ldots, z_{1,i}, \theta) = z_{k,i} - H\hat{x}_{k|k-1,i} = \nu_{k,i}, \tag{3.1}$$

is independent of $\nu_{\ell,i}$ for $\ell \neq k$ and has mean value zero and variance p_k. Hence we can compute the likelihood function for a sample of N trajectories as

$$f((z_1^T)_1^N|\theta) = \prod_{i=1}^{N} \prod_{k=1}^{T} \frac{1}{\sqrt{2\pi p_k}} \exp - \frac{\nu_{k,i}^2}{2p_k} ,$$

and the scaled negative log-likelihood function is given (up to an additive constant term) by

$$L(\theta|(z_1^T)_1^N) = \frac{1}{N} \sum_{i=1}^{N} \left(\sum_{k=1}^{T} \left[\nu_{k,i}^2/2p_k + \frac{N}{2} \log p_k \right] \right). \tag{3.2}$$

Definition 3.1 Given any compact subset Ω of the parameter space $\mathbb{R}^n$ where η is the dimension of the parameter vector θ, the ML (maximum likelihood) estimate $\theta_N = \theta_N((z_1^T)_1^N)$ is defined as that point in Ω which minimizes $L(\theta|(z_1^T)_1^N)$ or, in the case of multiple global minima, as a ($(z_1^T)_1^N$-measurable) specification of a member of the minimizing set. □

The scaled negative log-likelihood function can be rearranged in a form that makes clear its dependence on second-order sample-statistics:

$$L(\theta|(z_1^T)_1^N) = \frac{1}{2} \left[\log|\Sigma| + (\bar{z}^N - \mu)' \Sigma^{-1} (\bar{z}^N - \mu) \right.$$

$$\left. + \frac{1}{N} \sum_{i=1}^{N} ((z_1^T)_i - \bar{z}^N) \Sigma^{-1} ((z_1^T)_i - \bar{z}^N) \right] \tag{3.3}$$

$$= \frac{1}{2} \left[\log|\Sigma| + \mathrm{tr}\, \Sigma^{-1} S^N + (\bar{z}^N - \mu)' \Sigma^{-1} (\bar{z}^N - \mu) \right],$$

where $\mu = \mu(\theta)$ and $\Sigma = \Sigma(\theta)$ are the first and second moments of the vector random variable z_1^T and where

$$\bar{z}^N \triangleq \frac{1}{N} \sum_{i=1}^{N} (z_1^T)_i$$

and

$$S^N \triangleq \frac{1}{N} \sum_{i=1}^{N} ((z_1^T)_i - \bar{z}^N)((z_1^T)_i - \bar{z}^N)' ,$$

are the corresponding sample values. This identity permits us to write $L(\mu,\Sigma,\bar{z}^N,S^N)$ for $L(\theta|(z_1^T)_1^N)$.

Definition 3.2 Let $\theta_N \in \Omega$, $N = 1,2,...,$ where Ω is a compact set. Then the sequence $\{\theta_N;\ N \leq 1\}$ is a ML(T,Ω) sequence if

$$(\forall\ \theta \in \Omega)\ L(\mu(\theta),\ \Sigma(\theta),\ \bar{z}^N,S^N) \geq L(\mu(\theta_N),\Sigma(\theta_N),\bar{z}^N,S^N),\ \forall\ N \geq 1. \square$$

The problem of identifiability is a standard issue in time-series analysis and (longitudinal) system-identification theory, but the present case requires the new formulation which we give below.

Definition 3.3 Let $\{z_k;\ k = 1,2,...,T\}$ be generated by the model (2.3), with $\theta = \mathring{\theta} \in \Omega$, where Ω is any prescribed subset of the parameter space $\mathbb{R}^\eta$. Let us denote the first and second moments by $\mu_T(\mathring{\theta})$ and $\Sigma_T(\mathring{\theta})$. Then $\mathring{\theta}$ is (second order) identifiable (T,Ω) if and only if

$$(\forall\theta\in\Omega)\ \mu_T(\theta) = \mu_T(\mathring{\theta}),\ \Sigma_T(\theta) = \Sigma_T(\mathring{\theta}) \Rightarrow \theta = \mathring{\theta}, \qquad (3.4)$$

i.e., if and only if $\mathring{\theta}$ is the unique parameter in Ω with image $(\mu_T,\ \Sigma_T)$. We say that $\mathring{\theta}$ is (second order) locally identifiable (T) if and only if there exists a neighbourhood $N_{\mathring{\theta}}$ of $\mathring{\theta}$ such that $\mathring{\theta}$ is second order identifiable $(T,N_{\mathring{\theta}})$. Finally we say the set Ω is (second order) identifiable (T) if and only if the parameter θ is identifiable (T,Ω) for all $\theta \in \Omega$. □

Our definition is structural, distinguishing it from those that involve the existence of consistent sequences of estimators (see, e.g., Tse [8] or Ljung [3]). It leads immediately to the following simple nesting theorem.

Theorem 3.1 Let the parameter $\mathring{\theta}$ be identifiable (T, Ω). Then $\mathring{\theta}$ is identifiable $(T+1, \Omega)$. □

Before we prove our main consistency result we observe the following facts, where the subscript T on μ and Σ is dropped for convenience:

(i) $L(.,.,.,.)$ is jointly continuous in all its arguments (3.5a)

(ii) $$\min_{\mu,\Sigma} L(\mu,\Sigma,\bar{z};S) = L(\bar{z},S,\bar{z},S), \tag{3.5b}$$

(iii) $$L(\mu,\Sigma,\bar{z},S) = L(\bar{z},S,\bar{z},S) \Rightarrow \mu = \bar{z},\ \Sigma = S. \tag{3.5c}$$

Statements (ii) and (iii) together yield the uniqueness of the minimizing arguments of $L(.,.,\bar{z},S)$.

Theorem 3.2 (Consistency and Asymptotic Normality). [9]. Let $\{\theta_N;\ N \geq 1\}$ be a $ML(T,\Omega)$ sequence for data generated by the system (2.2) with parameter $\mathring{\theta}$ in the compact set Ω. Let $\mathring{\theta} \in \Omega$ be identifiable (T,Ω) and let $\Sigma_T(\mathring{\theta}) > 0$. Then $\theta_N \xrightarrow{a.s.} \mathring{\theta}$ as $N \to \infty$, i.e., the $ML(T,\Omega)$ estimator is strongly consistent. In addition, let $\mathring{\Omega}$ denote the interior of Ω, assume $\mathring{\theta} \in \mathring{\Omega}$, and assume that the map $\theta \to (\mu(\theta),\Sigma(\theta))$ has a Jacobian of full rank at $\mathring{\theta}$. Then $\{\theta_N;\ N \geq 1\}$ is asymptotically normal with

$$\sqrt{N}(\theta - \mathring{\theta}) \xrightarrow{dist.} N(0,M^{-1})$$

where M is given by the formula

$$(M)_{p,q} = \sum_{k=1}^{T} \left\{ \frac{1}{2} p_k^{-2}(\theta) \frac{\partial p_k(\theta)}{\partial \theta_p} \frac{\partial p_k(\theta)}{\partial \theta_p} + p_k^{-1}(\theta)\, E \frac{\partial \nu_k(\theta)}{\partial \theta_p} \frac{\partial \nu_k(\theta)}{\partial \theta_q} \right\} \Bigg|_{\theta=\mathring{\theta}}, \tag{3.6}$$

$$1 \leq p,\ q \leq \eta.$$

and M^{-1} exists. □

We remark that the Hessian matrix of the log-likelihood function (used in most numerical minimization algorithms) converges to M as $N \to \infty$. It is shown in [10] that M is non-singular if and only if θ is locally identifiable (T), subject to a regularity condition.

4 SUFFICIENT CONDITIONS FOR IDENTIFIABILITY

One of the principal conditions for Theorem 3.2 to hold was called (second order) identifiability (T,Ω), where T denotes the length of the (longitudinal) sequence of observations and Ω denotes a subset of the parameter space (see Definition 3.3). In this section we present a theorem which provides sufficient conditions for identifiability (T,Ω) of systems of the form (2.2). In particular, this result yields a lower bound for the "magic number" T of (longitudinal) samples required for identifiability in terms of the state dimension d of the model (2.2).

Consider the set of η-component parameter vectors of ordered lists of the distinct entries of the matrices $(F,G,\sigma_w^2,\sigma_{wv},\sigma_v^2,V_{1|0},d_r,\sigma_r^2)$ in (2.2). We shall only consider the subset $\Theta_1 \subset \mathbb{R}^\eta$ for which $V_{1|0} = V'_{1|0} \geq 0$.

<u>Definition 4.1</u> The system (2.2) parameterized by $\theta \in \Theta_1 \subset \mathbb{R}^\eta$ is said to be in full rank disequilibrium if and only if

$$|V_{1|0} - FV_{1|0}F' - G\Sigma_w G'| \neq 0. \qquad \square$$

Note that the system (2.2) in steady state satisfies

$$\Delta V_{1|0} \triangleq V_{1|0} - FV_{1|0}F' - G\Sigma_w G' = 0,$$

while systems that are <u>not</u> in full rank disequilibrium lie on the manifold defined by $|\Delta V_{1|0}| = 0$, i.e., on a proper algebraic variety in $\mathbb{R}^\eta$. Hence, the system (2.2) is generically in full rank disequilibrium. We also remark that the rank of $\Delta V_{1|0}$ is related to the rank α of a matrix which plays an important rôle in the so-called Chandrasekhar algorithms [11] for recursive estimation in linear time-invariant systems.

The following theorem specializes the main result of [2] to the case of scalar z,w,v,r.

Theorem 4.1 [2] Consider the map $\Phi:\Theta_1 \rightarrow \mathbb{R}^{T(T+1)/2}$ given by $\theta \mapsto \Sigma_T(\theta)$.

Let $\Theta \subset \mathbb{R}^{\eta}$ denote the subset of Θ_1, such that $|F| \neq 0$ and let $\sigma_w^2(\mathring{\theta}) > 0$.

We consider two cases:

Case I: $(\sigma_{wv}, \sigma_r^2, d_r) = 0$

Case II: $\sigma_{wv} = 0$ and $|I-F| \neq 0$.

Then the following implications hold:

(i) If (2.2) is in full rank disequilibrium at $\mathring{\theta} \in \Theta$, then (Case I) $\mathring{\theta}$ is locally identifiable (d + 2), or (Case II) $\mathring{\theta}$ is locally identifiable (d + 3).

(ii) If $(F, V_{1|0}H')$ is controllable at $\mathring{\theta} \in \Theta$, then (Case I) $\mathring{\theta}$ is locally identifiable (2d + 1), or (Case II) $\mathring{\theta}$ is locally identifiable (2d + 2).

(iii) Suppose, in addition, that F is asymptotically stable and $W(z) \triangleq \sum_{k=0}^{\infty} HF^kGz^{-k}$ is of full rank outside the closed unit disc $\bar{\mathbb{D}}$ i.e., the factor W(z) of the spectral density matrix of the process Hx in (2.2) is inverse stable. Then (i) and (ii) hold globally. □

The standard result concerning the stochastic realization of stationary processes generated by finite-dimensional state space models, i.e. processes for which $\Delta V_{1|0} = 0$, is given in [12]. The following lemma yields the result as a special case of Theorem 4.1.

Lemma 4.1 [2] Let $\Theta \subset \mathbb{R}^{\eta}$ denote the subset of Θ_1, such that $|F| \neq 0$, $\sigma_w^2(\mathring{\theta}) > 0$ and let $\Delta V_{1|0} = 0$. Then, for the model (2.2), $(F, V_{1|0}H')$ is controllable if and only if (F,G) is controllable. □

Table I

Filter Equations and Gradient Algorithms

State Prediction Equations	$\hat{x}_{k+1\|k} = F\hat{x}_{k\|k-1} + FK_{k+1}(z_{k+1} - H\hat{x}_{k+1\|k})$
Filter Gain Equations	$K_{k+1} = V_{k+1\|k} H' (HV_{k+1\|k}H' + \sigma_v^2)^{-1}$
Variance Equations	$V_{k+1\|k} = FV_kF' + G\sigma_w^2G'$ $V_{k+1} = (1-K_{k+1}H)V_{k+1\|k}$
Innovation Variance Equation	$p_{k+1} = HV_{k+1\|k}H' + \sigma_v^2$

Kalman Filter Equations

State Prediction Gradient Equation	$\frac{\partial \hat{x}_{k+1\|k}}{\partial\theta} = F(I-K_kH)\frac{\partial \hat{x}_{k\|k-1}}{\partial\theta} + \frac{\partial F}{\partial\theta}(\hat{x}_{k\|k-1} + K_k\nu_k) + F\frac{\partial K_k}{\partial\theta}\nu_k$
Filter Gain Gradient Equation	$\frac{\partial K_{k+1}}{\partial\theta} = \frac{\partial V_{k+1\|k}}{\partial\theta}H' p_{k+1}^{-1} - V_{k+1\|k}H'p_{k+1}^{-2}\left\{H\frac{\partial V_{k+1\|k}}{\partial\theta}H' + \frac{\partial\sigma_v^2}{\partial\theta}\right\}$
Variance Gradient Equations	$\frac{\partial V_{k+1\|k}}{\partial\theta} = \frac{\partial F}{\partial\theta}V_kF' + FV_k\frac{\partial F'}{\partial\theta} + \left\{\frac{\partial G}{\partial\theta}G' + \frac{G\partial G'}{\partial\theta}\right\}\sigma_w^2$ $+ F\frac{\partial V_k}{\partial\theta}F' + GG'\frac{\partial\sigma_w^2}{\partial\theta}$ $\frac{\partial V_k}{\partial\theta} = \frac{\partial V_{k\|k-1}}{\partial\theta} - \left\{\frac{\partial V_{k\|k-1}}{\partial\theta}H'HV_{k\|k-1} + V_{k\|k-1}H'H\frac{\partial V'_{k\|k+1}}{\partial\theta}\right\}p_k^{-1}$ $+ V_{k\|k-1}H'HV'_{k\|k-1}p_k^{-2}\left\{H\frac{\partial V_{k\|k-1}}{\partial\theta}H' + \frac{\partial\sigma_v^2}{\partial\theta}\right\}$

Kalman Filter Gradient Equations

Innovation Variance Gradient Equation	$\frac{\partial p_{k+1}}{\partial\theta} = H\frac{\partial V_{k+1\|k}}{\partial\theta}H' + \frac{\partial\sigma_v^2}{\partial\theta}$
Likelihood Function Gradient	$\frac{\partial L}{\partial\theta} = \sum_{i=1}^{N}\sum_{k=1}^{T}\left\{-\nu_kH.\frac{\partial\hat{x}_{k\|k-1}}{\partial\theta}p_k^{-1} - \frac{1}{2}\nu_k^2p_k^{-2}\frac{\partial p_k}{\partial\theta} + \frac{1}{2}\frac{\partial p_k}{\partial\theta}p_k^{-1}\right\}$

Likelihood Function Gradient Equations

Hence, under the conditions of Lemma 4.1 we have, via Theorem 4.1 (ii) (Case I), local identifiability (2d + 1). If, in addition, F is asymptotically stable and W(z) is inverse stable, we obtain global identifiability. (The latter is, of course, the standard result [12]).

5. IMPLEMENTATION OF IDENTIFICATION ALGORITHM (Table I)

We compute the function $L(\theta|(z_1^T)_1^N)$ by recursively computing the innovations $\{\nu_k(\theta);\ 1 \leq k \leq T\}$ via the Kalman filter equations. Minimization of this function is achieved by use of a numerical optimization algorithm. We calculate the required gradients via recursive equations derived by straightforward differentiation of the Kalman filter equations. An approximation to the Hessian matrix is obtained by substituting a sample-mean for the expectation appearing in the formula (3.6). The FORTRAN computer program in [1] implements the equations in Table I and has been used in a number of applications. one of which is described in Section 6.

The computer program was constructed with the option of determining $(\sigma_{vw},\sigma_r^2,d_r)$ or assuming them to be zero. Table I presents the recursive equations for this latter case, but they are easily extended to the more general case. The computer program also allows identification of non-zero-mean systems by methods detailed in [1].

Structure determination is carried out by making sequences of likelihood maximization runs under different structural assumptions and then applying the asymptotic likelihood ratio or Akaike AIC criterion [13]. The computer program in [1] also provides sample covariances of lagged residuals. These are used in testing whiteness and, hence, goodness-of-fit and system order.

On a given run, one obtains estimates of the parameter covariance matrix by computing the approximate inverse Hessian (i.e., the information matrix), as discussed above. These estimates are useful not only in hypothesis tests involving parameter values, but also in examining identifiability. This is the case because, except possibly on a null set where a regularity condition fails, local identifiability and asymptotic nonsingularity of the Hessian matrix are equivalent conditions [10].

6 APPLICATION OF TECHNIQUE TO HUMAN DEVELOPMENT DATA

Our identification method was used in an analysis of non-stationary longitudinal and cross-sectional Stanford-Binet IQ data from the classical Fels longitudinal study described in detail by Sontag et al. [14] and in an important monograph by McCall et al. [15]. Our sample of 80 subjects, from a largely professional population in Southern Ohio, includes IQ measurements at 2 1/2, 3, 3 1/2, 4, 4 1/2, 5, 5 1/2, 6, 7, 8, 9, 10, 11, 12, 14, 15 and 17 years of age. This sample exhibits extreme longitudinal variability of IQ scores and argues against portrayal of IQ as a basically stable human trait. Subjects in the Fels sample changed (in absolute value) an average of 28.5 IQ points between 2 1/2 and 17 years of age, and one in seven changed by more than 40 points, while the average IQ increased by only about 3 points. One of the conclusions of McCall et al. is that a substantial fraction of longitudinal variation of IQ is caused by "idiosyncratic interaction of specific environmental events with the skills and motivational distinctions of the child", in other words, by inputs that resemble a random process. The main object of the analysis was to test this psychological hypothesis of stochastic behaviour and to determine the dynamic propagation of stochastic effects as the child grows.

Equation (2.1b) follows standard psychological test theory [16] in representing the measured IQ score as the sum of the "true score" $(y_{k,i}+r_i)$ and a measurement error $v_{k,i}$ characteristic of the Stanford-Binet instrument. In a departure from classical test theory, the true score is further decomposed into a dynamically varying component $y_{k,i}$ and a general level r_i characteristic of the individual. Evolution of the dynamic component y is described by the stochastic difference equation (2.1a). Equation (2.2) presents an equivalent representation of the same model. In this formulation all subjects are assumed to have the same dynamics (2.1a). Subjects differ in that both r_i and the initial state $y_{1,i}$ are random variables distributed across the population of subjects.

The unit of time for the algorithms in Table I was selected to be six months in order to correspond to the six monthly testing schedule from 2 1/2 to 6 years. Measurements are missing for ages 6 1/2, 7 1/2, etc. and for ages 13 and 16. The calculation of the likelihood function and its derivatives was adapted to this irregular data schedule by (effectively) setting the measurement error variance for the missing data to infinity. This is equivalent to propagating the state estimation error covariance via the Liapunov equation when data are missing while, as already described, it is propagated by the Riccati equation (see Table I) when data are present. It appears that this simple technique for the missing data problem is not yet used by statisticians.

Table II

Maximum Likelihood Parameter Estimates for First Order IQ Model with half-year time scale.

a_1	σ_w^2	σ_v^2	$V_{1\|0}$	d_r	σ_r^2
.914	11.0	21.3	170.2	12.6	3.2

Table III

Estimated Standard Errors (diagonal) and Correlations of Parameter Estimates for First-Order IQ Model with half-year Time Scale.

	a_1	σ_w^2	σ_v^2	$V_{1\|0}$	d_r	σ_r^2
a_1	.027	-.311	.285	-.008	-.785	-.926
σ_w^2	-.311	1.78	-.700	.012	.227	.244
σ_v^2	.285	-.700	1.99	-.025	-.209	-.224
$V_{1\|0}$	-.008	.012	-.025	29.0	.368	.077
d_r	-.785	.227	-.209	.368	5.25	.862
σ_r^2	-.926	.244	-.224	.077	.862	1.71

The structure of the model describing the data was identified as first order (d=1) with $\sigma_r^2 > 0$, on the basis of asymptotic likelihood ratio tests. The estimated model fits the data well in terms of second order statistics, whiteness of innovations, and the distribution of 80 simulated growth profiles, which resembled that of the original data. The resulting model was asymptotically stable. McCall's hypothesis of the essentially stochastic nature of the process was tested and accepted at the level of significance .01. The estimated value of the autoregressive parameter, .914 per half-year, predicts persistence of stochastic effects for long periods. Details of this analysis are given in [1] and are the subject of a separate paper [17]. Table II presents estimates for parameters of the accepted model. Table III presents corresponding estimates of the standard-errors and correlations of these estimates obtained by manipulation of the Hessian matrix.

7 ACKNOWLEDGEMENT

Partially supported by U.S. Office of Naval Research, Contract N00014-75-C-0648.

8 REFERENCES

1. Goodrich, R. L. "Stochastic Models of Human Growth", Ph.D. Thesis, Division of Applied Sciences, Harvard University, May, (1978).

2. Goodrich, R. L. and Caines, P. E. "On the Non-Stationary Covariance Realization Problem", to appear, *IEEE Trans. Automat. Contr.*, Oct., (1979).

3. Ljung, L. "On Consistency of Prediction Error Identification Methods", in Systems Identification: Advances and Case Studies, D. G. Lainiotis and R. K. Mehra, Eds., New York: Academic Press, (1977).

4. Caines, P. E. "Prediction Error Identification Methods for Stationary Stochastic Processes", *IEEE Trans. Automat. Contr.*, **AC-21**, 500-505, August, (1976).

5. Caines, P. E. and Ljung, L. "Asymptotic Accuracy and Normality of Prediction Error Estimates"; Control Systems Report No. 7602, Dept. E. E., U. of Toronto, Feb. 1976. Presented as "Prediction Error Estimators: Asymptotic Normality and Accuracy", Conference on Decision and Control, Clearwater Beach, Fla., December, (1976).

6. Caines, P. E. and Rissanen, J. "Maximum Likelihood Estimation of Parameters in Multivariate Gaussian Stochastic Processes", *IEEE Trans. Inform. Theory,* **IT–20**, 102-104, January, (1974). To appear in extended form in *Ann. Math. Statist.*

7. Rosenberg, B. "The Analysis of a Cross Section of Time Series by Stochastically Convergent Parameter Regression", *Annals of Economic and Social Measurement,* **2**, 399-428, October, (1973).

8. Tse, E. "Information Matrix and Local Identifiability of Parameters", Paper 20-3 in Proc. J.A.C.C., Columbus, (1973).

9. Goodrich, R. L. and Caines, P. E. "Non-Stationary Linear System Identification from Cross-Sectional Data", to appear, *IEEE Trans. Automat. Control,* **AC–24**, April, (1979).

10. Goodrich, R. L. and Caines, P. E. "Necessary and Sufficient Conditions for Local Identifiability", *IEEE Trans. Automat. Control,* **AC–24**, February, (1979).

11. Morf, M., Sidhu, G. S. and Kailath, T. "Some New Algorithms for Recursive Estimation in Constant, Linear, Discrete-Time Systems", *IEEE Trans. Automat. Contr.,* **AC–19**, 315-323, August, (1974).

12. Faurre, P. L. "Stochastic Realization Algorithms", in *System Identification: Advances and Case Studies,* D. G. Lainiotis and R. K. Mehra, Eds., New York: Academic Press, (1977).

13. Akaike, H. "Canonical Correlation Analysis of Time Series and the Use of an Information Criterion", in *System Identification: Advances and Case Studies,* D. G. Lainiotis and R. K. Mehra, Eds., New York: Academic Press, (1977).

14. Sontage, L. W., Baker, C. T. and Nelson, V. L. "Mental Growth and Personality Development: A Longitudinal Study", *Monographs of the Society for Research in Child Development,*

23, (2, Serial No. 68), (1958).

15. McCall, R. B., Apelbaum, M.I. and Hogarty, P. S. "Development Changes in Mental Performance", *Monographs of the Society for Research in Child Development*, **38**, (3, Serial No. 150), (1973).

16. Lord, F. M. and Novack, M. R. *Statistical Theories of Mental Test Scores,* Reading, MA: Addison-Wesley, (1968).

17. Goodrich, R. L. "A Stochastic Dynamic Analysis of the Fels Institute Longitudinal IQ data", 1979 Meeting of the Society for Research in Child Development, San Francisco, March, (1979).

CHAPTER 8

APPLICATIONS

8.1 OPTIMAL SCHEDULING FOR THE TREATMENT OF CANCER

N. Carmichael and A. J. Pritchard
(Control Theory Centre, University of Warwick, Coventry, CV4 7AL, UK)

1 INTRODUCTION

This paper provides a general framework within which one may pose problems of scheduling for treatment regimes with particular reference to optimality. At present schedules are derived in a purely heuristic fashion. Our mathematical formulation follows the work of Lions and Bensoussan [1] on stock control in which the system is described by a nonlinear stochastic differential equation and the controls are impulsive in nature. Such controls change the state of the system at instants and by amounts which are available for choice. The minimization of a performance index leads to optimality criteria which can be formulated in terms of quasi-variational inequalities for which numerical methods of solution have recently been developed, see [2] and [3].

In cancer the underlying dynamics are determined by the tumour growth. The mathematical model is given in terms of nonlinear differential equations representing the growth of oxic and hypoxic cell population.

Uncertainty in the dynamics is modelled by additive white noise. The performance index penalises the expected number of cancer cells at the end of and during a treatment period. The controls are the sizes and times at which doses are given, and the performance index includes terms which limit the level and frequency of the doses. Thus toxicity is taken into account. The numerical computation required to derive a complete schedule is not straight forward and for simple deterministic problems analytic solutions show a variety of features - cf. Shima [4]. Work continues on the construction of algorithms for the stochastic case - the intention being to evaluate schedules currently implemented by clinicians (see Wheldon and Kirk [5]).

2 THE DETERMINISTIC MODEL

The model of Fischer [6] describes the development and interplay between four different cell types; live oxygenated, dead oxygenated, live anoxic, and dead anoxic. These four cell types are emphasised because of their distinct responses to irradiation. The live oxygenated cells differ from the live anoxic cells in that they divide hence causing tumour growth. Whereas the dead cells are those that have been damaged by radiation, and although they may metabolise they no longer have the capacity to divide. It is possible for anoxic cells to become oxygenated and vice versa, and this interchange plays an important role in determining the effect of a treatment schedule.

We let S_O, S_1 be the surviving fraction of live oxygenated and live anoxic cells, respectively after a radiation dose of D rads; and assume

$$S_i(D) = 1 - (1 - e^{-D/D_i})^{n_i} \quad i = 0,1 \tag{2.1}$$

where D_i, n_i, i = 0, 1 are constants derived from empirical data, the structure of this formula having been derived from a multi-target single hit model of the effect of radiation.

We assume that doses of size d_i, i = 1,2... are given at times t_i, i = 1,2.... Constraints can be imposed on the treatment schedule - we have chosen, initially, to place direct bounds on the dose size

$$d_i \in U \text{ for some compact set} \tag{2.2}$$

A more realistic constraint would result from a modification of the cumulated radiation effect (or CRE) index. Inclusion of this constraint is among the objectives of further studies.

In order to model the fact that as the tumour size increases the fraction of live oxygenated cells decreases, Fischer used

$$\frac{N_{O(t)}}{N(t)} = e^{-\beta N(t)} \tag{2.3}$$

where $N_0(t)$ and $N(t)$ are the number of live oxygenated and total number of tumour cells at time t, respectively, and β is an empirically determined constant. Then since the growth of the tumour is determined by the number of live oxygenated cells, a simple law for an untreated tumour is

$$\frac{dN}{dt} = \alpha N_0, \quad \alpha \text{ constant.} \tag{2.4}$$

From (2.3) we obtain

$$\frac{dN}{dt} = \alpha N e^{-\beta N} \tag{2.5}$$

The general features of (2.5) concur with observed behaviour of tumours which initially grow exponentially, but then this growth slows as the tumour size increases.

Using (2.3) and (2.1) we see that if N is the number of cells before a dose of size d_i, then the number surviving immediately after the dose is

$$S(N,d_i) = S_0(d_i) N e^{-\beta N} + S_1(d_i) N (1 - e^{-\beta N}) \tag{2.6}$$

3 IMPULSE CONTROL

In this section we give a brief review of the theories of impulse control as developed by Lions et al.

Let N(t), a vector in R^1 be the state of the system at time t and assume N(t) satisfies the stochastic differential equation

$$\begin{aligned} dN &= g(N, t)dt + \sigma (N, t)\, dw(t) \\ N(s) &= \overline{N} \end{aligned} \tag{3.1}$$

where

$$g(N, t) = \alpha N e^{-\beta N}$$

Our reasons for using a stochastic model are twofold. Intuitively we may think of $\sigma(N,t)$ as a random perturbation of the deterministic growth function g. More precisely, if the system is in state N at time t, then between t and t + Δt it will change randomly with mean equal to N + g (N,t) Δt, and variance $\sigma^2(N,t)\ \Delta t$. On the other hand the presence of σ makes the problem, or at least part of it, easier, since it is equivalent to the introduction of a regularization - i.e. we are faced by the solution of equations of parabolic rather than hyperbolic type.

As usual we assume w to be a one dimensional Wiener process adapted to F_t, which is an increasing family of sigma algebras representing the information contained in the process. Then subject to some technical conditions on σ, [1], it is possible to obtain existence and uniqueness of the solution of (3.1) on a time interval which we denote by T.

Now we introduce our control variables - the impulses - in our case they represent the time at which a dose is given and the number of cells killed by that dose. There is assumed to be no delay between the administration of a dose and the occurrence of its full effect. The impulses are assumed to occur at times $t_1, t_2 \ldots$ with $0 \leq t_1 \leq t_2 \ldots$ and associated with these times there is a sequence of random variables d_i in U, a subset of R^1, such that each d_i is F_{t_i} -measurable. In our case U will not only be closed, but it will also be compact because of the constraints imposed by the dose toxicity.

Note that $\{d_i \in A\}\ \{t_i \leq t\}$ is a subset of F_t for all t, and A any Borel set in R^1. The set

$$V = \{t_i,\ d_1;\ \ldots;\ t_i,\ d_i\ ;\ \ldots\} \tag{3.2}$$

is referred to as an impulse control, and it is possible to describe the evolution of the dynamical system (3.1) subject to controls (3.2). First suppose that $N(t_i)$ is given, then we define N(t) for $t \in (t_i, t_{i+1})$ by

$$dN = g(N,t) + \sigma\ (N,t)\ dw(t)$$

We then set $N(t_{i+1}) = S(N(t^-_{i+1}),\ d_{i+1})$ where $S(N,d_i)$ is given by (2.6) and $N(t^-_{i+1})$ denotes the limit from the left, with existence ensured by continuity. Thus we define N(t) on $[0,T]$.

The cost functional given $\bar{N}$ cells at time s, is taken to be

$$E\{\sum_i c(d_i) + \int_s^T f(N(t), t)\, dt + h(N(T))\}$$

where f measures the cost of a cancer cell continuing to survive, h penalises the size of the final population, $c(d_i)$ is the cost of a dose of size d_i, and E denotes expectation. We set

$$\rho(\bar{N},s) = \inf_V E\{\sum c(d_i) + \int_s^T f(N(t),t)dt + h(N(T))\}. \quad (3.3)$$

We recall the philosophy behind the dynamic programming approach to optimal control. Suppose at time s there are $\bar{N}$ cells present, then there are two choices.

a) Immediately apply a dose d. If we do this then the state becomes $S(\bar{N},d)$, and hence

$$\rho(\bar{N},s) \leq \inf_{d\varepsilon U} (c(d) + \rho(S(\bar{N},d), s)) \quad (3.4)$$

We shall set

$$M\,\Phi = \inf_{d\varepsilon U} (c(d) + \Phi(S(\bar{N},d), s)) \quad (3.5)$$

Obviously M is non-linear, and non-local but it is monotonic.

b) Proceed for a small time δ without applying a dose; then

$$\rho(N,s) \leq E\{\int_s^{s+\delta} f(N(t),t)dt\} + E\{\rho\ (N(s+\delta), s+\delta)\}$$

By taking δ small enough, expanding, and ignoring all but the first order terms in δ, we have

$$\begin{aligned}\rho(\bar{N},s) &\leq \delta\ f(\bar{N},s) + E\{\rho(\bar{N} + g\delta + \sigma(w(s+\delta)-w(s)), s+\delta)\} \\ &\leq \delta\ f(\bar{N},s) + E\{\rho(\bar{N},s) + \delta \frac{\partial\rho}{\partial s}(\bar{N},s) + \delta g \frac{\partial\rho}{\partial N}(\bar{N},s) + \\ &\quad + \sigma(N(s+\delta)-w(s)) \frac{\partial\rho}{\partial\bar{N}} \bar{N},s) + \frac{\sigma^2}{2} (w(s+\delta)-w(s))^2 \frac{\partial^2\rho}{\partial\bar{N}^2}(\bar{N},s) \\ &\quad + \delta g\sigma\ (w(s+\delta) - w(s)) \frac{\partial^2\rho}{\partial\bar{N}^2}(\bar{N},s)\}\end{aligned} \quad (3.6)$$

The term $\frac{\sigma^2}{2}(w(s+\delta) - w(s))^2 \frac{\partial^2\rho}{\partial\bar{N}^2}(\bar{N},s)$ is derived from the formula of Itô. From (3.6) using standard properties of the Wiener process, we have

$$\rho(\bar{N},s) \leqslant \delta\, f(\bar{N},s) + \rho(\bar{N},s) + \delta \frac{\partial\rho}{\partial s}(\bar{N},s) + \delta g \frac{\partial\rho}{\partial\bar{N}}(\bar{N},s)$$

$$+ \delta\sigma^2 \frac{\partial^2\rho}{\partial\bar{N}^2}(\bar{N},s)$$

and so

$$-\frac{\partial\rho}{\partial s} - \sigma^2 \frac{\partial^2\rho}{\partial\bar{N}^2} - g\frac{\partial\rho}{\partial\bar{N}} - f \leqslant 0. \tag{3.7}$$

This equation together with (3.4) defines the value function ρ, but since the policy is optimal and either a) or b) must occur, we also have

$$\left(-\frac{\partial\rho}{\partial s} - \sigma^2 \frac{\partial^2\rho}{\partial\bar{N}^2} - g\frac{\partial\rho}{\partial\bar{N}} - f\right)(\rho - M\sigma) = 0. \tag{3.8}$$

We also have boundary conditions for a bounded state variable N, so that if $N \varepsilon \Sigma$, on $\partial\Sigma$

$$\frac{\partial\rho}{\partial n} \leqslant 0, \quad \rho - M\rho \leqslant 0 \tag{3.9}$$

and $\frac{\partial\rho}{\partial n}(\rho - M\rho) = 0$

where n is the normal to $\partial\Sigma$.

Finally when s = T no control action is taken so that

$$\rho(N,T) = h(N). \tag{3.10}$$

The above formal manipulations which result in (3.4), (3.7), (3.8), (3.9) and (3.10) can be made rigorous, and a useful tool in this analysis which is also important in computation is to reformulate the problem as quasi-variational inequality. In order to do this we define the operator A, by

$$- A\Phi = \sigma^2 \frac{\partial^2\Phi}{\partial N^2} + g \frac{\partial\Phi}{\partial N}$$

and introduce the Sobolev spaces $H^O(\Sigma)$, $H^1(\Sigma)$, where the inner product, and norm on $H^O(\Sigma)$ are given by

$$<f, g>_{H^O} = \int_\Sigma f(N) \; g(N) \; dN$$

$$|f|^2_{H^O} = \int_\Sigma f^2(N) \; dN$$

and

$$\|f\|^2_{H^1} = |f|^2_{H^O} + \left|\frac{\partial f}{\partial N}\right|^2_{H^O}$$

Then we define the bilinear form for $u, v \in H^1$,

$$a(u,v,t) = \int_\Sigma \sigma^2 (N,t) \frac{\partial u}{\partial N} (N) \frac{\partial v}{\partial N} dN + \int_\Sigma g(N,t) \frac{\partial u}{\partial N} v(N) \; dN$$

If $\sigma(N,t) > O$ for all N,t, there exist constants λ,μ such that

$$a(u,v,t) + \lambda \; |u|^2 \geqslant \mu \; \|u\|^2$$

where $\mu > O$. Moreover for $\Phi \in D(\Sigma)$, the space of test functions on Σ, and $u \in D(A)$

$$a(u,\Phi,t) = < Au, \Phi >$$

So for $f \in L^2|O,T,H|$, we may seek a function $u \in L^\infty(O,T,H^1) \cap L^\infty(O,T,L^\infty)$ with $\frac{du}{dt} \in L^2|O,T;H|$, such that

$$- < \frac{du}{dt}, v - u > + a(u,v-u) - <f, v-u> \geqslant O \qquad (3.11)$$

For all $v \in H_0'$, with $v \leqslant u$, $v \leqslant Mu$, $u(T) = h$ and the boundary conditions (3.9). To see that this solution will solve (3.4), (3.7) ... (3.10) we set $v = u - x$, then since $v \leqslant u$ we have $x \geqslant 0$. Then for $u \in D(A)$ we obtain from (3.11)

$$< -\frac{\partial u}{\partial t} + Au - f, x > \leqslant 0 \quad \Phi \geqslant 0$$

and hence

$$-\frac{\partial u}{\partial t} + Au - f \leqslant 0.$$

For $v = Mu$, we have

$$< -\frac{\partial u}{\partial t} + Au - f, Mu - u > \geqslant 0$$

But $-\frac{\partial u}{\partial t} + Au - f \leqslant 0$, and $u \leqslant Mu$, so

$$(-\frac{\partial u}{\partial t} + Au - f)\ (Mu - u) = 0.$$

The formulation of our problem as a quasi-variational inequality (3.11) ensures the existence of a solution by a theorem of Lions and Bensoussan [1].

4 NUMERICAL RESULTS

The solution of the quasi-variational inequality determines a boundary in the (N,t) space where $\rho = M\rho$. It is this surface which provides a solution to the impulse control problem. The quasi variational formulation is also advantageous here since we may use existing knowledge concerning numerical approximations for such systems. In the calculations for this paper we have used an approximate scheme to be found in [2].

So far we have computed the quasi-variational surface for a variety of different cost functionals. The following graphs show, as shaded, those regions of the (N,t) space for which $\rho = M\rho$. The data common to all the graphs are

Maximum dose at any one time = 500 rads

D_0 (oxic) = 100

N_0 (oxic) = 4

D_1 (hypoxic) = 250

N_1 (hypoxic) = 1

Oxic growth law α = 0.004

B = 2.5×10^{-10}

Note that the value of B gives the fraction of oxygenated cells in a tumour of total population 10^{11} to be about 0.1. Figs. 1 and 2 show the effect of increasing the function f in the cost function; f in Fig. 2 being a tenth of that in Fig. 1. Fig. 3 shows the effect of reducing the dose cost function, $c(d_i)$ to a hundredth of its value in Fig. 1; Figs. 1, 2 and 3 were all produced with the same terminal cost.

The doses are administered at those times when the observed state N lies on the boundary. The dose sizes are determined from the values of ρ locally.

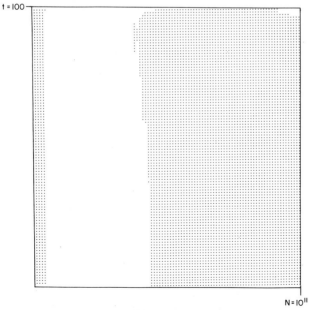

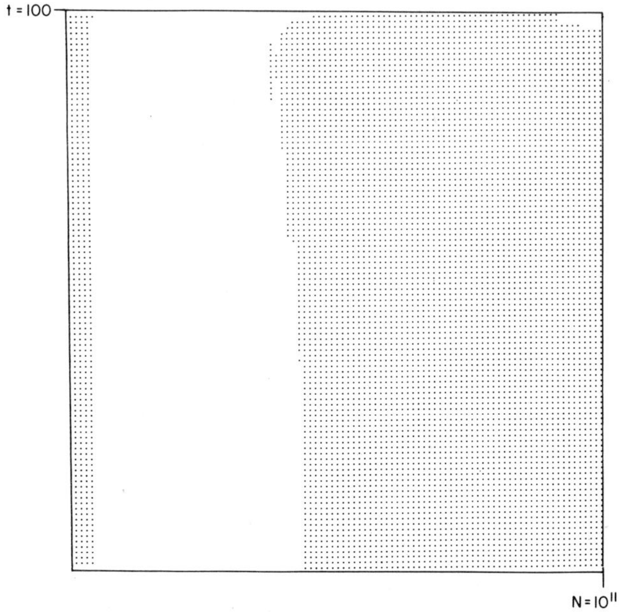

Fig. 1

Fig. 2

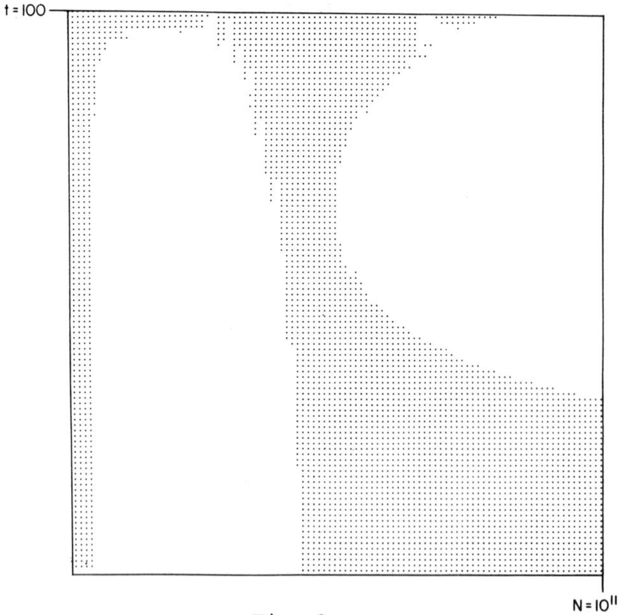

Fig. 3

5 CONCLUSIONS

Much work remains to be done to turn the theoretical considerations described here into operationally realistic schedules. The main areas of interest concern the numerical construction of solutions, choice of constraints and cost functionals and also the effect of large parameter uncertainties. Indications are, however, that the work could provide a useful tool for the evaluation of treatment scheduling and not only for cancer.

6 ACKNOWLEDGEMENTS

The authors would like to thank the Medical Research Council under whose support this work is continuing.

7 REFERENCES

1. Lions, J. and Bensoussan, A. "Nouvelles Methodes en Contrôle Impulsionnel", *Applied Mathematics and Optimisation*, **1**, 289-312, (1975).

2. Goursat, M. and Maarek, G. "Nouvelle Approche des problèmes de gestion de stocks", Rapport Laboria, No. 148, I.R.I.A., (1976).

3. Mosco and Scarpini "Complementarity Systems and Approximation of Variational Inequalities", R.A.I.R.O., 83-104, (1975).

4. Shima, M. "Analytical Construction of the Free Boundary", Rapport Laboria, No. 194, I.R.I.A., (1976).

5. Wheldon, T. and Krik, J. "Mathematical Derivation of Optimal Treatment Schedules", *British Journal of Radiology*, **49**, 441-449, (1976).

6. Fischer, J. "Mathematical Simulation of radiation therapy of solid tumours", *Acta. Radiol. Ther: Phys. Biol.*, **10**, 73-85, (1971).

7. Almquist, K. and Banks, H. "A Theoretical and Computational Method for determining optimal Treatment Schedules", *Mathematical Biosciences*, **29**, 159-179, (1976).

8.2 THE CONTROL TEST AND ITS APPLICATION TO THE CHOICE AND THE EVALUATION OF ALTERNATIVE STOCHASTIC ECONOMIC SYSTEMS

Y. Uchida

(Research Institute for Economics and Business Administration, Kobe University, Rokko, Kobe, Japan)

1 INTRODUCTION

Often in physical, socio-economic systems and elswhere, we have the problem of choosing among different classes of dynamic models. Most previous studies have been only based on the prediction test criteria. [1,2] On the other hand, another test criterion for choosing models must be formed when dynamic models are used to analyse the effects of policy control as developed in the econometric literature. This paper is concerned with an application of optimal control theory to the problem of choice of dynamic models frequently encountered in statistical identification.

The structure of the paper is as follows. An algorithm for evaluating the models is presented in Section 2. In Section 3 we introduce the new concept of the control test (XY-test) and discuss the choice of models by the use of this test. Also, the two other criteria (final test (Y-test) and inverse final test (X-test)) for choosing among models are presented from the viewpoint of both the prediction and the control. The results in Section 2 and Section 3 were also given by Uchida [3]. In Section 4 the proposed test criteria are applied to the well-known Klein Model I [4]. It is shown that the single estimation method is effective in the control, and that the simultaneous estimation is rather effective in the prediction.

2 AN ALGORITHM FOR EVALUATING THE MODELS

2.1 The Model

In general econometric model is represented in the structural form

$$y_t = f(s_t, x_t, y_t, z_t) + u_t \tag{2.1}$$

where s_t is an NS-dimensional vector of state variables; x_t, an NX-dimensional vector of control variables; y_t, an NY-dimensional vector of endogenous variables; z_t, an NZ-dimensional vector of exogenous variables; u_t, the disturbances. s_t is the lagged variable vector, which is interpreted as

$$s_t \in \{x_{t-1}, x_{t-2}, \ldots, y_{t-1}, y_{t-2}, \ldots\}$$

Accordingly, s_t can be written as

$$s_{t+1} = \Delta_1 s_t + \Delta_2 x_t + F y_t \tag{2.2}$$

(2.1) and (2.2) indicate the state-variable form of nonlinear econometric models.

2.2 The Objective Functional

Suppose the objective functional is described by

$$\phi = \Sigma_{t=1}^{T} \phi_t \tag{2.3}$$

$$\phi_t = \| y_t - \hat{y}_t \|_{Q_t}^2 + k \| x_t - \hat{x}_t \|_{R_t}^2 \tag{2.4}$$

where $\hat{x}_t$ and $\hat{y}_t$ are the target vectors for x_t and y_t, respectively; Q_t and R_t are nonnegative definite and positive definite weighted matrices, respectively; k is the positive scalar $\| x \|_A^2 = x' Ax$.

2.3 The Weighted Matrices

The weighted matrices are constructed for given $\rho \geq 0$, $W_x(i) > 0$ and $W_y(i) \geq 0$ as follows.

$$\hat{\rho}_t = \frac{(1 + \rho)^t}{\Sigma_{\tau=1}^{T}(1+\rho)^{\tau}}$$

$$\hat{W}_x(i) = \frac{W_x(i)}{\Sigma_{j=1}^{NX} W_x(j)}, \qquad \hat{W}_y(i) = \frac{W_y(i)}{\Sigma_{j=1}^{NY} W_y(j)} \tag{2.5}$$

$$Q_t(i, j) = \hat{\rho}_t \hat{W}_y(i) \hat{y}_t(i)^{-2} \delta_{ij} \tag{2.6}$$

$$R_t(i, j) = \hat{\rho}_t \hat{W}_x(i) \hat{x}_t(i)^{-2} \delta_{ij} \tag{2.7}$$

where δ_{ij} denotes the Kronecker delta. The new and old data and the relative importance of the variables are adjusted by a time discount factor ρ and by W_x and by W_y, respectively.

2.4 POEM Algorithm

We describe briefly the POEM algorithm (Policy Optimization using Econometric Models), which gives the approximate solution of minimizing the objective functional (2.3), with respect to the control policy, subject to the dynamic constraint (2.1). The algorithm is given as follows.

The prior value: $\mu(0) = \{s_0^d, s_t^0, x_t^0, y_t^0, z_t^0\}_{t=1}^{T}$

The posterior value: $\mu(*) = \{s_0^d, s_t^*, x_t^*, y_t^*, z_t^d\}_{t=1}^{T}$

Data: $\mu(d) = \{s_0^d, s_t^d, x_t^d, y_t^d, z_t^d\}_{t=1}^{T}$

Step 1. Start: determine ρ, W_x, W_y and the target values

Step 2. The model (3.1) is linearized in the neighbourhoods of the prior value $\mu(0)$.

Step 3. Solve the optimal control problem of minimizing the objective functional (2.3) subject to the linearized approximation model (2.2) to find the posterior value $\mu(*)$ and the value for (2.3).

Step 4. Go to the Step 5, when the computation converges. Otherwise, return to the Step 2, where the posterior value $\mu(*)$ is regarded as the prior value $\mu(0)$ at the next iteration.

Step 5. Compute various kinds of evaluation values in Section 3.

Stop:

where when

$$\left| \frac{\phi^{\ell} - \phi^{\ell-1}}{\phi^{\ell-1}} \right| < EPS \tag{2.8}$$

then the computation converges at the ℓth iteration.

3 THE EVALUATION AND CHOICE RULE FOR DYNAMIC MODELS

Define

$$X(i) = \Sigma_{t=1}^{T} \hat{\rho}_t \left(1 - \frac{x_t^*(i)}{x_t^d(i)}\right)^2 \tag{3.1}$$

$$X = \Sigma_{i=1}^{NX} \hat{W}_x(i) X(i) \tag{3.2}$$

$$Y(i) = \Sigma_{i=1}^{T} \hat{\rho}_t \left(1 - \frac{y_t^*(t)}{y_t^d(t)}\right)^2 \tag{3.3}$$

$$Y = \Sigma_{i=1}^{NY} \hat{W}_y(i) Y(i) \tag{3.4}$$

Final Test (Y-Test)

The final test is formulated as

$$x_t = x_t^* = x_t^d, \ \hat{y}_t = y_t^d \tag{3.5}$$

$$\beta = 100\sqrt{\frac{Y}{2}} \tag{3.6}$$

Y-Test is x = 0 and evaluated by the error ratio (%) of the endogenous variables. The parameter β stands for the result of Y-Test.

Inverse Final Test (X-Test)

When Y = 0, we call it inverse final test denoted by X-Test. X-Test can be expressed by

$$\hat{x}_t = x_t^0, \quad \hat{y}_t = y_t^d, \quad k=0.001 \tag{3.7}$$

$$\alpha = 100\sqrt{\frac{X + Y}{2}} \quad (\doteqdot 100\sqrt{\frac{X}{2}}) \tag{3.8}$$

Note that the posterior value x_t^* obtained from the current iteration ℓ is calculated as the posterior x_t^0 and the target value $\hat{x}_t$ at the next iteration $(\ell + 1)$. Thus we have

$$X^{\ell} < X^{\ell+1}, \quad Y^{\ell} > Y^{\ell+1}$$

and as $\ell \to \infty$, $Y^{\ell} \to 0$ when the model is perfectly controllable (with 100%). The controllability of the model is calculated as by-products of X-Test and is defined as follows. y(i) is $\gamma(i)$-controllable (%) if and only if

$$\gamma(i) = 100(1 - \sqrt{Y(i)}\ ) \tag{3.9}$$

and the model is strictly γ-controllable (%) if and only if

$$\gamma \geq \gamma(i) \quad \text{for } i = 1, 2, \ldots, NY \tag{3.10}$$

Furthermore the model is $\bar{\gamma}$-controllable (%), if and only if

$$\bar{\gamma} = 100(1 - \sqrt{Y}\ ) \tag{3.11}$$

The parameter α in X-Test becomes large when the model is ill-conditioned. Thus Y-Test implies the criterion of whether the model is fitted to the data.

Control Test (XY-Test)

XY-Test is formulated as

$$\hat{x}_t = x_t^d, \quad \hat{y}_t = y_t^d \tag{3.12}$$

$$\pi = 100\sqrt{\frac{X + Y}{2}} \tag{3.13}$$

The two methods are considered; the first one is that k is fixed, and the second one is to modify some procedures in the iteration process according to the following rules.

$$X \gtrless Y \Rightarrow k^{\ell} \gtrless k^{\ell+1} \tag{3.14}$$

$$X \doteq Y \Rightarrow \sigma = \frac{k}{1 + k}, \quad \text{(stop)} \tag{3.15}$$

where σ stands for the autonomy of the model. $\bar{\gamma} = 0$ when $\sigma = 1$. Inversely $\bar{\gamma} = 100$ when $\sigma = 0$.

X and Y are increasing function and decreasing function in k, respectively. The performance parameters are depicted as Fig. 1.

Decision Rule for the Choice of Models

As was seen in X-Test, Y-Test and XY-Test, we contend that the model is chosen based on the control test criterion, when $\pi < \beta$. XY-Test for comparing and choosing among different models is performed by integrating ρ, k, W_x and W_y that indicate the evaluation criteria. In such case, the parameter β is calculated in the process of implementation of XY-Test. Decision rule for the choice of models is best given, when π is less than β and minimized. Furthermore this rule can be also applied to the model selection of the simultaneous equations case where the size of each model is different.

4 EMPIRICAL STUDY ON THE EVALUATION AND CHOICE OF MODELS

4.1 The Model and the Computation Results

In this section we investigate empirically the three tests proposed in the preceding sections. The model used to perform our investigation is well-known Klein's Model I, which gives

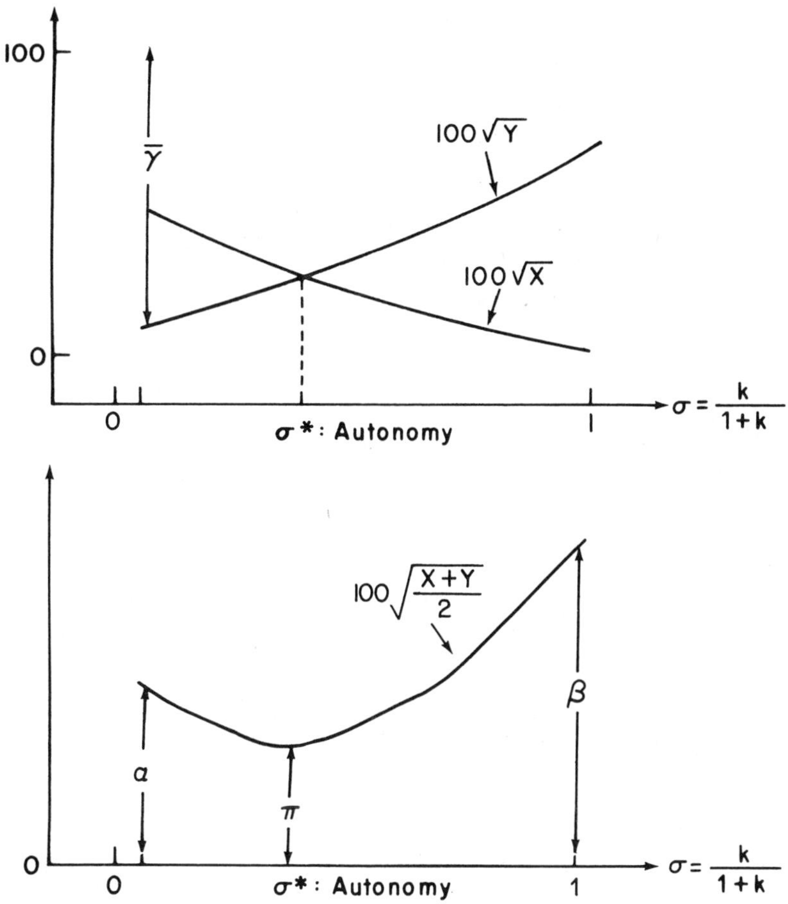

Fig. 1 The performance parameters of control test and final test

a simplified description of the U.S. economy (for the period 1921-41). The model is represented by

$$C = \theta_{11} + \theta_{12}P + \theta_{13}P_{-1} + \theta_{14}(WP + WG) \tag{4.1}$$

$$I = \theta_{21} + \theta_{22}P + \theta_{23}P_{-1} + \theta_{24}K_{-1} \tag{4.2}$$

$$WP = \theta_{31} + \theta_{32}E + \theta_{33}E_{-1} + \theta_{34}(YEAR - 1921) \tag{4.3}$$

$$Y = C + I + G - T \tag{4.4}$$

$$P = Y - (WG + WP) \tag{4.5}$$

$$K = K_{-1} + I \tag{4.6}$$

$$E = Y + T - WG \tag{4.7}$$

where C = Consumption, I = Investment, WP = Private wage bills, P = Profits, Y = National income, K = Capital stock at the beginning of the year, E = Private product, WG = Government wage bills, G = Government expenditure, T = Indirect Taxes, YEAR = The Calendar year.

In the original model (4.1) through (4.7), we consider the log-type and the lagged-type equations of private wage bills in (4.3) as follows:

$$\log WP = \theta_{41} + \theta_{42}\log E + \theta_{43}\log E_{-1} + \theta_{44}\log (YEAR - 1920) \tag{4.3$'$}$$

$$WP = \theta_{51} + \theta_{52}E + \theta_{53}WP_{-1} + \theta_{54} (YEAR - 1920) \tag{4.3$''$}$$

The estimates for each parameter are given in Table I. The log type model of (4.3)$'$ is non-linear only, but all other models are linear.

The estimation methods used are as follows. OLS (Ordinary Least Squares), LIML (Limited Information Maximum Likelihood), COMB (Combination of LIML and TSLS), TSLS (Two-Stage Least Squares). The estimations of log-type and lagged-type equations are performed by OLS.

Table I

Estimates of Parameters

	OLS	TSLS	LIML	COMB
θ_{11}	16.2366	16.5548	17.1477	16.6282
θ_{12}	0.1929	0.0173	-0.2225	-0.0232
θ_{13}	0.0899	0.2162	0.3960	0.2453
θ_{14}	0.7962	0.8102	0.8226	0.8134
θ_{21}	10.1258	20.2782	22.5908	22.6211
θ_{22}	0.4796	0.1502	0.0752	0.0742
θ_{23}	0.3330	0.6159	0.6804	0.6812
θ_{24}	-0.1118	-0.1578	-0.1683	-0.1684
θ_{31}	0.0637	-0.1134	-0.1381	-0.1137
θ_{32}	0.4395	0.4389	0.4339	0.4388
θ_{33}	0.1461	0.1467	0.1513	0.1468
θ_{34}	0.1303	0.1304	0.1316	0.1304
	θ_{41}	θ_{42}	θ_{43}	θ_{44}
LOG	-0.2900	0.7312	0.2003	0.0348
	θ_{51}	θ_{52}	θ_{53}	θ_{54}
LAG	-0.3575	0.4785	0.1948	0.1025

Table II

The Results of X-Test, Y-Test and XY-Test

			TSLS	LIML	COMB	OLS	LOG	LAG
$100\sqrt{Y(i)}$	C	Y-Test	7.577	7.509	7.457	9.692	9.899	9.399
		X-Test	3.757	5.795	4.405	1.395	1.398	1.398
		XY-Test	2.157	4.798	3.211	2.091	1.848	2.431
	I	Y-Test	185.1	201.6	203.6	126.8	129.3	123.2
		X-Test	50.15	121.6	123.3	5.033	4.993	5.037
		XY-Test	54.94	125.1	125.8	7.752	7.788	8.014
	WP	Y-Test	10.61	10.51	10.50	13.06	13.25	12.64
		X-Test	0.1322	0.1521	0.1471	4.041E-02	3.946E-02	4.111E-02
		XY-Test	3.757	5.498	5.034	3.313	3.084	4.016
	Y	Y-Test	12.18	12.25	12.07	15.06	15.55	14.56
		X-Test	3.360	4.943	3.794	1.221	1.223	1.219
		XY-Test	6.033	7.314	7.522	3.508	3.523	3.939
	P	Y-Test	26.71	28.36	26.79	28.57	29.11	28.07
		X-Test	20.65	23.80	23.89	9.302	9.343	9.300
		XY-Test	16.79	20.13	20.58	8.235	8.251	8.147
$100\sqrt{X(i)}$	WG	X-Test	51.37	43.72	56.50	27.08	27.10	27.08
		XY-Test	3.669	3.698	3.856	1.971	2.032	2.085
	G	X-Test	33.51	65.58	48.06	16.69	13.65	17.39
		XY-Test	31.27	38.19	34.19	11.48	11.61	11.41
	T	X-Test	64.67	62.98	63.11	33.18	33.70	35.96
		XY-Test	26.80	33.92	31.42	10.39	10.64	10.85
$100\sqrt{X}$		X-Test	24.36	55.54	56.23	4.802	4.810	4.802
		XY-Test	25.91	56.85	57.17	5.578	5.551	5.799
$100\sqrt{Y}$		X-Test	51.46	58.25	56.23	26.53	26.18	27.86
		XY-Test	23.87	29.57	27.22	9.012	9.166	9.171
β		Y-Test	59.40	64.63	65.17	41.70	42.53	40.52
α		X-Test	36.93	56.57	56.23	16.69	16.48	17.48
π		XY-Test	25.16	48.45	48.18	7.064	7.125	7.250

LOG (log-type Klein Model I, (4.1), (4.2), (4.3)$'$, and (4.4) through (4.7))

LAG (lagged-type Klein Model I, (4.1), (4.2) (4.3)", and (4.4) through (4.7))

In the model (4.1) through (4.7), let each variable be defined by

$s' = (P_{-1}, K_{-1}, E_{-1})$, (except for lagged-type equation)

$s' = (P_{-1}, K_{-1}, WP_{-1})$, (only lagged-type equation)

$x' = (WG, G, T)$

$y' = (C, I, WP, Y, P, K, E)$

Then the model can be represented by (2.1) in Section 2.

For the weighted matrices, we define

$\rho = 0,\ W_x' = (1, 1, 1),\ W_y' = (1, 1, 1, 1, 1, 0, 0)$

$k = 0.001$ (X-Test), $k = 0.3$ (XY-Test)

The convergent condition is assumed to be EPS = 0.01.

Table II indicates the results of Y-Test, X-Test and XY-Test. The degree of controllability of the model is given as by-product of X-Test. These results are described by Table III.

4.2 The Results and Interpretations

(i) The results of Y-Test become worse in order: LAG, OLS, TSLS, LIML, COMB

(ii) The results of X-Test become worse in order: LOG, OLS, LAG, TSLS, COMB, LIML.

(iii) The results of XY-Test become worse in order: OLS, LOG, LAG, TSLS, COMB, LIML.

(See Table II)

Comment 1.

TSLS is selected as the best in the simultaneous estimation method. LIML is better than COMB, from the viewpoint of the prediction, but the result is in reverse, from the viewpoint

Table III

The Degree of Controllability (%)

		TSLS	LIML	COMB	OLS	LOG	LAG
$\gamma(i)$	C	96.24	94.21	95.60	98.61	98.60	98.60
	I	49.85	-21.6	-23.3	94.97	95.01	94.96
	WP	99.87	99.85	99.85	99.96	99.96	99.96
	Y	96.64	95.06	96.21	98.78	98.78	98.78
	P	79.35	76.20	76.11	90.70	90.66	90.70
$\bar{\gamma}$		75.64	44.46	43.77	95.20	95.20	95.20
γ		49.85	——	——	90.70	90.66	90.70

of the control. As a result, COMB is better than LIML, when considering both the prediction and the control.

Comment 2.

OLS (in the original model) is selected as the best in the single estimation method. The goodness of fitting the data is in order of LAG, OLS and LOG, but the ranking is placed in order of LOG, OLS and LAG, from the viewpoint of the control. On the other hand, the result seems to be preferable in order of OLS, LOG and LAG, when considering both the prediction and the control.

Comment 3.

Let us consider two groups classified as the single estimation group A = {OLS, LOG, LAG} and the simultaneous group B = {TSLS, LIML, COMB}. The group A, with respect to each of the parameters α and π, is better than the group B. However, the group B in Y-Test is better rather than the group A, when the investment equation (I) is eliminated. On the other hand, the group A, with respect to all variables, is better than the group B. Consequently, it is verified that the stimultaneous estimation method is preferable in the prediction problem, and that the single estimation method is preferable in the control problem. In addition, for XY-Test, the group A is better than the group B.

Comment 4.

As is seen in Table III, the degree of controllability in the group A is about 95%, but the degree of controllability in the group B is about 50%. An Econometric model in the structural form is estimated by using either the single estimation method or the simultaneous estimation method. Thus the computations of stabilization policies implemented by the use of a given econometric model are strongly affected by the estimation method used in the estimation of the econometric model in the structural form.

5 REFERENCES

1. Mallows, C. L. "Some Comments on C_p", *Technometrics*, **15**, (1973).

2. Akaike, H. "A New-Look at Statistical Model Identification", *IEEE Trans. Automatic Control*, **AC-19**, (1974).

3. Uchida, Y. "A Comparison of Dynamic Regression Models Based on the Control Test Criterion", Proc. in Computational Statistics, Physica-Verlag, (1978).

4. Klein, L. R. "Economic Fluctuations in the United States, 1921 - 1941", John Wiley, (1950).

8.3 NUCLEAR REACTOR OPERATION

G. Tunnicliffe Wilson

(University of Lancaster, UK)

and

R. S. Overton

(CEGB Berkeley Nuclear Laboratories, UK)

1 INTRODUCTION

This paper concerns the methods developed to maximise the financial return from a nuclear reactor, the operational life of which is regarded as divided into a number of two-year stages. Such methods are not problem-specific, but can be used for any multistage decision problem of a broadly similar structure.

The problem upon which the method was developed is described in its original deterministic formulation, together with a constraint of a probabilistic nature. A solution of the deterministic problem is a desirable basis for solving the stochastic problem.

Treatment of the deterministic version is effected in two ways. Firstly, a Lagrange Multiplier method is used to provide a quick solution, and this is followed by the use of a constrained differential dynamic programming approach, which, as a general FORTRAN computer program, gives a means of checking the Lagrangian solution.

The use of more conventional dynamic programming techniques is found to be necessary for solving the stochastic problem. The standard method is modified in a way which allows the probabilistic constraint to be handled easily. Nevertheless, some innovative techniques have to be used to ensure that computation time is not prohibitive because of the "curse of dimensionality".

A simulation confirms that the introduction of the stochastic features into the problem results in a value of the maximum expected return smaller than the maximum return attainable in a deterministic case.

2 DESCRIPTION OF THE REACTOR PROBLEM

The first stage in the production of electricity from a nuclear reactor is the transfer of energy from its core.

2.1 The Control Variables

A coolant gas is used to remove the heat energy from the core, and there are essentially three properties of this coolant - the controls (u) of the system-which determine how much electricity is produced. The first is the pressure of the gas. A higher pressure results in a greater mass and energy flow and, consequently, in more electricity. The second is the gas temperature which, if increased, also increases energy transfer. The third property is the water content of the coolant. Although not directly affecting electricity production, it plays an important part in the two constraints on the system.

2.2 The State Variables

For technical reasons two parameters (S) of the reactor have to be monitored during its lifetime. Both these state parameters increase in level as the reactor operates, but their rates of increase can be changed by varying the control variables. A non-linear model describes the transformation of state from one stage to the next as a function of the controls. Initially, this model was considered to be deterministic. In the stochastic version, random variables were introduced as an additive "error" to the model.

2.3 Objective Function

An assumption is made that the reactor will produce power for a period up to a specified maximum lifetime, which consists of a number of two-year stages during each of which the control variables are kept at a constant level.

The amount of electricity produced is valued by time-weighting on a present worth basis in accordance with common economic practice. The objective is then to maximise the total value of the electricity produced over all the stages.

A non-linear model describes the rate of electricity production and its value as a function of the three control variables.

2.4 *Constraints*

Both the state parameters have to be kept below certain limits during the reactor lifetime, with a given probability (close to unity) in the stochastic case. Simple upper and lower limits on the control variables must also be satisfied. One would expect all these constraints to be active at some stage during the reactor lifetime.

2.5 *Variable End Point*

Another important aspect of the problem was that the number of stages of reactor operation was also regarded as a parameter to be optimised. Several approaches to this aspect were employed.

2.6 *Mathematical Formulation*

The deterministic problem can be summarised as:

Maximise: $$\sum_{i=1}^{K} f_i\,(u_i) \quad (2.1)$$

subject to: $$S_K \leqslant S' \quad (2.2)$$

$$u'_{i2} \leqslant u_i \leqslant u'_{i1}, \; i = 1, K \quad (2.3)$$

$$K \in \mathbb{N}, \; K \leqslant 15 \quad (2.4)$$

where u_i are the control variables (pressure, temperature, water) during stage i

S_i are the state variables at the end of stage i ($S_0 = 0$)

and $$S_i = S_{i-1} + h_i(u_i), \; i = 1, K \quad (2.5)$$

defines the transformation of state.

Also S', (u'_{i1}, u'_{i2}, $i = 1, K$) are specified limits on state and control variables respectively, and the inequalities apply to all corresponding vector components.

For the stochastic problem, this formulation was modified by,
(a) maximising the expected value of the above objective (2.1)
(b) rewriting the state model (2.5) as

$$S_i = S_{i-1} + h_i(u_i) + \nu_i, \quad i = 1, K \tag{2.6}$$

with ν_i, $i = 1, K$, as independent random variables with uniform distributions of zero mean and known variance.
(c) using a new probabilistic state constraint:

$$P(S_K \leqslant S') > 1-\varepsilon \tag{2.7}$$

The precise formulations are, with $u_i = (R_i, T_i, W_i)$ = (pressure, temperature, water) in stage i:

$$f_i = a(i)\ (R_i/215)^{2/3}\ (y(T_i)\ (T_i - 643) - 955)$$

$$\text{where } y(T_i) = \begin{cases} 6.5 & \text{if } T_i \leqslant 643 \\ 7.5 & \text{if } T_i > 643 \end{cases}$$

and the discount factor a(i) is given by Table I below.

Table I

Objective Function Discount Factor

i	a(i)	i	a(i)	i	a(i)
1	4.0	6	1.8	11	0.5
2	2.9	7	1.4	12	0.4
3	3.6	8	1.2	13	0.3
4	2.9	9	0.8	14	0.25
5	2.2	10	0.6	15	0.2

$$h_{i1}(R_i, T_i, W_i)$$

$$= 2\ (W_i/3)^{\frac{1}{4}}\ (R_i/215)^{\frac{1}{2}}\ \exp\left\{-\ 15500\ \left(\frac{1}{T_i} - \frac{1}{633}\right)\right\} \quad i = 1, K$$

$h_{i2}(R_i, W_i)$

$$= 0.44\ (R_i/215)^{5/3}(1.451 - .53\log_{10} W_i), \quad i = 1, K$$

$$\left.\begin{aligned} u'_{1i} &= (400,\ 663,\ 30)^T \\ u'_{2i} &= (215,\ 603,\ 3)^T \end{aligned}\right\} \quad i = 1, K$$

Also $S_0 = (0, 0)^T$

and $S' = (50, 15.66)^T$

3 A LAGRANGE MULTIPLIER APPROACH

A solution to the deterministic problem, which also gives some insight into the effect of introducing random variables into the formulation, uses Lagrange Multipliers to handle the state variable constraints. Since the state model (2.5) is additive, the constraint (2.2) can be rewritten as:

$$S_K = \sum_{i=1}^{k} h_i(u_i) \leqslant S' \tag{3.1}$$

A Lagrangian function

$$L(\lambda) = \sum_{i=1}^{K} f_i(u_i) - \lambda_1 \sum_{i=1}^{K} h_{i1}(u_i) - \lambda_2 \sum_{i=1}^{K} h_{i2}(u_i)$$

was defined to deal with the state inequality constraints. (It could be shown that these would be active at the optimum.)

3.1 Decoupling the Stages

Rewriting

$$L(\lambda) = \sum_{i=1}^{K} (f_i(u_i) - \lambda_1 h_{i1}(u_i) - \lambda_2 h_{i2}(u_i)) \tag{3.2}$$

the optimum of $L(\lambda)$ could be found, for a fixed λ, by maximising with respect to the controls for each stage separately, taking into account the simple limits on their ranges. By regarding the Lagrangian as net profit in an economic context, and by requiring the reactor to shut down at any stage at which the future total Lagrangian was negative, it was possible to optimise with respect to K, the number of stages, (for each fixed λ).

3.2 *Determining the Multipliers*

By repeating this optimisation over a grid of values of λ, contour plots, given in Fig. 1, of

(i) the (unpenalised) optimum return $F(\lambda) = \Sigma f_i$

(ii) the final level of each of the state parameters

$H_1(\lambda) = \Sigma h_{i1}$, $H_2(\lambda) = \Sigma h_{i2}$

were obtained as functions of the "prices" λ.

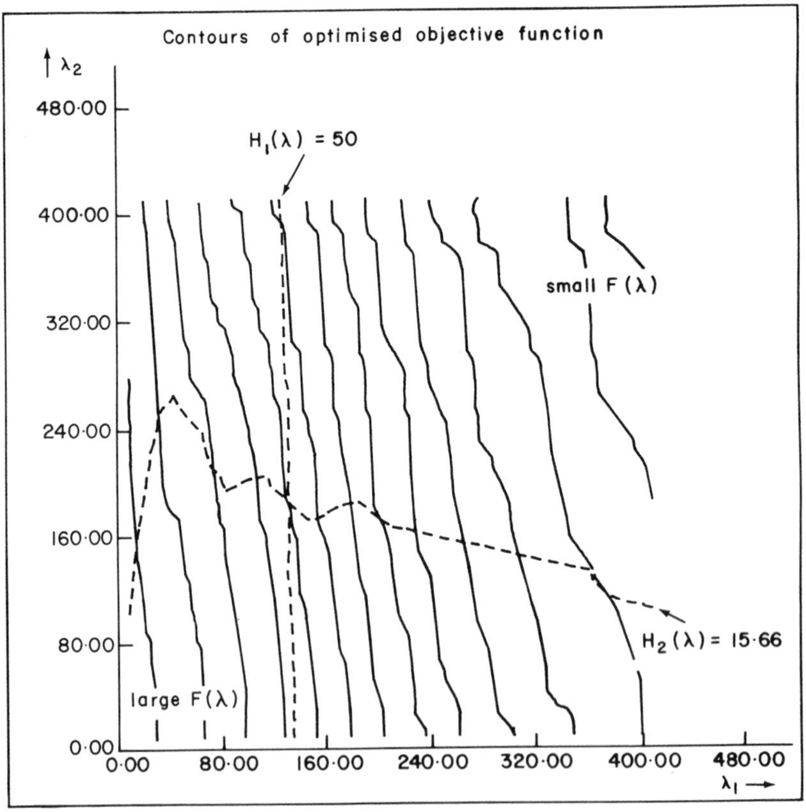

Fig. 1 Contours of the objective $F(\lambda)$ and of $H_1(\lambda) = 50$ and $H_2(\lambda) = 15.66$.

The required value λ of λ and the corresponding return $\mathcal{F}(\lambda)$ were given by the intersection of the contours $H_1(\lambda) = S'_1$ $H_2(\lambda = S'_2$.

The sensitivity of the optimum to variations in the limits S'_1, S'_2, could easily be seen from this solution, and the maximum adverse effect of the perturbations of the state variables due to stochastic effects could be inferred. The return from the deterministic problem with $S' = (50, 15.66)^T$ was 34200. Assuming that all the random variables had maximum realised values (case (b) in Table II), we would reduce S' to (46.228, 13.503) and the return would be 33000. Assuming the cumulative effect of the random variables to be Normal and taking the lower 0.01% point of this distribution, S' would reduce to (47.845, 14.39), and the return to 33400.

4 DIFFERENTIAL DYNAMIC PROGRAMMING

As an alternative to the use of Lagrange multipliers, a method of differential dynamic programming (DDP) has been implemented as a general FORTRAN program. The original DDP method [1] was modified to handle state inequality constraints such as (2.2). Lagrange multipliers with slack variables were recommended in [1] to deal with the control constraints (2.3). We have introduced exterior quadratic penalty functions to deal with all these constraints.

DDP combines the recursive structure of standard dynamic programming with general hill-climbing techniques. A trajectory in the state and control spaces is regarded as a "point" about which local linear and quadratic approximations are made to the state equations and objective function. The local approximation has a well-known recursive optimal solution which supplies the trajectory for the next iteration.

The three-dimensional graph in Fig. 2 shows how the pressure control variable changes with the iterations for each time stage, from a constant starting value to the final optimum.

We recommend DDP as a cheap and effective way of solving the deterministic problem.

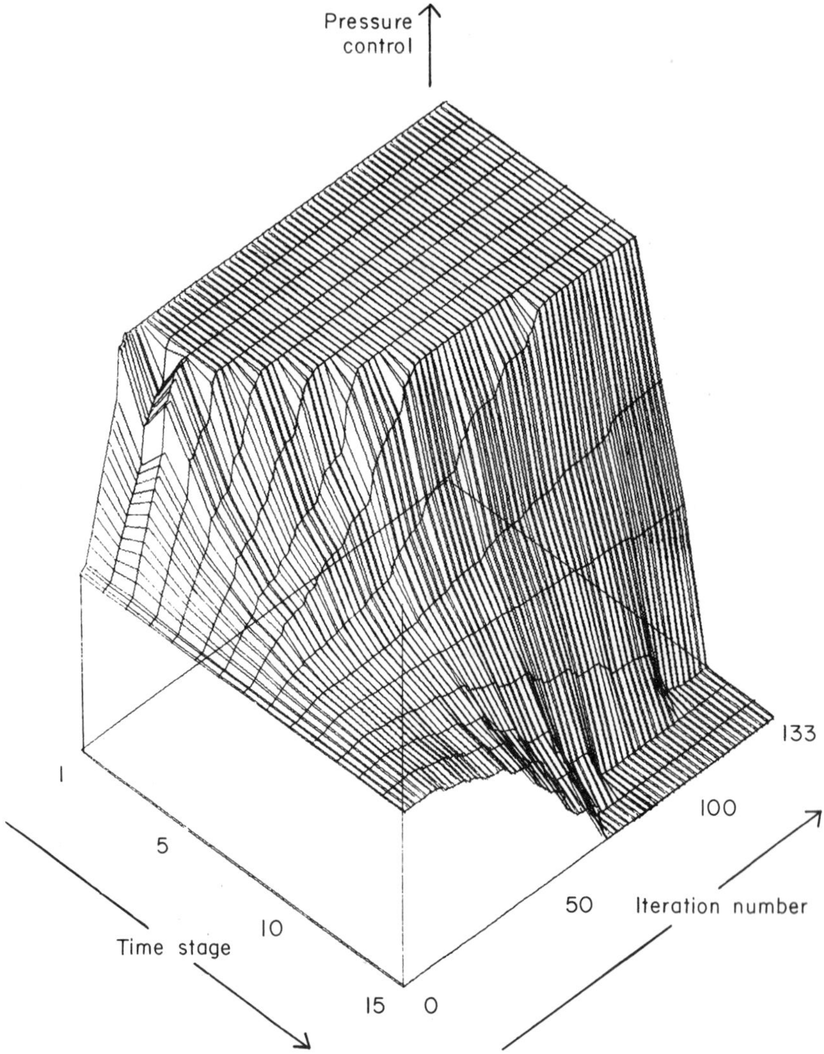

Fig. 2 Progress of the pressure variable (R_i) iterations under DDP.

5 A PROBABILITY STATE VARIABLE

For solving the reactor operation problem with random variables, uniformly distributed, as formulated in (2.6) and (2.7), it was necessary to revert to a modified version of standard dynamic programming, since just one sequence of controls and the associated state trajectory could not be optimal for all possible realisations of the random variables. For each time stage i, and for each combination of states S_{i-1} at the start of that stage, it was necessary to find the maximum, over u_i, of

$$E\left\{f_i(u_i) + V_{i+1}\,(S_i(S_{i-1},\ u_i))\right\} \tag{5.1}$$

where V_{i+1} is the total future maximised expected objective from the end of stage i onwards.

In order to accommodate the probabilistic constraint (2.7) we have defined a <u>probability state variable</u> (psv) for each stage i:

P_i = probability that either reactor state parameter exceeds its limit by the end of the process, given that neither value has been exceeded during the first i stages.

Taking

p_i = probability that one or more constraints are exceeded <u>during</u> stage i (a function of the state and control variables),

it can be shown that

$$P_i = (P_{i-1} - p_i)/(1 - p_i) \tag{5.2}$$

This provides an updating formula for the psv which neatly conforms with a DP structure (cf 2.5). The new state variable P_i has to be constrained to lie below a given value ε at each stage i in order to satisfy constraint (2.7).

However, since P_i is decreasing, we need only consider a range of P in $[0,\varepsilon]$, and the most important constraint to be maintained is that $p_i \leqslant P_{i-1}$.

6 PERFORMING THE OPTIMISATIONS

For each discrete state variable combination at the start of each time stage, dynamic programming requires an optimisation of the expected value of $f_i(u_i) + V_{i+1}(S_i)$ to be carried out over the next stage control variables u_i. Bellman [2] envisaged this as being done by discretising the control variables if necessary, evaluating the objective function at all combinations of controls resulting, and then comparing (subject to state constraints), to find the optimum. However, if the objective is a differentiable function of the controls, this can be a wasteful method. We have used a hill-climbing method with first derivatives.

We retain a discretisation of the state variables, but treat control variables as continuous. Constraints on the control variables - simple limits - are dealt with by projection methods. Within any time stage, with specified initial state S_{i-1}, the present return $f_i(u_i)$ is an analytic function of the controls u_i. The future return is to be derived from the table of $V_{i+1}(S_i)$, but is modified to take into account the stochastic effects, the probabilistic constraint, and the shutdown policy. Even so, it was possible to obtain by interpolation sufficiently precise values of the future return and its first derivatives to exploit successfully hill-climbing methods over the control variable space. This was much more efficient than a grid search for the optimum control.

7 EXPECTED FUTURE RETURN INCORPORATING PROBABLISTIC CONSTRAINTS

With prescribed values of S_{i-1}, P_{i-1}, we wish to determine the expected future return as a function of u_i. A given choice of u_i leads to a uniform distribution of S_i (see Fig. 3). There are two possibilities.

(i) If, from the distribution of S_i, $P(S_i > S') = p_i \leqslant P_{i-1}$ i.e. the constraint is satisfied, then the expected future return, conditional on the constraint being satisfied, is approximated by taking the mid-point S* of the portion of the distribution lying within the constraint and interpolating

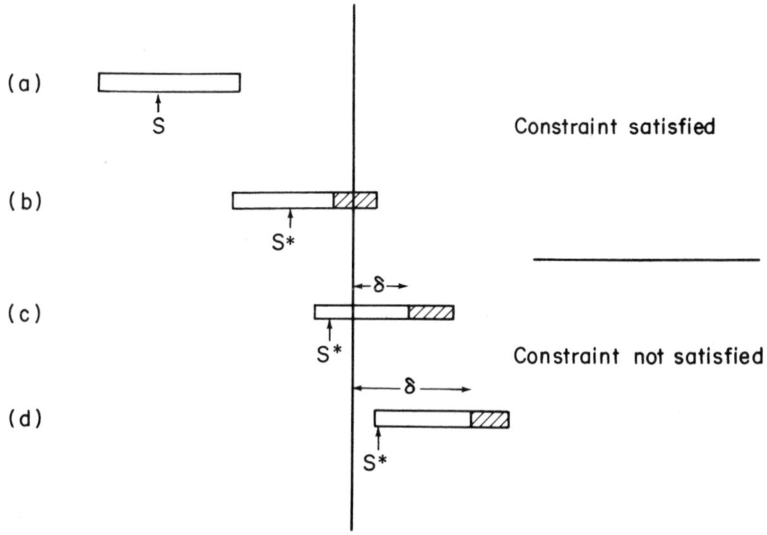

Fig. 3 Possibilities for the distribution of state variables

for $V_{i+1}(S^*)$.

(ii) If the constraint is not satisfied, the theoretical policy is to shutdown and take the present plus future return as zero, leading to a discontinuity in total return as a function of the controls. See Fig. 4.

In order that a small step over the constraint will be led back, this discontinuity is removed. $V_{i+1}(S^*)$ is evaluated as before and then the total is multiplied by $1/(1 + K_j\delta^2)$, where δ is the distance by which the distribution is shifted beyond the point at which the constraint becomes active. A suitable large value of the weight K_j was found by experiment for this penalisation.

8 CONTROL-LED STATE DEFINITION

The major drawback of DP cited in [2] is the "curse of dimensionality", referring to the exponential increase in calculation with the number of state variables. Since calculations must be performed at each combination of a number of discrete state levels, it is important to keep this number of

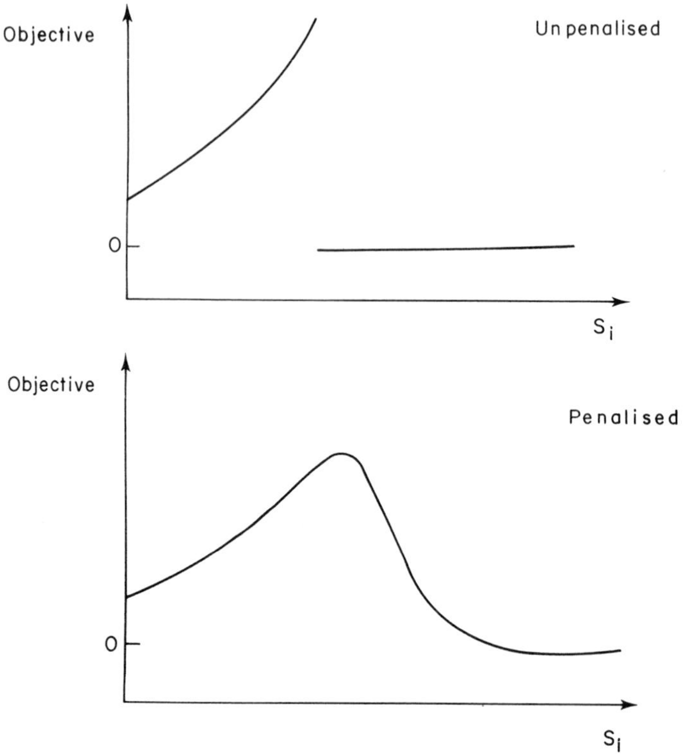

Fig. 4 Penalisation of the objective function

levels small, but large enough to secure sufficient accuracy in interpolation. This can be achieved by considering ranges of the state variables which will just contain any possible state trajectory but no others.

If the state model (2.5) is monotonic in the control variables then the minimum and maximum possible state levels at each time stage (allowing for <u>any</u> control sequence) can be computed by considering the limits of the control variables together with the extrema of the uniform distributions. The resulting state ranges can be reduced further because it is not necessary to consider the state variables above the constraint limits S'. See Fig. 5.

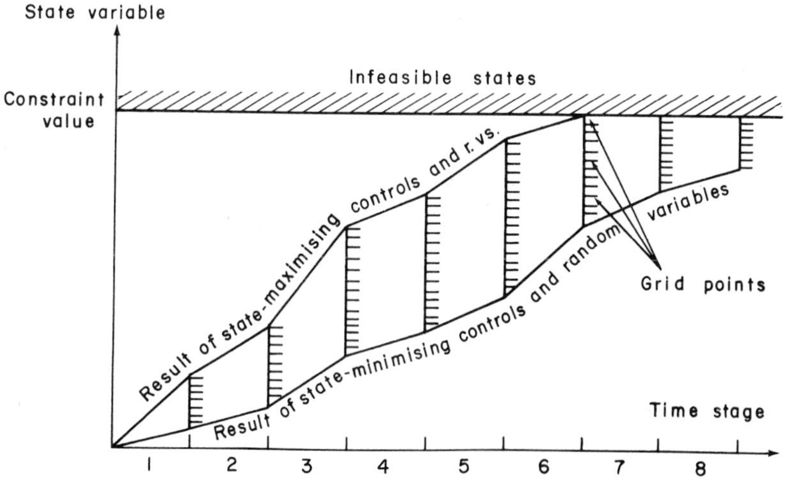

Fig. 5 A method of calculating a suitable state variable range

The state ranges thus obtained were found to be too wide to give sufficient accuracy for the reactor problem. Instead we have devised a method of control-led state definition which has proved much more effective. Rather than allow for any control sequence, we need only consider those which might arise from an optimal control policy. As a first step, the solution to the deterministic problem is taken, and used to create new limits on the control variables by placing a small interval about the optimal control trajectory. Some of these limits will coincide with the original ones, and some will be new "false" limits. From these limits and the random variable extrema, the corresponding possible state range is recomputed. Then standard DP tabulation of optimal controls is performed as usual in reverse time.

The adequacy of the new limits is tested by simulation, working forward from the initial state and sampling from the distributions of the random variables. Optimal controls are calculated in this part of the algorithm by interpolation in the state-space tabulation of control policy. If any control value required in the interpolation is a "false" limit, then the whole algorithm is reworked but with a slightly extended "false" limit in that particular control variable. This procedure is repeated until no further extension of the control limits is necessary, thus yielding the most sensitive optimal control table possible.

9 RESULTS

The stochastic reactor operation problem was solved using the above methods. Ten levels of each of the "physical" state variables and seven levels of the psv were used at each time stage. Tabulation of optimal controls took of the order of 350 seconds (IBM 370/168). To perform the simulation, forward working through the tables was carried out 200 times, sampling from the appropriate uniform distributions. A histogram of the total return obtained is given in Figs. 6 for two cases. In case (b) the variance of each random variable has been increased by 50% over the variance of each corresponding random variable in case (a).

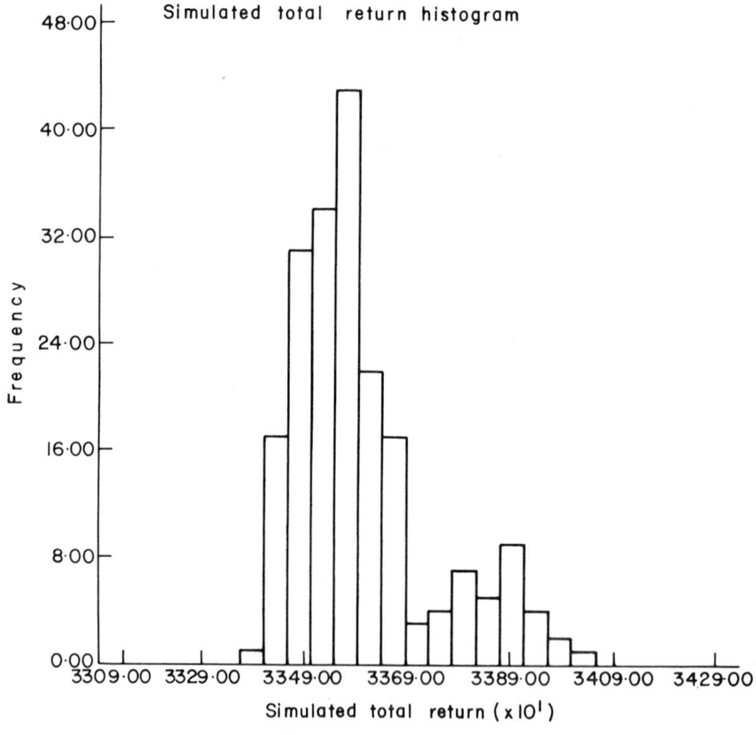

Fig. 6(a) Histogram of the simulated total return for case (a)

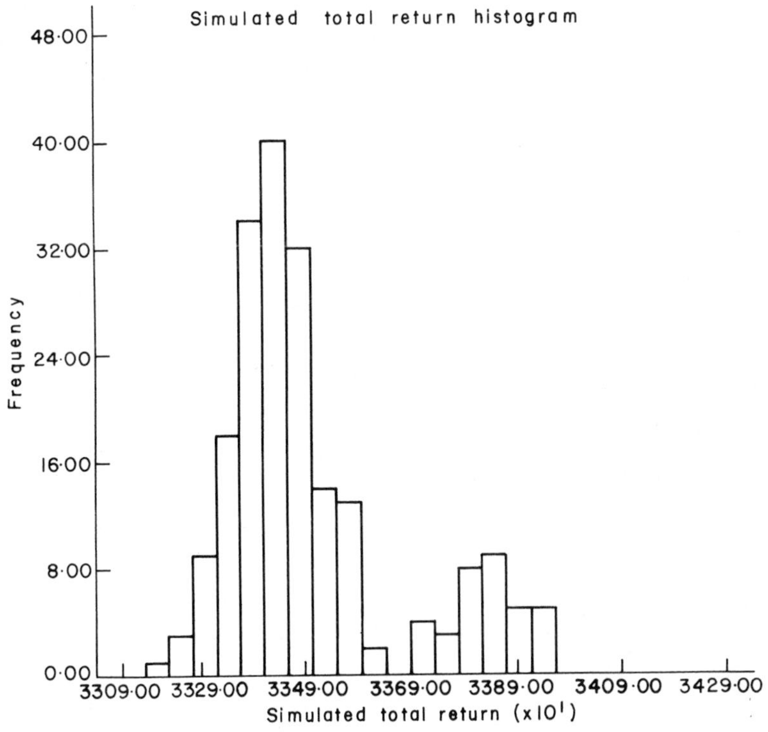

Fig. 6(b) Histogram of the simulated total return for case (b)

The following table shows that the larger variance in case (b) results in a lower mean simulated total return than in (a), reflecting the greater uncertainty in the system. A greater spread of the distribution of simulated total return also results.

There is a small difference between the expected return from the DP tables and the simulated mean return, which is most likely accounted for by bias in the approximation used in computing the expected return. This approximation can be improved by using a finer discretisation of states. However, we have good reason to believe that the resulting control policies are acceptable.

Table II

Results of Stochastic DP Applied to the Reactor Problem

	case (a)	case (b)
Deterministic model total return	34446.4	34446.4
Maximum simulated total return	34041.6	33959.7
Minimum simulated total return	33374.7	33213.2
Mean simulated total return	33600.0	33502.9
Standard deviation of simulated total return	160.8	183.1

10 ACKNOWLEDGEMENTS

This paper is published by permission of the Central Electricity Generating Board.

11 REFERENCES

1. Jacobson, D. M. and Mayne, D. Q. "Differential Dynamic Programming", Elsevier, N.Y., (1970).

2. Bellman, R. "Dynamic Programming, Princeton U.P., N.J., (1957).

8.4 PROBLEMS IN THE STOCHASTIC STORAGE OF SOLAR THERMAL ENERGY

J. Haslett

(Statistics Department, Trinity College, Dublin)

1 INTRODUCTION

Solar energy is a variable energy resource. Its exploitation requires a controlled system for collecting, storing and distributing the energy. Such systems are amenable to analysis by stochastic storage theory. Systems involving the storage of energy as sensible heat, by raising the temperature of a storage tank, lead to new and interesting problems in stochastic storage. It is the purpose of this paper to bring some of these problems, many of which are as yet unsolved, to the attention of others involved in applied stochastic systems research.

In section 2 therefore, we shall briefly motivate a general model of a solar thermal energy system. This model will then be used to highlight two important aspects of the system that arise from the way it is controlled. These will be examined in some detail. In section 3 we imagine both controls to be trivial and immediately obtain a solution from the resulting linear model. In sections 4 and 5 we separately examine the controls by imagining each in turn to be trivially operated. In section 6 we discuss the unresolved difficulties and indicate areas where similar research is needed.

2 SOLAR ENERGY SYSTEMS

Solar thermal energy systems operate by allowing solar radiation to heat a solar panel, which is continuously cooled by a circulating fluid [4]. This fluid gives up its energy to a storage unit, which is tapped as required by the user. The simplest storage unit is a tank of water to which the fluid transmits its energy by mixing. The user is not entirely dependent on solar energy for his energy requirements, and has access to an auxiliary (conventional) energy source should the solar system fail to provide all his energy requirements.

The criterion by which the system is judged is, therefore, the long term average rate at which the system provides energy to the user. Continuity of supply is *not* the overriding criterion.

Such a system, intended to reduce the conventional energy requirements for domestic water heating, is shown in Fig. 1. This is achieved by drawing one's water not from the mains cold water supply, but from the solar heated storage tank and refilling the store with an equal amount of cold water. One important control is exercised by the user at this stage, for if the temperature of the stored water is in fact *greater* than the temperature at which the water is required, the user's total energy requirements can be met by withdrawing a lesser amount of water at this greater temperature, and subsequently cooling it by dilution with cold water. If the desired temperature is not exceeded, then the full volume requirements are withdrawn and heated at the auxiliary heater.

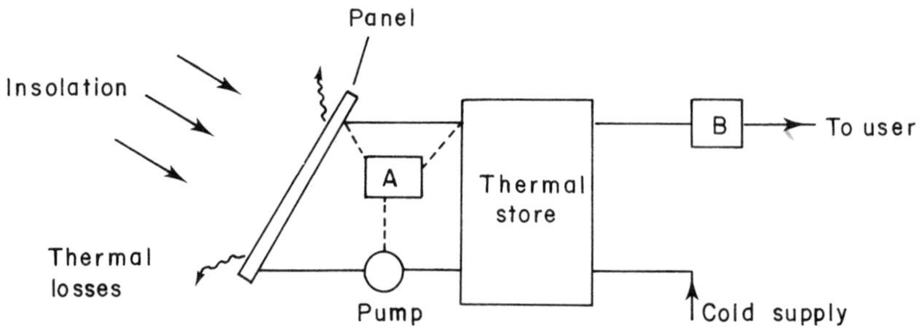

Fig. 1 A Schematic of a Solar Water Heating System

The units A and B represent a differential thermostat controlling the circulation of liquid through the panel, and an auxiliary heater, respectively.

The other critical control is due to the operation of a differential thermostat between the panel and the store. For it is only advantageous to operate the panel when to do so would result in a *net* energy gain at the store. This is because the panel is in two-way thermal communication with the environment-gaining by radiation, and losing by convection/conduction. Thus to circulate fluid through the panel at night could only cool the store. Further to operate it on a relatively overcast day will only be advantageous if, as a result of its previous history of operation, the store is already relatively cool. Since insolation is highly variable - seasonally, diurnally and randomly - this aspect of system feedback is of the greatest importance. It also gives rise to considerable modelling difficulties.

For certain simple cases [6,7] concerning the diurnal variation in insolation it is sufficient to model such a system on a day-by-day basis. We may also imagine that the user makes all his demands for energy in the evening, when the panel is not operating. Then if X_{n-1} denotes the useful energy content of the store just before the demand D_n is placed on day n, it may be shown that the energy W_n actually supplied to the user is

$$W_n = \min\{(1 - \beta_n)X_{n-1}, D_n\} \tag{2.1}$$

where β_n $(0 < \beta_n < 1)$ is, in general, a function of D_n. This represents symbolically the first control above.

The energy content then becomes $X_{n-1} - W_n$. This characterises the state of the store when it next faces the insolation. The useful energy Q_n gained on day n depends on the insolation available, and on $(X_{n-1} - W_n)$. In fact

$$Q_n = \max\{Y_n - (1 - \alpha)(X_{n-1} - W_n), 0\} \tag{2.2}$$

where Y_n is a system independent variable characterising the insolation and ambient termperature on day n, and α is a system constant $(0 < \alpha < 1)$. This represents the second control above. Clearly

$$X_n = X_{n-1} - W_n + Q_n \tag{2.3}$$

This cycle is shown schematically in Fig. 2. Y_n represents the maximum available energy on day n.

The objective of the analysis is now to determine μ_W, the mean of W_n, as a function of the system parameters and of the known distributions of Y_n and D_n, the system independent

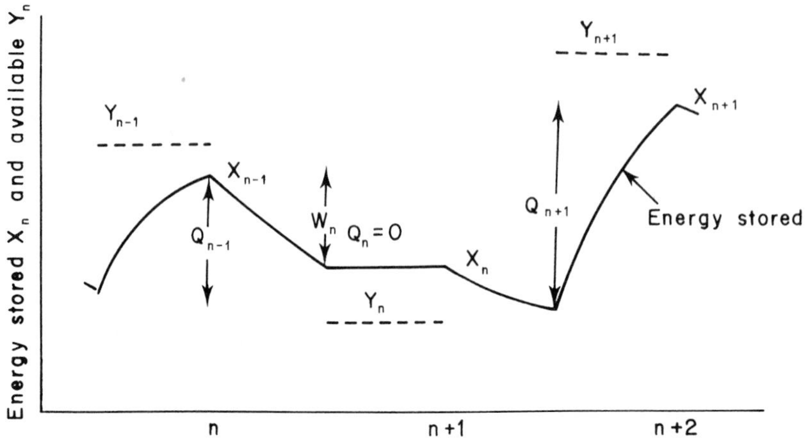

Fig. 2 An idealisation of the heating and cooling of the thermal store

variables. Note that $\mu_Q = \mu_W$ in equilibrium.

The problems associated with this analysis will be illustrated in the following sections by an examination of some very special cases. We shall regard D_n (and hence β_n) as constant, and $\{Y_n\}$ as independent and identically distributed. The special cases will be chosen to highlight the control problems represented by (2.2) and (2.3). Firstly, however, we examine an almost trivial case where neither control is operative.

3 TRIVIAL CONTROLS

Suppose in fact that rather than (2.1) and (2.2) the system were defined by

$$W_n = (1 - \beta) X_{n-1} \tag{3.1}$$

$$Q_n = Y_n - (1 - \alpha)(X_{n-1} - W_n) \tag{3.2}$$

It follows immediately from (2.3) that

$$\mu_W = (1 - \beta)(1 - \alpha\beta)^{-1} \mu_Y \tag{3.3}$$

where μ_Y denotes the mean of Y.

This simple result is of relevance to undersized systems where the system rarely if ever heats sufficiently either to supply the full requirements of the user or to cause insolation to be rejected. It has also been used with some success as a variance reduction device in computer simulation models [9].

4 NO THERMOSTAT CONTROL

Suppose now that we have

$$W_n = \min\{(1-\beta)X_{n-1}, D\} \tag{4.1}$$

$$Q_n = Y_n - (1-\alpha)(X_{n-1} - W_n) \tag{4.2}$$

It is convenient to write $Z_n = (1-\beta)X_n - D$, from which it follows via (2.3) that

$$\Delta Z_n = Z_n - Z_{n-1} = Y'_n - \gamma_1 Z_{n-1} \quad \text{if } Z_{n-1} \geq 0 \tag{4.3}$$

$$= Y'_n - \gamma_2 Z_{n-1} \quad \text{if } Z_{n-1} < 0 \tag{4.4}$$

where $\gamma_1 = 1-\alpha$, $\gamma_2 = 1-\alpha\beta$ and $Y'_n = (1-\beta)Y_n - \gamma_2 D$. This model is very close to classical storage models. One useful technique for such models has been the use of diffusion approximations [10]. This approach has also been used here with some success [8]. In general terms it is likely to be a particularly useful approach when the changes ΔZ_n in Z_n are small, which corresponds to small γ_1, γ_2 and Var (Y'). To illustrate the technique we shall only consider the very special case of Y' having zero mean and unit variance.

In this case (4.3) and (4.4) may be identified as analogous to an Ornstein-Uhlenback process with different centralising parameters γ_1 and γ_2 according as Z is positive or negative [3, sec. 5.3]. The equilibrium probability density function is readily shown to be

$$f(z) = k_1 f_1(z) \qquad z > 0$$

$$\qquad = k_2 f_2(z) \qquad z < 0 \tag{4.5}$$

where $f_i(z) = \exp\{-\gamma_i z^2\}$, $i = 1, 2$. k_1 and k_2 are constants such that $f(z)$ is a density function, and that the long term mean of ΔZ_n is zero. This in fact requires that $k_1 = k_2$

$$k_1 = k_2 = \frac{2}{\sqrt{\pi}} \{1/\sqrt{\gamma_1} + 1/\sqrt{\gamma_2}\}$$

It will therefore be noted that Z has, in equilibrium, a distribution the right and left hand tails of which are the tails of Normal distributions with zero means but different variances. It follows that

$$\mu_z = \{1/\sqrt{\gamma_1} - 1/\sqrt{\gamma_2}\}/\sqrt{\pi} \tag{4.6}$$

whence, since $\mu_Q = \mu_W$ it can be seen that

$$\mu_W = \{(1-\alpha)/(1-\beta)\}(\mu_z/\alpha) \tag{4.7}$$

which is of course less than D and less than μ_W identified by (3.3).

A simple extension is required to meet the case where Y' does not have zero mean or unit variance. When D is variable but β is constant the approach is useful but breaks down for certain system parameters, for the reason that $\mathrm{Var}(Y')$ can no longer be guaranteed to be small. However, the solution methodology appears to be reasonably robust even when γ_1 and γ_2 cannot be guaranteed to be small. No solution has been found when β is variable, except that implicit in replacing β by its mean μ_β and using the approach above.

5   THERMOSTAT CONTROL DOMINANT

Suppose now that our defining equations are

$$W_n = (1 - \beta) X_{n-1} \tag{5.1}$$

$$Q_n = \max\{Y_n - (1 - \alpha)(X_{n-1} - W_n), 0\} \tag{5.2}$$

This is equivalent, with (2.3) to

$$X_n = \max\{Y_n + \alpha\beta X_{n-1}, \beta X_{n-1}\} \quad (5.3)$$

This is in fact a remarkably difficult process to analyse, and little progress can be reported. To illustrate the difficulties we can consider the very special case of $\alpha = 0$, when we have

$$X_n = \max\{Y_n, \beta X_{n-1}\} \quad (5.4)$$

This process is extremely simple in form, and has an extremely simple formal solution. For if we denote by F(.) and $\phi_n(.)$ the distribution functions for Y and X_n respectively, we immediately have

$$\phi_n(x) = F(x)\ \phi_{n-1}(x/\beta) \quad (5.5)$$

If $\phi(x) = \lim_{n\to\infty} \phi_n(x)$ exists, then we have

$$\phi(x) = F(x)\ \phi(x/\beta) \quad (5.6)$$

or alternatively

$$\phi(x) = \prod_{n=0}^{\infty} F(x/\beta)^n \quad (5.7)$$

Simple though (5.4) and (5.6) are, it seems frustratingly difficult to obtain explicit solutions even for μ_x as a function of the parameter β and the distribution of Y. Explicit solutions are, however, available for two very special distributions F(.).

5.1 Case 1

$$F(y) = \exp\{-y^{-p}\} \text{ for } y > 0,\ p > 1. \quad (5.8)$$

It follows from 5.7 that

$$\phi(x) \doteq \exp\{-(1 - \beta^p)^{-1}\ x^{-p}\} \quad (5.9)$$

whence

$$\mu_x = (1 - \beta^p)^{-1/p}\ \mu_y \quad (5.10)$$

Clearly $\mu_x > \mu_y$ as $\beta \to 0$, and as $\beta \to \infty$, $\mu_x \to \infty$ also.

Hence also

$$\mu_W = (1 - \beta)(1 - \beta^p)^{-1/p} \mu_y \tag{5.11}$$

which may usefully be compared to the corresponding version of (3.3). It is clearly seen that μ_W is increased by a factor $(1 - \beta^p)^{-1/p}$ by the imposition of a thermostat control.

5.2 Case 2

$$F(y) = 1 - y^{-p} \text{ for } y > 1,\ p > 1 \tag{5.12}$$

It follows from (5.7) that

$$\phi(x) = \prod_{0}^{\infty} (1 - \beta^{np} x^{-p}) \tag{5.13}$$

which may be expressed [1, p.19] as a power series in x^{-p}.

$$\phi(x) = 1 + \sum_{1}^{\infty} (-1)^n A_n x^{-pn} \tag{5.14}$$

where $A_n = \beta^{np}(1 - \beta^{np})^{-1} A_{n-1}$ and $A_0 = \beta^{-p}$

Hence we find

$$\mu_x = \{\sum_{1}^{\infty} (-1)^{n+1} A_n (p - 1)/(np - 1)\} \mu_y \tag{5.15}$$

and hence μ_W follows.

The two cases above are the only non-trivial cases for which the author has been able to obtain a solution. Further, these bear little relation to the original problem. Yet the process (5.4) represents in a very real sense the essence of the thermostat control exercised by the system.

6 DISCUSSION

Remarks of two kinds may be made on the foregoing. The first, at the technical level, is an earnest plea for advice in solving some of the technical problems posed. The methods used (see [7]) to overcome the completely unsolved problems of section 5 are inelegant, and the solution methodology of section 4 could be strengthened.

The second is to bring the subject as a whole - that is, the modelling of renewable, and hence variable, energy systems - to the attention of applied probabilists. It is a large and increasingly important area, littered with applied and theoretical problems, essentially concerned with the concept of variability, and yet, with rare exceptions [2,5], completely ignored by statisticians and probabilists.

7 ACKNOWLEDGEMENT

This work has arisen from research conducted by the Statistics and Operations Research Laboratory (SORL) of Trinity College for the Energy R & D program of the European Communities and for the National Board for Science and Technology, Dublin. Their support is acknowledged.

8 REFERENCES

1. Andrews, G. E. "The Theory of Partitions", Addison Wesley, (1976).

2. Bae, H. M. and Devine, M. D. "Optimisation Models for the Economic Design of Wind Power Systems", *J. Solar Energy*, **20**, 469-482, (1978).

3. Cox, D. R. and Miller M. D. "The Theory of Stochastic Processes", Methuen, (1965).

4. Duffie, J. A. and Beckman, W. A. "Solar Energy Thermal Processes", Wiley, (1974).

5. Goldstein, L. H. "A Fokker Planck Analysis of Photo-voltaic Systems", *Energy*, **3**, 51-62, (1978).

6. Haslett, J. "The Stochastic Modelling of Solar Thermal Energy Systems", IMA Conf. Proc. 15, (1977).

7. Haslett, J. "The Analysis by Stochastic Modelling of Solar Systems for Space and Water Heating" in Sun: Mankinds Future Source of Energy, Proceedings of International Solar Energy Congress, New Delhi, Pergamon, (1978).

8. Haslett, J. "A Diffusion Model for the Storage of Solar Thermal Energy", *J. Opl. Res. Soc.*, **30**, 433-438, (1979).

9. Haslett, J. and Hand, F. P. "Models in Solar Water Heating - a Critical Comparison", *Euro. J. Op. Res.*, to appear.

10. Pliska, S. R. "Diffusion Model for the Optimal Operation of a Reservoir System", *J. Appl. Prob.*, **12**, 859-863, (1975).

LIST OF PARTICIPANTS

Mr. B.L. Anand	Middlesex Polytechnic, UK
Mr. J. Ansell	Sheffield City Polytechnic, UK
Mr. M. Armstrong	University of Leicester, UK
Dr. A. Bagchi	Twente University of Technology, Netherlands
Professor A.V. Balakrishnan	University of California, USA
Mr. A. Bendell	Sheffield City Polytechnic, UK
Dr. V.E. Beneš	Bell Laboratories, USA
Dr. A. Bensoussan	IRIA, France
Mr. E. Bertsch	TU Berlin, FB Mathematik, Germany
Dr. J.M. Bismut	University of Paris, France
Dr. S. Bittanti	Polytechnic of Milan, Italy
Dr. R. Boel	Bell Laboratories, USA
Professor R.W. Brockett	Harvard University, USA
Professor P.E. Caines	Harvard University, USA
Mr. N. Carmichael	University of Warwick, UK
Miss H.K. Chadha	Leicester Polytechnic, UK
Dr. C.W. Chan	Unilever Research, UK
Dr. N. Christopeit	University of Bonn, Germany
Mr. Norman Clarke	The Institute of Mathematics and its Applications, UK
Dr. R.F. Curtain	University of Groningen, Netherlands
Dr. I. Dancs	OT S.K. Coordination and Scientific Secretariat, Hungary
Mr. B. Darbyshire	British Aerospace, UK
Dr. M.H.A. Davis	Imperial College of Science and Technology, UK
Dr. L. de Biase	University of Milan, Italy
Mr. Ph. de Larminat	University of Nantes, France
Dr. M.A.H. Dempster	Balliol College, Oxford, UK
Dr. L.C.W. Dixon	Hatfield Polytechnic, UK
Mr. C. Doncarli	University of Nantes, France
Dr. D.C. Dowson	University of Salford, UK
Professor R.J. Elliott	University of Hull, UK
Dr. F. Faruqi	Hunting Engineering Limited, UK
Professor W.H. Fleming	Brown University, USA
Dr. P.J. Gawthrop	University of Oxford, UK
Mr. R.L. Goodrich	Abt Associates and Harvard University, USA
Dr. M.J. Grimble	Sheffield City Polytechnic, UK
Dr. P.A. Hall	British Steel Corporation, UK
Professor A. Halme	University of Oulu, Finland
Dr. C.J. Harris	University of Oxford, UK
Mr. J. Haslett	Trinity College, Dublin, UK
Professor U.G. Haussmann	The University of British Columbia, Canada

Mrs. C. Hinds	The Institute of Mathematics and its Applications, UK
Professor Y.C. Ho	Harvard University, USA
Dr. S. Humble	Sheffield City Polytechnic, UK
Dr. A. Ichikawa	University of Warwick, UK
Dr. G.W. Irwin	Loughborough University of Technology, UK
Dr. O. Jacobs	University of Oxford, UK
Mr. L. James	Hatfield Polytechnic, UK
Dr. E. Klafszky	OT S.K. Coordination and Scientific Secretariat, Hungary
Dr. M. Kohlmann	University of Bonn, Germany
Professor D.G. Lainiotis	State University of New York at Buffalo, USA
Dr. J.P. Lepeltier	University of Maine, USA
Dr. R.M. Lewis	University of Bath, UK
Professor L. Ljung	University of Linkoping, Sweden
Professor F. Louveaux	Facultes Universitaires Notre-Dame de la Paix, Belgium
Mr. U.E. Makov	Chelsea College, London, UK
Dr. B. Marchal	University of Paris, France
Professor S.I. Marcus	The University of Texas at Austin, USA
Dr. J. Marshall	University of Bath , UK
Professor Dr. H.H. Martens	Norwegian Institute of Technology, Norway
Mr. D. McElhenny	Imperial College of Science and Technology, UK
Professor H. Myoken	Nagoya City University, Japan
Dr. M.M. Newmann	Queen's University of Belfast, UK
Professor G.R. Nicoll	Heriot-Watt University, UK
Dr. M. Nisio	Kobe University, Japan
Mr. R.S. Overton	CEGB, Berkeley, UK
Dr. D.H. Owens	University of Sheffield, UK
Professor P.C. Parks	Royal Military College of Science, Shrivenham, UK
Mr. P. Pawsey	Hunting Engineering Limited, UK
Mr. J.D. Perkins	Imperial College of Science and Technology, UK
Mr. F. Pew	Hunting Engineering Limited, UK
Mr. P.H. Phillipson	University of Leicester, UK
Mr. J.H. Powell	Health and Safety Executive, Sheffield, UK
Dr. A.J. Pritchard	University of Warwick, UK
Professor U. Pursiheimo	University of Turku, Finland
Miss C. Richards	The Institute of Mathematics and its Applications, UK

Dr. J. Rissanen	IBM Research Laboratories, USA
Professor A.P. Roberts	Queen's University of Belfast, UK
Miss I. Robertson	University of Strathclyde, UK
Professor W.L. Root	University of Michigan, USA
Dr. M. Ruohonen	University of Turku, Finland
Professor R.W.H Sargent	Imperial College of Science and Technology, UK
Dr. L.A. Shepp	Bell Laboratories, USA
Professor A.F.M. Smith	University of Nottingham, UK
Mr. M.W.A. Smith	The Northern Ireland Polytechnic, UK
Dr. D. Sprevak	The Queen's University of Belfast, UK
Mr. W.R.S. Sutherland	University of Dalhousie, Canada
Dr. G. Tunnicliffe Wilson	University of Lancaster, UK
Mr. Y. Uchida	Kobe University, Japan
Dr. M. Ullrich	Institute of Information Theory and Automation, Prague, Czechoslovakia
Mr. B. Uhrin	OT S.K. Coordination and Scientific Secretariat, Hungary
Dr. D. Vermes	University of Szeged, Hungary
Mr. N. Virani	Imperial College of Science and Technology, UK
Mr. C.B. Wan	Imperial College of Science and Technology, UK
Mr. W. Watson	Leicester Polytechnic, UK
Professor R.J.B. Wets	University of Kentucky, USA
Professor J.C. Willems	University of Groningen, Netherlands
Dr. H.S. Witsenhausen	Bell Laboratories, USA
Dr. J.H. Wright	University of Bristol, UK
Mr. M.B. Zarrop	Imperial College of Science, UK and Technology, UK

SUBJECT INDEX